Applied Seismic Geostructures

Zonghu Liao

Applied Seismic Geostructures

Characterizing Fault Damage Zones Quantitatively

Springer

Zonghu Liao
College of Geosciences and College
of Neutrality Future Technology
China University of Petroleum
Beijing, China

ISBN 978-981-97-7078-6 ISBN 978-981-97-7079-3 (eBook)
https://doi.org/10.1007/978-981-97-7079-3

This Springer imprint is published by the registered company Springer Nature Singapore Pte Ltd.
The registered company address is: 152 Beach Road, #21-01/04 Gateway East, Singapore 189721, Singapore

Preface

Within the esteemed halls of the University of Oklahoma, our journey into the depths of seismic interpretation began an endeavor that has since blossomed into a comprehensive exploration of fault damage zones through the lens of seismic attributes. Under the guidance of Prof. Kurt J. Marfurt and with the invaluable insights of Prof. Ze'ev Reches, we embarked on a project that not only sought to characterize the strike-slip fault in the Anadarko Basin but also to lay the groundwork for a methodical, quantitative analysis of subsurface structures—a task that was both ambitious and pioneering in 2010.

Over fourteen years, our scholarly pursuits have unfolded along two parallel paths. One path led us through the intricate world of fault dynamics using ROGA (Rotary Gouge Apparatus) designed by Prof. Ze'ev Reches, engaging in fault experiments and rotary shears to decode the mechanics of faulting. The other path navigated the complexities of fault geometries, interpreting the Earth's cryptic subsurface patterns using seismic attributes. This dual approach underscored a pressing need for advancements in seismic resolution, to ensure our interpretations and models closely mirror the real geological constructs that lie beneath.

Our work of the fault geometries to the field, detailed within the pages of this book, includes rigorous delineation of fault damage zones across various types, e.g., normal faults, thrust faults, strike-slip faults, etc., and extending our analysis to a systemic level. Through meticulous evaluation, we established a reliable resolution within 50–150 m, thus enhancing our understanding of subsurface structures significantly.

This volume, *Applied Seismic Geostructures: Characterizing Fault Damage Zones Quantitatively*, encapsulates our dedication and the collective insights garnered through years of research. It unfolds across three pivotal parts: (1) Fault Damage Zones: This part presents our initial foray into the characterization of fault types, offering a foundation for understanding the complex dynamics and geometries of fault damage zones. (2) Structure Systems: Elevating our analysis of faults as a system, we explore the interconnectedness of fault damage zones and their broader geological implications in the oil industry and earthquake hazards, utilizing seismic attributes to unearth new insights. (3) New Technologies and Applications: Looking toward the future, we delve into the integration of cutting-edge technologies, including artificial intelligence (AI), to push the boundaries of seismic interpretation and fault analysis.

Through this narrative, we bridge the theoretical and practical aspects of seismic science, aiming to enrich the scientific community's understanding of the fault structures in Earth's subsurface. Our work, grounded in the scholarly environment of the University of Oklahoma and China University of Petroleum (Beijing), extends an invitation to readers to explore the fascinating interplay between seismic survey and fault geometries, heralding a new era in the field of seismic interpretation.

The journey of creating this book has been both challenging and rewarding, and it would not have been possible without the generous support and assistance from Luyao Hu, Brett M. Carpenter, Saiyyna Vasileva, Yun Zhou, Lin Zhang, Hui Liu, Wei Li, Huayao Zou, Fang Hao, Xiaofeng Chen, Mengni Wu, Lianbo Zeng, Xiaojuan Huang, Yangpu Chen, Li-Yun Fu, Shikai Jian, Jiarun Yang, Xinming Wu, Zhengfa Bi, Zhicheng Geng. I must acknowledge Prof. Xinming Wu for contributing to Chap. 11 "Artificial Intelligence Model for 3D Seismic Fault Analysis." I am deeply grateful to my colleagues and mentors at the University of Oklahoma and the China University of Petroleum for their invaluable insights and encouragement. Special thanks are extended to Profs. Ze'ev Reches and Kurt J. Marfurt for their guidance and inspiration that sparked the genesis of this work.

I must also acknowledge the unwavering support of my parents Dexing Liao and Xiuzhi Cao, my wife Kun Yang, and my son Darwin (Haoxian) Liao, whose patience and understanding provided the necessary foundation for my research and writing. This book is for Kun and our new baby Haoru Liao. The contributions of the numerous graduate students and research assistants who engaged with the project, offering their expertise and dedicating countless hours to data analysis and interpretation, have been indispensable.

Financial support from various funding bodies, e.g., the Israel Science Foundation-Natural Science Foundation of China (ISF-NSFC) Joint Scientific Research Program 42161144005, has been crucial in facilitating our research endeavors and is hereby gratefully acknowledged.

Lastly, I wish to express my appreciation to the editorial team of Springer, my master's student Shu Yuan, and reviewers whose constructive feedback helped shape the final manuscript, ensuring that the book meets the highest academic standards. Their commitment to excellence has significantly enhanced the quality of this work.

Beijing, China Zonghu Liao

Competing Interests The authors have no conflicts of interest to declare that are relevant to the content of this book.

Ethics Approval The research conducted for this book/chapter has no human participants and does not involve animals.

About This Book

Applied Seismic Geostructures: Characterizing Fault Damage Zones Quantitatively meticulously unravels the intricate world of seismic geostructures, employing cutting-edge methodologies to decode the mysteries beneath the Earth's crust. This volume is artfully divided into three comprehensive parts, each tailored to impart a deep understanding of the seismic attributes and their pivotal role in geoscience.

Part I: Fault Damage Zones lays the foundational knowledge on characterizing various fault damage zones, including normal, thrust, and strike-slip faults, across diverse geological settings like in eastern China, the Longmen Shan Range in Sichuan, and the Anadarko Basin in Oklahoma, USA. It follows a structured approach to introduce seismic variance methodologies, attribute analysis, and the critical implications of fault damage zones on reservoir compartmentalization and fluid flow.

Part II: Structure Systems progresses into the realms of multiscale fracture analysis within fault damage zones and lithologically controlled fault networks. This part not only uncovers the geological and seismic attributes that delineate fault networks and their control on hydrocarbon accumulation but also delves into the integrity of caprock in carbonate and evaporite settings, crucial for sealing performance in reservoirs.

Part III: New Technologies and New Applications introduces forward modeling for characterizing fault damage zones, discusses the elastic characteristics of these zones within superdeep carbonates, and ventures into the promising field of artificial intelligence (AI) for 3D seismic fault analysis. This final part culminates with innovative research on new mechanisms for understanding earthquake distribution and fault patterns, indicating the forward path for seismic research and its application in understanding geohazards. This part also highlights the development of an AI model for 3D seismic fault analysis, including multi-task learning frameworks that significantly enhance fault detection and characterization capabilities. This part presents a forward-looking perspective on leveraging computational advancements to refine our understanding of fault mechanisms, earthquake distributions, and fault patterns.

Throughout *Applied Seismic Geostructures*, didactic elements such as comprehensive methodologies, seismic attribute analysis, AI methods, and case studies are strategically employed to foster a bridge between theoretical concepts and their practical applications. These educational components are pivotal for empowering readers—ranging from geologists and seismologists to petroleum engineers and students—with the knowledge and tools necessary for advancing their research and professional endeavors within the realm of seismic geostructures.

Contents

About the Author

Dr. Zonghu Liao Professor of Geology and Geophysics at the China University of Petroleum, Beijing, is renowned for his pioneering work in seismic interpretation and fault dynamics. Earning his Ph.D. from the University of Oklahoma under Profs. Ze'ev Reches and Kurt J. Marfurt, he has contributed over 35 publications to the advancement of geology and geophysics. His expertise in fault experiments and seismic data interpretation has been showcased in leading journals including *Earth and Planetary Science Letters* (EPSL), *Tectonophysics, Journal of Petroleum Science and Engineering* (JPSE), and *Interpretation.* He has made significant contributions as an associate editor for *Interpretation,* curating special sections on fractures in the subsurface and fault damage zones. Additionally, he serves as a Professor of Finance, underscoring his interest in the development of interpretation methods and their interdisciplinary applications in finance. He has received accolades from the China Ministry of Natural Resources and the China Ministry of Education in recognition of his significant contributions to science and technology, earning him the distinguished title of University Professor. He is dedicated to mentoring the next generation, having advised and mentored over 20 graduate students and numerous undergraduates.

Part I
Fault Damage Zones

Chapter 1
Normal Fault Damage Zones in Eastern China

Abstract We investigate the distribution and thickness of damage-zones for a system of secondary normal faults in the subsurface of the Wangxuzhuang Oilfield, China. Based on seismic variance analysis, we find: (1) Four isolated faults with approximately 2 km length and approximately 200 m damage-zone thickness; (2) The damage-zones of these isolated faults reveal a decaying intensity of deformation from the fault core to the protolith, which fits a power-law form, $y = ax^{-b}$, similar to that observed in the field; (3) A merged fault with approximately 400 m thickness; (4) A bifurcated fault with approximately 400 m thickness and three linked segments. Damage-zones that consist of several subsidiary faults are thicker than those of isolated faults. The displacement-length analyses of the four isolated faults suggest the constant-length growth of the limestone in this case. We demonstrate the potential to apply seismic variance to systematically characterize damage-zones as potential fluid migration conduits on a basin scale.

1.1 Introduction

Fault zones are commonly described to consist of a fault core and surrounding damage-zone, which differs structurally, mechanically, and hydrologically from the undeformed host rock (protolith) (Caine et al., 1996; Chester & Logan, 1986; Katz et al., 2003; Sagy et al., 2001; Savage & Brodsky, 2011). The damage-zone, with structures that formed as a result of the faulting process, likely consists of complex fracture networks, owing to fault-related diagenesis, segmentation, and evolution (Laubach et al., 2014). With limited data and access, it is challenging to evaluate the structure of damage-zones in the subsurface and to determine how the damage-zone affects the migration, accumulation, and leakage of subsurface fluids (e.g., Busetti et al., 2012; Caine et al., 1996; Ellis et al., 2012; Faulkner et al., 2010).

Previous quantitative methods for characterizing fault damage-zones have been based on measured fracture density, seismic event data, and geophysical logging data. (1) Using outcrop measurements to map the fracture density as a function of distance from the fault core. The damage-zone thickness of the fault is determined

Z. Liao, *Applied Seismic Geostructures*, https://doi.org/10.1007/978-981-97-7079-3_1

by determining the distance at which the damage falls below the background level on both sides of the fault (e.g. Mitchell & Faulkner, 2009; Vermilye & Scholz, 1998). (2) Via the analysis of the density and distribution of seismic events recorded during the 1992 Landers earthquake in California, the damage-zone width was roughly estimated (Peng et al., 2003). (3) Zoback et al. (2011) interpreted structural characteristics of the San Andreas fault zone using borehole logging methods. The damage-zone of the fault was identified by identifying a reduction in Vp and Vs. Although these three methods are widely accepted, particularly for isolated fault strands, it lacks feasibility and proper resolution for characterizing faults in the subsurface on a basin-scale. Such faults have been shown to significantly impact reservoir connectivity and petroleum migration (e.g., Faulkner et al., 2010; Rossetti et al., 2010; Yan et al., 2016).

With the advantage of seismic surveys, the detection of fault zones in the subsurface can be done indirectly by using seismic analyses, including seismic simulation (Botter et al., 2016) and seismic attributes (Iacopini and Butler, 2011). Seismic attributes such as azimuth, dip, and coherence, can not only be used to enhance the recognition of a fault network but also to characterize fault zone architecture and properties (Botter et al., 2016; Chopra & Marfurt, 2007; Liao et al., 2017). The attribute of seismic coherence measures changes in the waveform from the seismic reflection and provides a potential quantitative measure of damage-zone thickness. This approach has been used to analyze the wide damage-zone of a 20-km strike-slip fault in Oklahoma's subsurface (Liao et al., 2019). It has been demonstrated that this approach can reveal the dimensions and shapes of damage-zones with some indications of the intensity of its deformation. However, the seismic expression of fault damage is usually combined with seismic noise and unknown complex features from the fault itself. With the exception of a few cases, it is challenging to obtain a realistic, conceptual understanding of the fault system in the subsurface. In this study, we apply the seismic attribute of variance (Liao et al., 2017; Pigott et al., 2013) to describe a system of small-size secondary normal faults in the Wangxuzhuang Basin, China.

1.2 Geological Setting

The Wangxuzhuang oilfield is located in the western part of the Bohai Bay Basin (Fig. 1.1a; Jiang et al., 2019), in eastern China. It is a Cenozoic lacustrine basin formed upon the basement of the North China Craton. Our analyses utilize a three-dimensional seismic survey from the Wangxuzhuang Oilfield, covering the southern portion of the Nandagang fault zone (Fig. 1.1b). The basin has experienced an active tectonic history, including: East–West striking granitic intrusions into the upper block of the Nandagang fault in the middle and late Proterozoic, and the development of a large number of North-East subsidiary faults due to the formation of the Luxi Uplift and Taihang Mountains in the Cenozoic and Late Jurassic. The strata revealed by drilling of the Wangxuzhuang Oilfield from top to bottom are: The Quaternary plain group, the Neocene (Minghuazhen and Guantao Formations), the Paleogene (Dongying and Shahejie Formations), and the Mesozoic formations (Fig. 1.1c). The Shahejie Formation is one of the primary hydrocarbon-bearing strata in this area. The lower Shahejie Formation (first member) is about 200–600 m thick and consists of micritic limestone, calcareous shale, and argillaceous dolomite, which is different from the middle and upper members that are composed entirely of sandstone and mudstone. This study investigates the normal faults developed within the lower member of Shahejie Formation at about 2–2.5 km in depth.

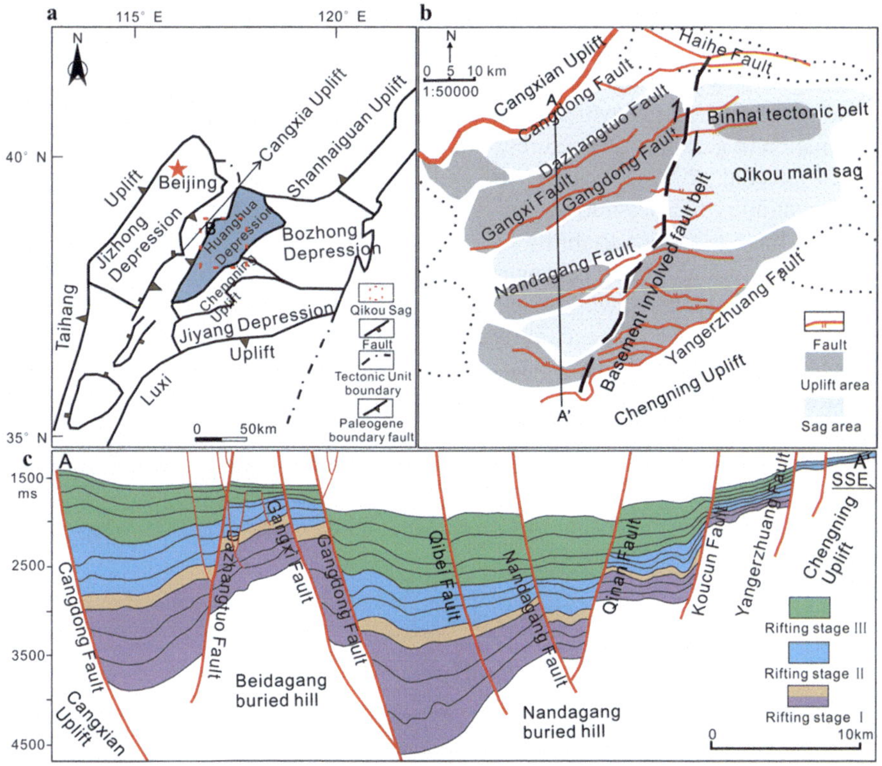

Fig. 1.1 **a** Location of study area in Huanghua Depression of eastern China (*red-dashed rectangle*), **b** Nadagang uplift and the Nandagang fault within, **c** the Shahejie formation (*lower green*). Red lines are generalized faults, and not all faults are shown. Three main rifting stages are presented (adapted from Jiang et al., 2019)

1.3 Methodology: Fault Damage-Zone and Seismic Variance

A fault damage-zone is composed of deformed rocks with reduced acoustic impedance compared to the undeformed host rock (protolith) (Caine et al., 1996; Chester & Logan, 1986). Permeability within the damage-zone is intrinsically determined by the architecture and lithology of the host rock, the displacement along the fault, and any diagenetic reactions that occurred within it (Billi et al., 2003; Caine et al., 1996). The distribution of damage around a fault, and thus the damage zone itself, is often asymmetrical, e.g., the hanging-wall damage-zone of normal faults (Fig. 1.2) tends to be larger than the footwall damage-zone. To detect the geometry of damage-zones in the subsurface, we used the attribute of seismic variance with a window of −5 to +5 ms, which integrates the cumulative seismic effects caused by deformation (e.g., faults or fractures; Chopra & Marfurt, 2007; Iacopini and Butler, 2011; Liao et al., 2019). Variance (the opposite of coherency) is a measurement of the discontinuity of a seismic section. It's calculated in three dimensions and represents the trace-to-trace variability over a particular sample interval and therefore produces interpretable lateral changes in acoustic impedance. Similar traces produce low variance coefficients, while discontinuities have high variance coefficients (fractures/faults). The area of increased variance near the fault is regarded as the fault damage-zone for the purpose of this study. As an analog to the intensity of fractures and structural discontinuities measured in outcrop, the high variance zone is interpreted as the fault core, the one-order wider zone of intermediate variance level is the damage-zone, and the low variance zone is the country rock (protolith zone) away from the fault. This approach (Chap. 3 of this book) has been validated to conform with observations in field outcrops (e.g., Mitchell & Faulkner, 2009; Sagy et al., 2001; Savage & Brodsky, 2011).

1.4 Attribute Results and Analysis

1.4.1 Distribution of Damage-Zones

The basin in this study is comprised of a system of normal faults located on the upper block of Nandagang fault (Fig. 1.3). As observed in core samples and image logs, the distribution of fracture networks is considered as significant conduits for the migration of fluids on a basin-scale (Yan et al., 2016). It is expected that the damage-zones around faults are densely fractured due to the uplift and bending of anticlinal formations (Reches, 1988; Staples, 2011) and associated stress localization around the secondary normal faults. We use the seismic attribute of variance, calculated from the three-dimensional seismic reflection data to image damage-zone widths, associated with faults in the Shahejie Formation.

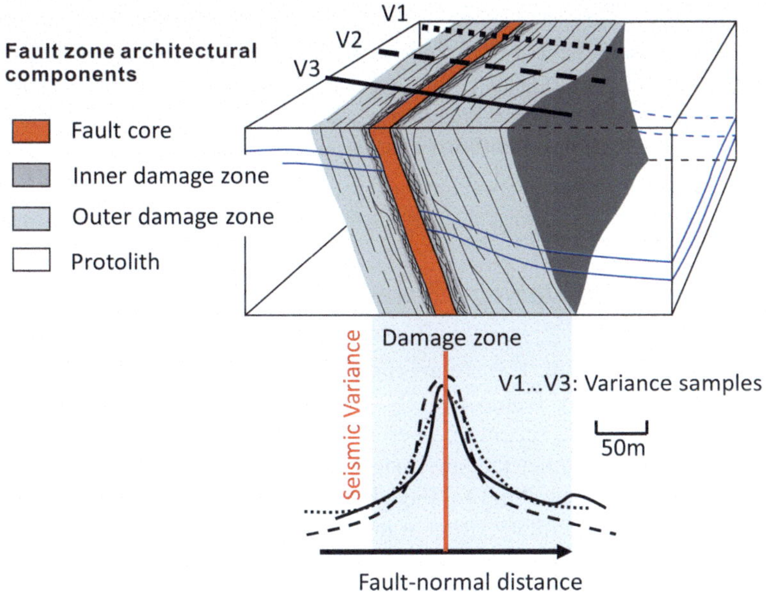

Fig. 1.2 Conceptual models for the damage-zone of a normal fault. Upper: Schematic diagram showing the fault-zone architectural components. Red indicates fault core, dark grey for inner damage-zone, grey for outer damage-zone, and the protolith is removed; Seismic variance profiles are show perpendicular to the fault. Lower: Seismic variance profiles versus fault distance corresponding to the schematic fault zone

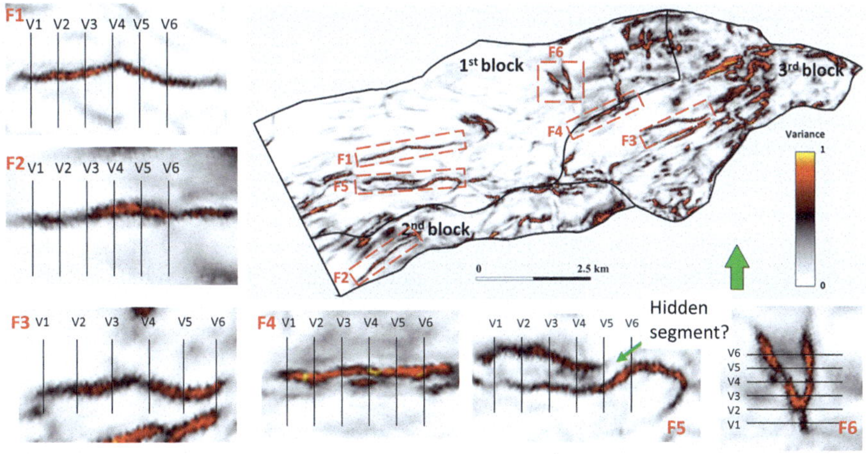

Fig. 1.3 Map of the calculated variance of the three-dimensional seismic survey in the lower member of Shahejie Formation. (Nandagang uplift area in Fig. 1.1b). This map is on the upper block of Nandagang fault. Strips of variance attribute within red boxes are secondary faults with zoom-in maps. F1: Fault 1; F2: Fault 2; F3: Fault 3; F4: Fault 4; F5: Fault 5; F6: Fault 6

Figure 1.3 shows the variance image computed from a patented algorithm. Significant observations include: (1) Distinguishable strips of high variance values in red–black, approximately trending east–west, parallel to the controlling Nandagang fault; (2) A few red–black strips trending north–south in the 1st block; (3) Red–black patches scattering in the 2nd and 3rd blocks; and (4) Large areas of low variance values in white–grey distributed throughout the map.

1.4.2 Thickness of Damage-Zones

The red–black patches (Fig. 1.3) are interpreted as structural discontinuities and are not the focus of this study. Comparatively, the white–grey area is relatively continuous, but with changes is seismic properties. The red–black strips are interpreted as normal fault damage-zones that are systematically formed due to extension associated with Nandagang anticlinal folding (Fig. 1.1b) during the Cenozoic and Late Jurassic. We picked six primary strips of normal faults (red–black stripes; Fig. 1.3) in this study. For each strip, we prepared 6 vertical profiles of the seismic variance. The six profiles in each section display distribution of the seismic variance that we generalized as the characteristics of normal fault damage-zone. These sections reveal two groups of damage-zones:

(1) Isolated normal faults: a typical damage-zone with a well-developed single, central, fault core, represented by a zone of high variance, >0.6, and two zones of intermediate variance, 0.2–0.6 (grey in Fig. 1.4), that decay to the background variance, 0–0.2 (Fig. 1.4). The thicknesses of these damage-zones range from 100 to 200 m. The fault lengths range from 1 to 3 km. The seismic variance pattern of faults 1–3 appears slightly asymmetrically, while fault 4 presents a strong asymmetrical shape.
(2) Bifurcated faults: larger damage-zones with a banded section of approximately 400–450 m thickness (Fig. 1.4e and Fig. 1.4f). Fault 5 is approximately 3 km long and accompanied by a 1.5 km sub-fault, and Fault 6 is composed of two approximately 1.3 km sub-faults (trending north–south). These faults include different linked segments and several fault cores to be discussed later.

1.4.3 Synthesis of Isolated Faults

Previous studies of fault damage-zones have used different methods to characterize fracture distribution for exposed faults. The mathematical modeling of fracture intensity decay with respect to off fault distance yields model fits that include lognormal, power-law, gamma, and exponential laws, among which the power-law distribution is the preferred model for damage-zones (e.g., Liao et al., 2019; Savage & Brodsky, 2011). Figure 1.5 shows seismic variance (deformation) as a function of distance from the fault core for the aforementioned faults. We found that the variance value

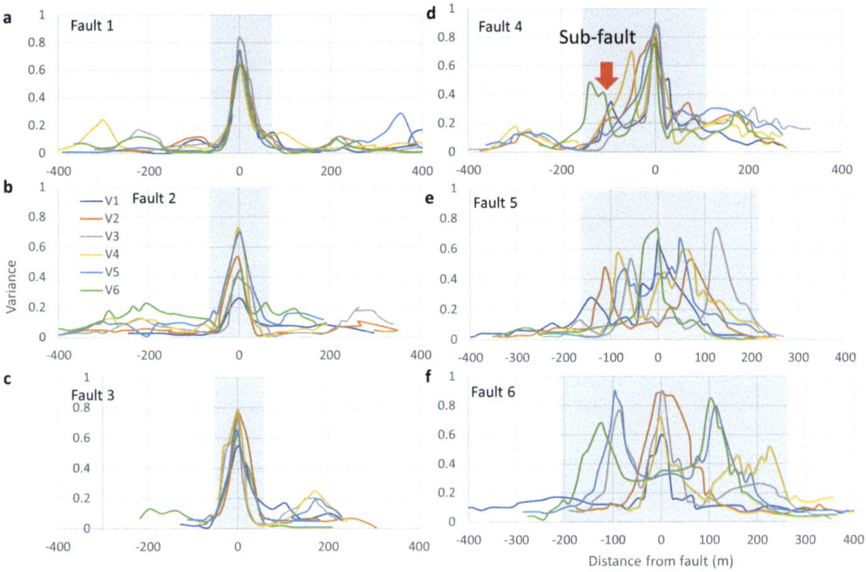

Fig. 1.4 Profiles of the seismic variance values across the damage-zones of selected normal faults; profiles locations in Fig. 1.3; variance values above background are interpreted as damage-zones. Isolated faults: F1, F2, F3, F4; Bifurcated faults: F5, F6. Note the variance asymmetry of Fault 4 that is associated with a subsidiary fault

decay is well fit by the power-law function: $y = ax^{-b}$, where y is variance values, x is the distance from the fault core, b is an exponent describing the decay, and a is a constant that is fault-specific. a and b are two constants that may reflect the physical properties related to the layer thickness or brittleness of the rock with slight differences. Such typical characteristics of the dimensions and shapes of the deformation zones in Fig. 1.5a–c (Faults 1–3) are observed in the field cases (Mitchell & Faulkner, 2009; Sagy et al., 2001; Vermilye & Scholz, 1998). For high values of seismic variance in the fault core, the damage is likely related to intense damage in the form of gouge zones and pulverized rock in the fault-core (e.g., Marone & Scholz, 1989). At great distances from the fault, the damage and fractures decrease to merge with the protolith values (Liao et al., 2019).

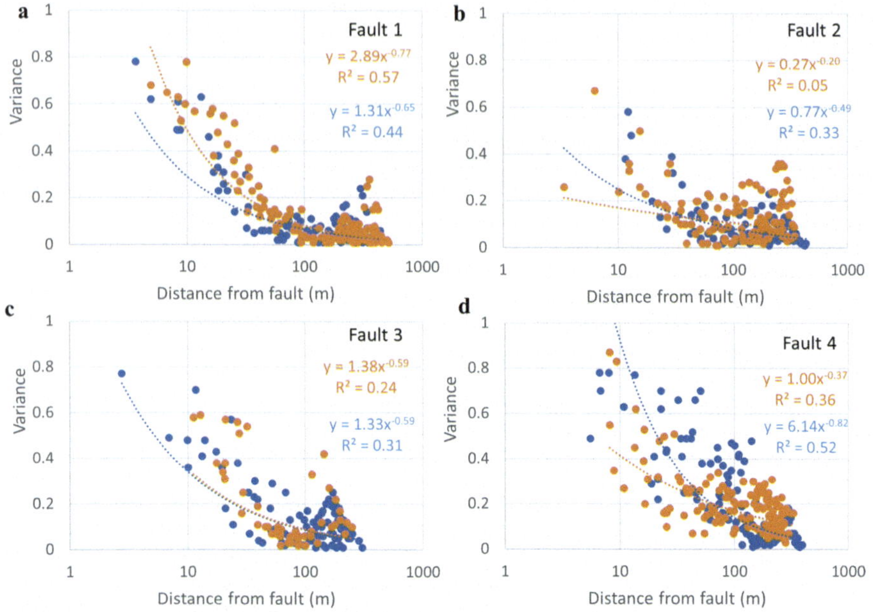

Fig. 1.5 Seismic variance as a function of fault distance from the fault. All data (**a**, fault F1; **b**, fault F2; **c**, fault F3; **d**, fault F4) are well fit by the model $y = ax^{-b}$ where coefficients a and b are fault-related. Orange is the hanging wall, blue is the footwall

1.5 Discussions

1.5.1 Fault Asymmetry and Bifurcation

Fault asymmetry is commonly observed in the field. Berg and Skar (2005) used outcrop data to examine the spatial arrangement of fractures in the damage-zones of a segment of the large-scale Moab Fault (45 km in length), SE Utah. They found that the hanging wall damage-zone is more than three times wider than the footwall damage-zone, and there was a higher abundance of antithetic fractures and deformation bands in the hanging wall than in the footwall. We do not observe such block differences for Faults 1–3 in this study. The slight asymmetry might be associated with slight preferential rupture propagation that is commonly assumed along normal faults (Ben-Zion & Shi, 2005; Ampuero and Mao, 2017).

Another possibility is that fault asymmetry is a consequence of structures associated with multiple secondary faults (Aydin & Johnson, 1978). Such secondary discontinuities impact the seismic variance values, as observed in the aforementioned faults. As such, the more damaged blocks are expected to have more secondary structures. It is suggested that the damage-zones of Faults 4–6 are formed by the superposition of several sub-faults.

We examine each profile of the Faults 5 and 6 that clearly present systems of fault bifurcations relating to such sub-faults. Two simple, idealized systems are discussed here: a merged system and a bifurcated system. For both systems, each profile of seismic variance perpendicular to the fault shows the internal structural components at a specific location. The interpreted damage-zones in Fault 5 is shown by two connected red lines in Fig. 1.6a. From V1–V5, one small damage-zone (left one with lower variance values) and a large one (right one with a higher value) are recognizable with two peaks of seismic variances. These two peaks merged with each other in profile V6. The variance values between the two peaks illustrate an overprinted increase, which reveals the increasing deformation of the mixed zone heterogeneity when the two faults merge. The profile of V6 becomes a typical peak of seismic variance for an isolated fault. For the missing variance strip in Fig. 1.3 (green arrow in Fault 5), we proposed a hidden (sub-seismic) fault segment with little displacement between profiles of V5 and V6. In the case of little fault displacement, the seismic attribute of variance is not sensitive enough to detect the discontinuity.

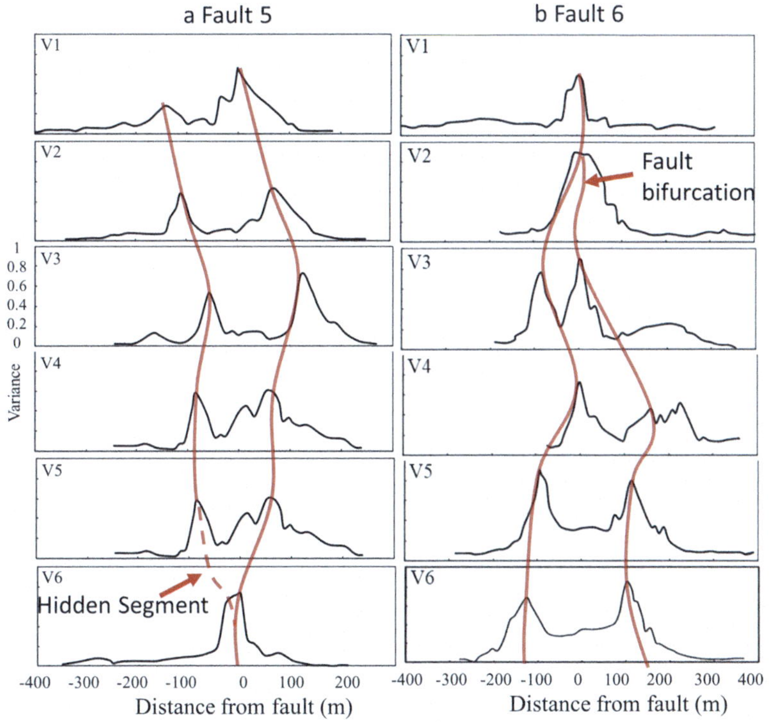

Fig. 1.6 Attribute variations of damage-zones along with strike intervals. **a** A merged fault (Fault 5) with one hidden segment, **b** A bifurcated fault (Fault 6) with two sub-segments. Note the variance value between the pulses is higher than background values, which indicates the superposition effect of the two sub-segments

Fault 6 in Fig. 1.3 is treated as a bifurcated system. Profile of V2 in Fig. 1.6b illustrates the exaggeration of seismic variances before bifurcated into two faults in the profile of V3. The seismic variances express the interacting deformation of the two damage-zones of sub-faults. The damage-zones thickness is about 400 m composed of overprinted bifurcated-segments. This cumulative effect is expected to disappear if the spaced distance between the two bifurcated-segments is larger than the thickness of an isolated fault.

1.5.2 Fault Length and Fault Displacement

To investigate the relationship between the total fault length and displacement, we combined our results with published displacement-length (D-L) profiles (Fig. 1.7). Assuming the limestone velocity is 3.5 km/s, we calculated the displacements for Faults 1–4 by multiplying time differences between fault blocks from the 3-D seismic data (Table 1.1). Figure 1.7 shows that the amount of displacement on a fault is proportional to the fault length. Our data (green dots) of these four faults fit well into such a general relation; however, the displacements increase at a faster rate in this study. Jackson and Rotevatn (2013) proposed two stages of D-L growth paths for normal faults: The first stage is lengthening rapidly with various displacements, e.g., tip propagation, relay formation, and breakdown, segment linkage etc.; The second stage is large displacement without much fault lengthening, e.g., sub-vertical constant-length growth. In this study, it is likely that the growth of normal faults in limestone demonstrates characteristics mostly in the second stage. Additionally, most normal faults in this study grow with damage-zones of approximately 200 m width, irrespective of the fault displacement and length, which we tentatively argue as potentially resultant of the systematical and abrupt deformation and the brittleness of limestone.

1.5.3 Reservoir Compartmentalization and Fluid Flow

Figure 1.3 demonstrates that the distribution of deformation (red strips and patches) in limestone is highly heterogeneous spatially. The fault damage-zones (red strips) are expected to provide high fault-parallel permeability pathways to transport fluids. Nevertheless, most hydrocarbon traps in this study have been found near the damage-zones. Estimation of the fault-zone permeability is difficult as faults are usually avoided by borehole drilling. The observation of tracer trajectories in boreholes indicates the significant channeling of flow within the damage-zones along the fault strike. The favored flow direction can be explained by the macro-scale fracture networks within the damage-zones. We inferred the variation of permeability exists horizontally as we found fault-bounded hydrocarbon traps. The fault sealing is due to gouge accumulation in the fault cores, which may lead to the compartmentalization of

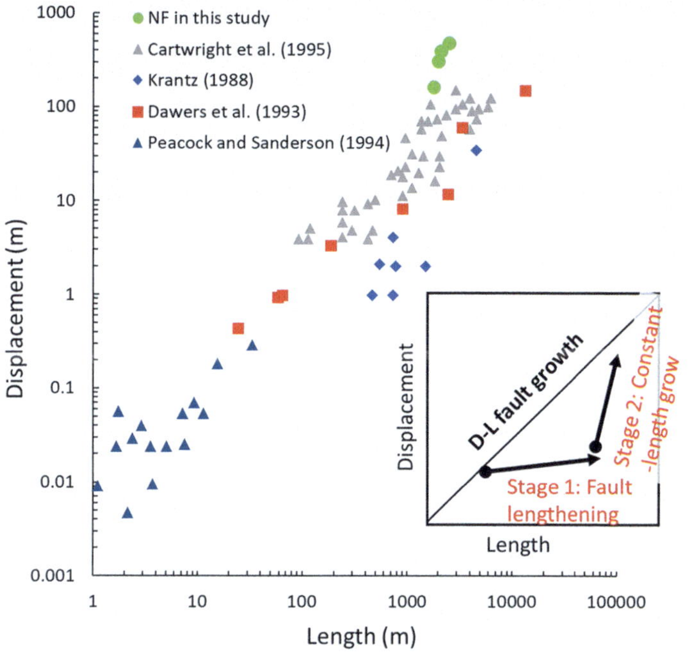

Fig. 1.7 Displacement versus length for normal faults plotted in log–log scales (NF: Normal fault; NF data extracted from Cartwright et al., 1995; Dawers et al., 1993; Krantz, 1988; Peacock & Sanderson, 1994). Inset is a schematic model for the two stages of fault growth. Stage 1: Fault lengthening with various displacement; Stage 2: Constant-length growth

Table 1.1 Parameters of fault geometries

	Displacement in time (ms)	Length (km)	Hanging wall		Footwall	
			a	b	a	B
Fault 1	131	2.5	2.89	0.77	1.31	0.65
Fault 2	104	2.1	0.27	0.2	0.77	0.49
Fault 3	35	1.8	1.38	0.59	1.33	0.59
Fault 4	65	2.0	1	0.37	6.41	0.82

the basin. Thus, the spatial distribution of the fault system is a highly interrelated approach to understand the fluid flow in the subsurface.

1.6 Conclusion

In this study, we display an approach to apply seismic variance to characterize the structure and distribution of a system of normal faults in eastern China. We have demonstrated characteristics of the damage-zones and within limestone in the subsurface, and recognize 4 isolated faults of approximately 200 m damage-zone thickness and two bifurcated faults of approximately 400 m thickness. We infer the overprinting of faults and rupture preference leads to the structure asymmetry. This approach and data present potential for evaluating fault damage-zones on a regional scale that impacts the migration and accumulations of petroleum.

References

Ampuero, J. P., and X. Mao (2017) Upper limit on damage zone thickness controlled by seismogenic depth. In: M. Y. Thomas, T. M. Mitchell, & H. S. Bhat (Eds.) Fault zone dynamic processes: Evolution of fault properties during seismic rupture (1st edn, pp. 243–253). Wiley, New York

Aydin, A., & Johnson, A. M. (1978). Development of faults as zones of deformation bands and as slip surfaces in sandstone. *Pure and Applied Geophysics, 116*(4–5), 931–942. https://doi.org/10.1007/BF00876547

Ben-Zion, Y., & Shi, Z. (2005). Dynamic rupture on a material interface with spontaneous generation of plastic strain in the bulk. *Earth and Planetary Science Letters, 236*(1–2), 486–496. https://doi.org/10.1016/j.epsl.2005.03.025

Berg, S. S., & Skar, T. (2005). Controls on damage zone asymmetry of a normal fault zone: Outcrop analyses of a segment of the Moab fault, SE Utah. *Journal of Structural Geology, 27*(10), 1803–1822. https://doi.org/10.1016/j.jsg.2005.04.012

Billi, A., Salvini, F., & Storti, F. (2003). The damage zone-fault core transition in carbonate rocks: Implications for fault growth, structure and permeability. *Journal of Structural Geology, 25*(11), 1779–1794. https://doi.org/10.1016/S0191-8141(03)00037-3

Botter, C., Cardozo, N., Hardy, S., Lecomte, I., Paton, G., & Escalona, A. (2016). Seismic characterisation of fault damage in 3D using mechanical and seismic modelling. *Marine and Petroleum Geology, 77*, 973–990. https://doi.org/10.1016/j.marpetgeo.2016.08.002

Busetti, S., Mish, K., Hennings, P., & Reches, Z. E. (2012). Damage and plastic deformation of reservoir rocks: Part 2. Propagation of a hydraulic fracture. *AAPG Bulletin, 96*(9), 1711–1732. https://doi.org/10.1306/02011211011

Caine, J. S., Evans, J. P., & Forster, C. B. (1996). Fault zone architecture and permeability structure. *Geology, 24*(11), 1025–1028. https://doi.org/10.1130/0091-7613(1996)024%3c1025:FZAAPS%3e2.3.CO;2

Cartwright, J. A., Trudgill, B. D., & Mansfield, C. S. (1995). Fault growth by segment linkage: An explanation for scatter in maximum displacement and trace length data from the Canyonlands Grabens of SE Utah. *Journal of Structural Geology, 17*(9), 1319–1326. https://doi.org/10.1016/0191-8141(95)00033-A

Chester, F. M., & Logan, J. M. (1986). Implications for mechanical properties of brittle faults from observations of the Punchbowl fault zone, California. *Pure and Applied Geophysics, 124*, 79–106. https://doi.org/10.1007/BF00875720

Chopra, S., & Marfurt, K. J. (2007). *Seismic attributes for prospect identification and reservoir characterization*. SEG publisher.

Dawers, N. H., Anders, M. H., & Scholz, C. H. (1993). Growth of normal faults: Displacement-length scaling. *Geology, 21*(12), 1107–1110. https://doi.org/10.1130/0091-7613(1993)021%3c1107:GONFDL%3e2.3.CO;2

Ellis, M. A., Laubach, S. E., Eichhubl, P., Olson, J. E., & Hargrove, P. (2012). Fracture development and diagenesis of Torridon Group Applecross Formation, near An Teallach, NW Scotland: Millennia of brittle deformation resilience? *Journal of the Geological Society, 169*(3), 297–310. https://doi.org/10.1144/0016-76492011-086

Faulkner, D. R., Jackson, C. A. L., Lunn, R. J., Schlische, R. W., Shipton, Z. K., Wibberley, C. A. J., & Withjack, M. O. (2010). A review of recent developments concerning the structure, mechanics and fluid flow properties of fault zones. *Journal of Structural Geology, 32*(11), 1557–1575. https://doi.org/10.1016/j.jsg.2010.06.009

Iacopini, D., & Butler, R. W. H. (2011). Imaging deformation in submarine thrust belts using seismic attributes. *Earth and Planetary Science Letters, 302*, 414–422. https://doi.org/10.1016/j.epsl.2010.12.041

Jackson, C. A.-L., & Rotevatn, A. (2013). 3D seismic analysis of the structure and evolution of a salt-influenced normal fault zone: A test of competing fault growth models. *Journal of Structural Geology, 54*, 215–234. https://doi.org/10.1016/j.jsg.2013.06.012

Jiang, S., Wang, W., Zhang, A., & Zhou, W. (2019). Genetic mechanism and evolution of the covert fault zone and its oil-controlling mode in Qikou Sag, eastern China. *Energies, 12*(1), 98. https://doi.org/10.3390/en12010098

Katz, O., Reches, Z., & Baer, G. (2003). Faults and their associated host rock deformation: Part I. Structure of small faults in a quartz-syenite body, southern Israel. *Journal of Structural Geology, 25*, 1675–1689. https://doi.org/10.1016/S0191-8141(03)00011-7

Krantz, R. W. (1988). Multiple fault sets and three-dimensional strain: Theory and application. *Journal of Structural Geology, 10*, 225–237. https://doi.org/10.1016/0191-8141(88)90056-9

Laubach, S. E., Eichhubl, P., Hargrove, P., Ellis, M. A., & Hooker, J. N. (2014). Fault core and damage-zone fracture attributes vary along strike owing to interaction of fracture growth, quartz accumulation, and differing sandstone composition. *Journal of Structural Geology, 68*, 207–226. https://doi.org/10.1016/j.jsg.2014.08.007

Liao, Z., Liu, H., Carpenter, B. M., Marfurt, K. J., & Reches, Z. (2019). Analysis of fault damage-zones by using 3D seismic coherence in Anadarko Basin, Oklahoma. *AAPG Bulletin, 103*(8), 1771–1785. https://doi.org/10.1306/1219181413417207

Liao, Z., Liu, H., Jiang, Z., Marfurt, K. J., & Reches, Z. (2017). Fault damage zone at subsurface: A case study using 3D seismic attributes and a clay model analogue for the Anadarko Basin, Oklahoma. *Interpretation, 5*(2), T143–T150. https://doi.org/10.1190/INT-2016-0033.1

Marone, C., & Scholz, C. H. (1989). Particle-size distribution and microstructure within simulated fault gouge. *Journal of Structural Geology, 11*(7), 799–814. https://doi.org/10.1016/0191-8141(89)90099-0

Mitchell, T. M., & Faulkner, D. R. (2009). The nature and origin of off-fault damage surrounding strike-slip fault zones with a wide range of displacements: A field study from the Atacama fault system, northern Chile. *Journal of Structural Geology, 31*, 802–816. https://doi.org/10.1016/j.jsg.2009.05.002

Peacock, D. C. P., & Sanderson, D. J. (1994). Geometry and development of relay ramps in normal fault systems. *AAPG Bulletin, 78*, 147–165. https://doi.org/10.1306/BDFF9046-1718-11D7-8645000102C1865D

Peng, Z., Ben-Zion, Y., Michael, A. J., & Zhu, L. (2003). Quantitative analysis of seismic fault zone waves in the rupture zone of the 1992 Landers, California, earthquake: Evidence for a shallow trapping structure. *Geophysical Journal International, 155*, 1021–1041. https://doi.org/10.1111/j.1365-246X.2003.02109.x

Pigott, J. D., Kang, M. H., & Han, H. C. (2013). First-order seismic attributes for clastic seismic facies interpretation: Examples from the East China Sea. *Journal of Asian Earth Sciences, 66*, 34–54. https://doi.org/10.1016/j.jseaes.2012.11.043

Reches, Z. (1988). Evolution of fault patterns in clay experiments. *Tectonophysics, 145*(1–2), 141–156. https://doi.org/10.1016/0040-1951(88)90322-8

Rossetti, F., Aldega, L., Trcce, F., Balsamo, F., Billi, A., & Brilli, M. (2010). Fluid flow within the damage zone of the Boccheggiano extensional fault (Larderello–Travale geothermal field, central Italy): Structures, alteration and implications for hydrothermal mineralization in extensional settings. *Geological Magazine, 148*(4), 558–579. https://doi.org/10.1017/S001675681000097X

Sagy, A., Reches, Z., & Roman, I. (2001). Dynamic fracturing: Field and experimental observations. *Journal of Structural Geology, 23*, 1223–1239. https://doi.org/10.1016/S0191-8141(00)00190-5

Savage, H. M., & Brodsky, E. E. (2011). Collateral damage: Evolution with displacement of fracture distribution and secondary fault strands in fault damage-zones. *Journal of Geophysical Research, 116*(B03405), 428–452. https://doi.org/10.1029/2010JB007665

Staples, E. (2011). Subsurface and experimental analyses of fractures and curvature. Dissertation, University of Oklahoma.

Vermilye, J. M., & Scholz, C. H. (1998). The process zone: A microstructural view of fault growth. *Journal of Geophysical Research, 103*(B6), 12233–12237. https://doi.org/10.1029/98JB00957

Yan, R., Zeng, L., Zhao, X., Liao, Z., & Chen, M. (2016). Fracture development characteristics and its formation mechanism in Lower Member 1 Shahejie biolithite reservoir, Z oilfield, Bohai Bay. *Chinese Journal of Geology, 51*(2), 484–493.

Zoback, M. D., Hickman, S., Ellsworth, W., et al. (2011). Scientific drilling into the San Andreas Fault Zone: An overview of SAFOD's first 5 years. *Scientific Drilling, 11*(14–28), 2011. https://doi.org/10.2204/iodp.sd.11.02.2011

Chapter 2
Thrust Fault Damage Zones in Eastern Longmen Shan Range, China

Abstract We analyze here the subsurface damage zone of a thrust fault, named the Pengxian fault (PXF), at the eastern margin of the Longmen Shan Range, Sichuan, China. By using a 3-D seismic reflection survey and the seismic attributes, we find: (1) The PXF is ~55 km long, intra-basinal, frontal thrust fault that dips 40°–50° with a displacement of 100–800 m; (2) The thickness of the composite damage zones varies along the fault, and it ranges from 1 km to more than 2 km; (3) The PXF is segmented, and its damage zones terminate against the Triassic detachment zone from above and below. The thickness of PXF damage zones and their internal structure are interrelated to the local displacement and the regional carbonate lithology. We argue that the observed composite damage zones increased in thickness during the evolution from the fault's initial stage to its current mature state. This study presents the methodology for identifying the fault damage zones in carbonate sequences.

2.1 Introduction

Field analyses of fault zones recognized three primary structural components: fault core, damage zone, and protolith (Fig. 2.1; Caine et al., 1996; Katz et al., 2003; Kim et al., 2004; Savage & Brodsky, 2011). The fault core usually comprises single or multiple slip surfaces of fault rocks, e.g., gouge and cataclasite, and it is commonly assumed that most of the fault displacement was accommodated in the fault core (e.g., Chester et al., 1993; Faulkner et al., 2003; Wibberley et al., 2008). The fault core could be a sealing zone with thick clay bodies (e.g., Billi et al., 2003), a permeable conduit (e.g., Caine et al., 1996), or both, depending on the fault's state in its seismic cycle (e.g., Sibson, 1990). The damage zone, which is located between the fault core and the protolith, is constituted of fractured, brecciated, and pulverized rocks (Billi et al., 2003; Caine et al., 1996). The fracture networks within the damage zone and their structural complexity can significantly affect the migration, accumulation, and leakage of subsurface fluids (Caine et al., 1996; Faulkner et al., 2010) and earthquake rupture characteristics (e.g., Liu et al., 2003; Savage et al., 2017; Weng et al., 2016). Therefore, documenting damage zones and their scaling with host fault properties,

Z. Liao, *Applied Seismic Geostructures*, https://doi.org/10.1007/978-981-97-7079-3_2

Fig. 2.1 Large-scale fault zone structure. **a** Panoramic view of a large-scale fault zone within the Triassic limestones in Xuanhan, Sichuan (Photo courtesy of Dr. Chunguo Du), **b** Line drawing of the fault zone shown in Fig. 2.1a

e.g., displacement and length, is central for understanding the mechanical, hydraulic, and seismic behavior of fault systems (Peacock, 2002; Powers & Jordan, 2010). Most analyses of fault damage zones are based on field observation or numerical simulation (Faulkner et al., 2003; Wibberley et al., 2008), but only a few studies were conducted on subsurface systems (e.g., Liao et al., 2017), and even fewer analyses covered subsurface damage zones in seismic active regions (e.g., Liao et al., 2020b).

The present subsurface analysis employs recent advances in acquiring and processing seismic reflection data and the availability of high-quality 3-D seismic surveys. These advances have allowed the study of the geometry of seismic-scale faults in the subsurface (Hale, 2013; Iacopini & Butler, 2011; Liao et al., 2023). We also employed seismic attributes of the seismic signal corresponding to the rock discontinuities (Chopra & Marfurt, 2007; Iacopini & Butler, 2011) to identify and interpret the seismic structure directly. For example, the coherence seismic quantifies changes in the waveforms and thus provides a potential quantitative tool to identify fault damage zones. This approach has been used to analyze the damage zone of a 20 km strike-slip fault in Oklahoma (Liao et al., 2019) and a 55 km thrust fault in the Longmen Shan range (Li et al., 2016; Lu et al., 2019). Although the seismic expression of a fault damage zone may include seismic noise and unknown complex features from the fault zone itself, it has been demonstrated that this approach can reveal the dimensions and shapes of the damage zone (Liao et al., 2023). In this study, we apply the volumetric seismic attributes (e.g., variance, curvature, and dip azimuth) to describe the fault damage zones in a system of Pengxian buried thrust in the Longmen Shan Range abutting the Sichuan Basin (Fig. 2.1).

2.2 Geological Setting

The Longmen Shan range (LMS) is located between the Songpan-Ganze Block and the Sichuan Basin (SCB). The LMS thrust belt comprises a series of NE-trending thrust faults. In general, this thrust range is divided (from west to east) into the Qingchuan-Maowen Fault, Beichuan-Yingxiu Fault, Pengguan Fault, Guankou Fault, and Pengxian Fault (PXF) (Burchfiel et al., 1995; Hubbard et al., 2010; Jia et al., 2006). In terms of seismicity, structural framework, and topographical relief, the LMS thrust belt can be subdivided (from north to south) into three longitudinal regimes with the internal boundaries of Dujiangyan and Anxian: northern, central, and southern (Jin et al., 2010; Liu et al., 2009).

It is generally accepted that since the beginning of the Mesozoic, the LMS developed during two major orogenic periods: a Late Triassic compressional event and a Late Cenozoic phase related to the India-Asia collision (Burchfiel et al., 1995; Jia et al., 2006). The central LMS outcrops are primarily Mesozoic strata (Chen et al., 1995; Liu et al., 1996), and the shallow structural deformation is characterized by imbricate thrusts and fault-related folds (Hubbard & Shaw, 2009; Lu et al., 2014, 2019). The surface geological and drilling cores observations indicate that the shallow sedimentary sequence of central LMS is primarily composed of Sinian-Middle Triassic shallow marine clastic and carbonate rocks and Late Triassic terrestrial clastic rocks (Li et al., 2012; Lu et al., 2019). Toward the east, the sequence is dominated by clastic rocks with a thickness of 5 km or more (Chen et al., 1995). An early Proterozoic crystalline basement lies under this sedimentary cover (Fig. 2.2c), and the SCB has been a foreland basin since the Late Triassic (Burchfiel et al., 1995; Jia et al., 2006).

We focus on the Pengxian area with the Pengxian buried anticline and the PXF is located between the central LMS and the western SCB (Fig. 2.2a, b). The PXF intersects the principal gas-bearing Triassic Leikoupo dolomite at a depth interval of 6.0–6.2 km, and its associated damage zones probably control the high local gas production. This area's subsurface structure was analyzed using a few 2D/3D seismic sections. These analyses led to four contradicting structural models of the PXF system: (i) The PXF as a deep-seated and deep-angle thrust fault that cuts the strata from the pre-Cambrian basement through the Quaternary (Yan et al., 2011); (ii) The PXF as a listric-reverse fault with a steep dip in the shallow strata and a gentle dip in greater depths (Liu et al., 2009); (iii) The PXF is a basement thrust cut through the Paleozoic layer and merged with the Middle-lower Triassic detachment fault, and did not cross the post-Upper Triassic sequence (Jin et al., 2010); and (iv) The PXF is a shallow, steep, thrust fault developed within the Middle-Lower Triassic stratigraphic sequence (Lu et al., 2019).

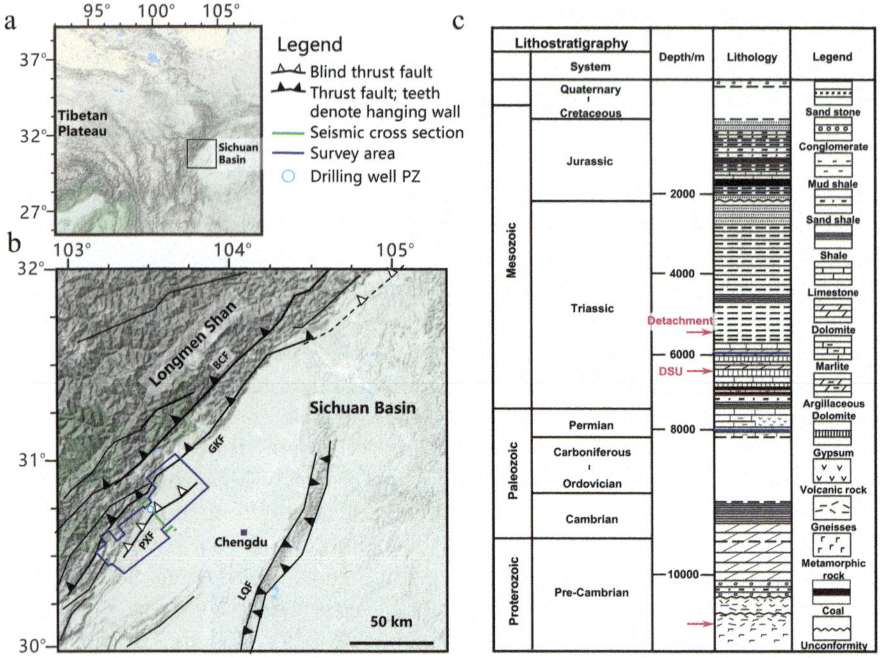

Fig. 2.2 General region of the study area **a** and the tectonic setting of the study area, **b** Liao et al. (2023), Lu et al. (2019) (Modified from). The green line is a seismic cross-section, and the blue polygon is a three-dimensional seismic survey area of 1370 km^2. PXF, Pengxian fault; GKF, Guankou fault; BCF, Beichuan fault; LQF, Longquan fault. **c** Sedimentary stratigraphy of Pengzhou area (modified from). Reddish arrows depict the locations of three detachments. Blue lines show the locations of stratigraphy in Figs. 2.3, 2.4. The velocity model is derived from Li et al. (2016)

2.3 Data and Methods

In the present analysis of the PXF system, we utilize a set of high-resolution 3-D seismic data collected from 2010 to 2012 (Fig. 2.3b). It was processed using post-stack time migration with a dominant frequency of 25 Hz. The seismic survey was recorded with two-way travel time to depth, which covers an area of 1370 km^2.

In the analysis, we employed primarily the volumetric seismic attributes of variance, curvature, and dip azimuth (Chopra & Marfurt, 2007). The variance attribute is based on changes of the wave amplitude, frequency, and phase, that depend on acoustic-impedance contrast (Chopra & Marfurt, 2007). It is defined as:

$$D^2 = \frac{\sum_i^n (x_i - \overline{x})^2}{n} \tag{2.1}$$

In which x indicates the energy value of the simple point, the $\overline{x}$ is the mean energy value of the sample points, and n indicates the number of samples. The variance

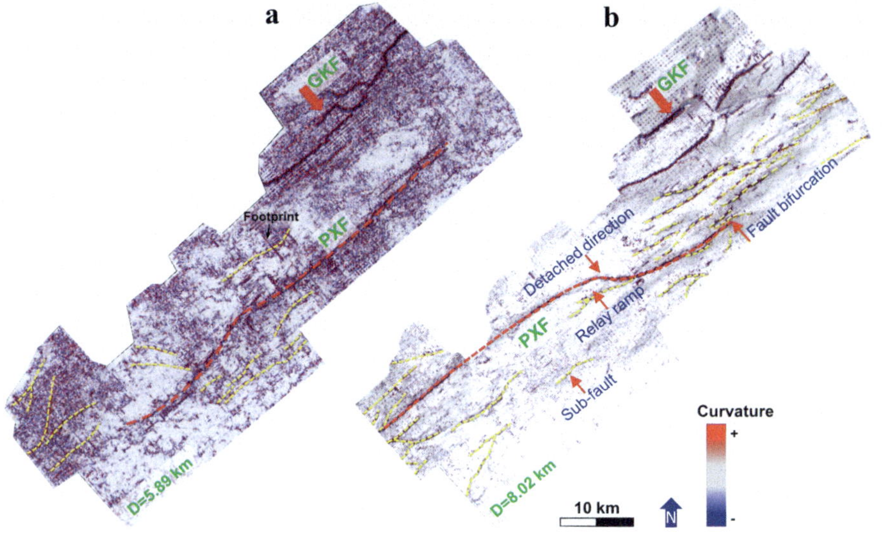

Fig. 2.3 **a** Stratigraphy composed of dolomite above the Triassic detachment, **b** Stratigraphy composed of limestone below the Triassic detachment. The seismic attribute of curvature illustrates PXF and its abutting structures. The locations of the two stratigraphic slices are in Fig. 2.2c. D, Depth

attribute quantifies the discontinuities of waveforms and traces (D, Eq. 2.1), and its high values indicate discontinuities due to strata deformation, e.g., faults, fractures, or flexures.

The curvature attribute is computed as the second derivative of a seismic interface (Eq. 2.2, see Thomas, 1972; pp. 473–475),

$$K = \frac{\frac{d^2y}{d^2x}}{[1 + (\frac{dy}{dx})^2]^{\frac{3}{2}}} \tag{2.2}$$

The local most positive curvatures indicate ridges, and the most negative ones indicate valleys. Dip azimuth is the strike measured toward the maximum downward dip (Chopra & Marfurt, 2007). If the apparent angle on the inline (longitudinal line) is θi and in the direction of the crossline (transverse line) is θc, then θi and θc are related to the dip azimuth (φ) as follows:

$$\varphi = atan(\theta i, \theta c) \tag{2.3}$$

2.4 Results

The tectonics of the study area is dominated by the presence of damaged stratigraphic unit which serves as a regional decollement zone in the Sichuan Basin, as we recently demonstrated (Liao et al., 2023). We documented a 1 km sequence of Triassic evaporites and carbonates at a 5.8–6.8 km depth interval that is pervasively damaged, and which we defined as a Damaged Stratigraphic Unit (DSU). It was found that the large, intra-basin system of PXF terminates from above and below against this relatively weak, damaged DSU (Liao et al., 2023). Based on these observations, we separate here the documentation of the PXF structure into two parts above and below the Middle-Lower Triassic DSU detachment.

2.4.1 Structural Features Above the Detachment

The main features above the detachment include (Fig. 2.3a): (1) the Pengxian fault (PXF) with quasi-linear trace trending 050° and length of ~55 km is located in the center of the study area. The damage zone of this fault is about 2 km thick (small red tangles in Fig. 2.4a). (2) the Guankou fault (GKF) striking 44° is partly located in the northwestern portion of the study area. (3) a few smaller faults distributing parallel or at an acute angle to the PXF with a length range of 3–15 km (Fig. 2.3a, yellow dashed lines). (4) the open Pengxian anticline (interlimb angle of ~162°) with NW limb dipping towards 320°–330°, and SE limb dipping to the southeast, 135°–145° (Fig. 2.4a).

2.4.2 Structural Features Below the Detachment

The main features below the detachment include (Fig. 2.3b): (1) the PXF trend displays a near open "z" shape and it is more than 55 km long. Here, the PXF is displaced about 8 km towards SE, relatively to the same fault above the detachment (Fig. 2.3a). (2) tens of small faults subparallel to the PXF with lengths from a few kilometers to tens of kilometers (Fig. 2.3b, yellow dashed lines). The density of these faults increases towards the northeastern portion of the study area, probably due to lithological heterogeneity (i.e., more brittle materials (shale) occurrence towards the northeastern part, see Fig. 2.2c, and Liao et al., 2023 for details). (3) the PXF damage zones about 1 km thickness (Fig. 2.4b, small red tangles). (4) the Pengxian anticline with the interlimb angle from 165° to 170° (Fig. 2.4b), including a relay ramp structure that can be inferred near the central part of the study area (Fig. 2.3b). Figure 2.4 also shows the shallow Pengxian back thrust, which constrains the formation of the Pengxian anticline combined with the PXF (Lu et al., 2019).

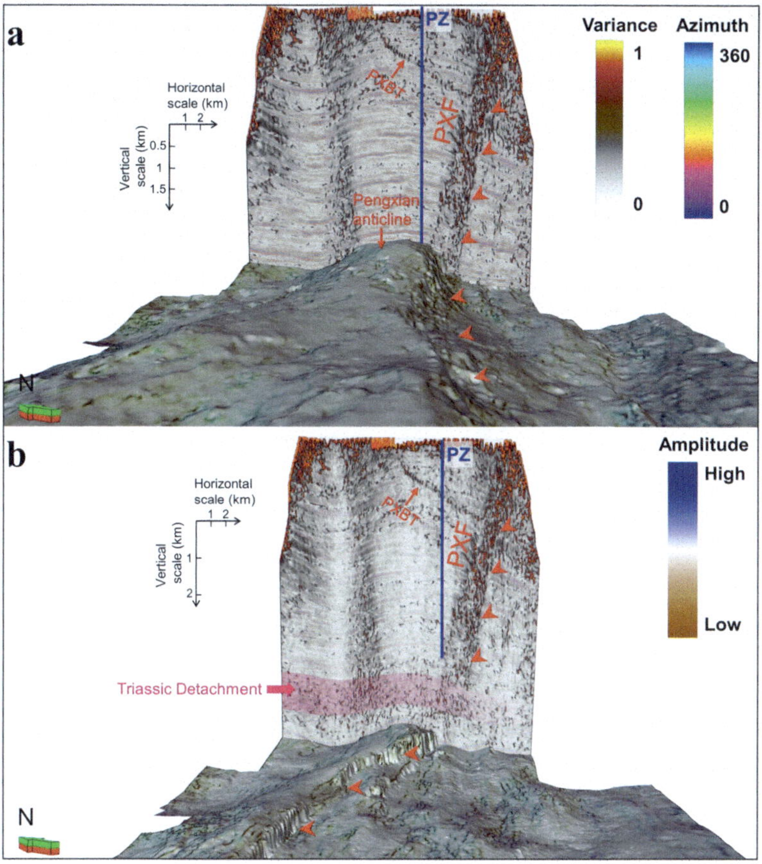

Fig. 2.4 Portions of chair chart consisting of the variance cube (co-rendered with azimuth) and a seismic profile (co-rendered with variance). Note that the small red tangles indicate the trace of PXF, and the red arrows indicate the location of Pengxian back thrust (PXBT)

2.4.3 Seismic Damage Zone Thickness-Displacement Relationships

The dimensions and distribution of the Pengxian fault damage zones can be recognized in maps of the variance attribute (Fig. 2.3). This attribute provides an estimate of the dissimilarity between waveforms or traces in 3-D seismic volume, and a local increase of its intensity near a fault is interpreted as a fault damage zone (e.g., Chopra & Marfurt, 2007; Liao et al., 2017). Following the procedures of Liao et al., (2017, 2019), the intensities of the variance attribute were calculated and plotted along six horizons are plotted in Fig. 2.5a. The six selected horizons, which are at a depth range of 5.70–8.38 km, define the damage zone thicknesses of the PXF.

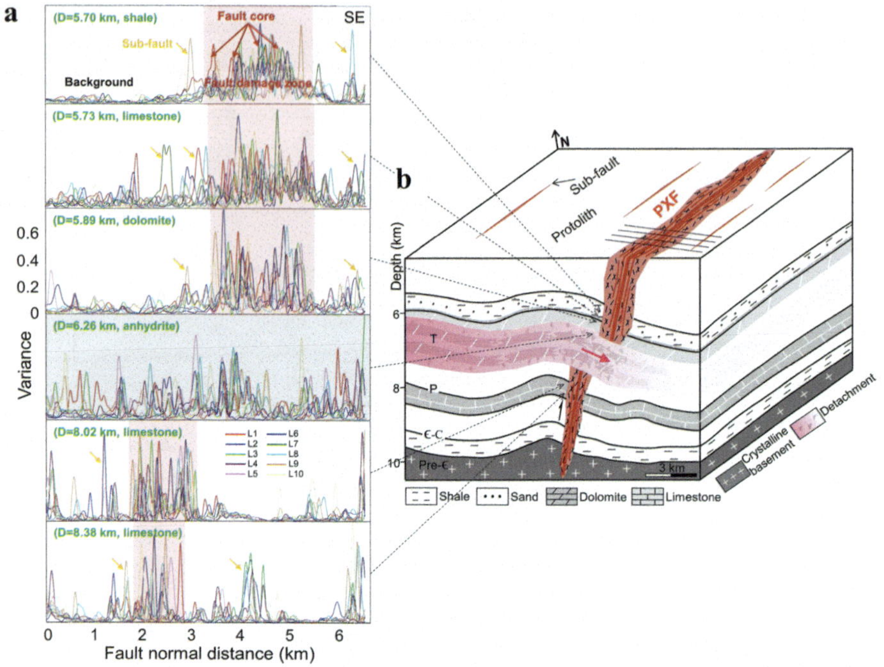

Fig. 2.5 **a** Variance composite damage zone variations of PXF with depth intervals from 5.70 to 8.38 km. The zone of high variance values above background variance indicates the damage zone (*colored in transparent red*). Note that the transparent gray zone indicates an anhydrite layer. Yellow arrows indicate sub-faults. **b** Conceptual model of the architecture of PXF. The arrow in reddish indicates the detached direction

2.5 Discussion

Fault-zones are deformed bodies that commonly include fault-cores of gouge and ultracataclasite (e.g., Caine et al., 1996; Chester et al., 1993), damage-zones of dense networks of fractures (e.g., Billi et al., 2003; Wilson et al., 2003), or distributed drag (e.g., Katz et al., 2003; Reches & Eidelman, 1995). The intensity within the damage zones typically exhibits power-law decay with distance from the fault core to merge with the surrounding protolith (Katz et al., 2003; Kim et al., 2004; Liao et al., 2019; Sagy et al., 2001; Savage & Brodsky, 2011; Wilson et al., 2003). We discuss three aspects of the present documentation of the subsurface damage zones and compare them to field-observed damage zones (Figs. 2.6 and 2.7).

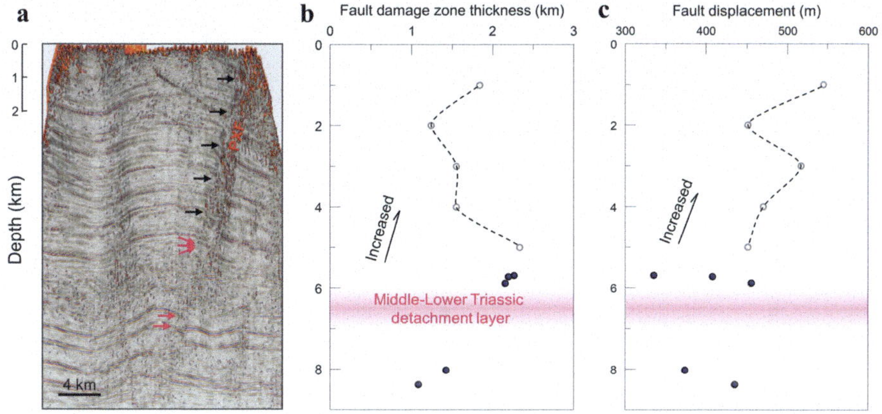

Fig. 2.6 Damage zone structure of the Pengxian fault on a cross-section view. **a** A seismic reflection profile (see Fig. 2.2b) is co-rendered with variance; arrows indicate positions of thickness measurements. **b, c** Damage zone thickness (**b**) and displacements (**c**) as a function of depth; solid black dots- accurate data at red arrows in Fig. 2.6a; open circles-estimated data at black arrows in Fig. 2.6a. Note the position of the Middle-Lower Triassic detachment layer (Liao et al., 2023)

2.5.1 Thickness of Fault Damage-Zones

The thickness of damage zones varies significantly between fault systems and along a single fault, and the scaling of these thickness variations also depends on the timing of the damage formation. For example, Vermilye and Scholz (1998) attributed the fault damage to microfracturing at the process zone during fault propagation and suggested that due to the finite dimension of the process zone, the damage reaches "…a constant maximum density that is independent of fault length." Yet, field analyses documented systematic dependency between damage zone thickness and fault length (Fig. 2.8a) (e.g., Vermilye & Scholz, 1998; Anders & Wiltschko, 1994; Katz et al., 2003). As it is difficult to assess the active length of a fault during its propagation stage, it is more common to relate the damage zone thickness to fault displacement, and next, we discussed these relationships.

Many studies indicate that damage zone thickness (T) seems to evolve with the net fault displacement (D) (Evans, 1990; Manighetti et al., 2001; Shipton & Cowie, 2001) and some of them even reveal a nearly linear relationship between T and D for D < ~150 m (Mitchell & Faulkner, 2009, 2012; Savage & Brodsky, 2011). To examine our observations of the PXF system, we first consider the analyses of Scholz (1987) and Savage and Brodsky (2011). These analyses differ in their compilation approaches. Scholz (1987) focused on the thickness of fault-cores as the product of wear and gouge formation during fault slip. His compilation indicates a linear TD = [thickness (T)/displacement(D)] ranging from 10–3 to 10–1 with most data centered around TD ~ 10–2 (block dots in Fig. 2.8b). Scholz applied a wear model, which is based on experimental observations, to explain the T/D ratio, and suggested that increase of normal stress, e.g., by depth increase, is expected to increase the

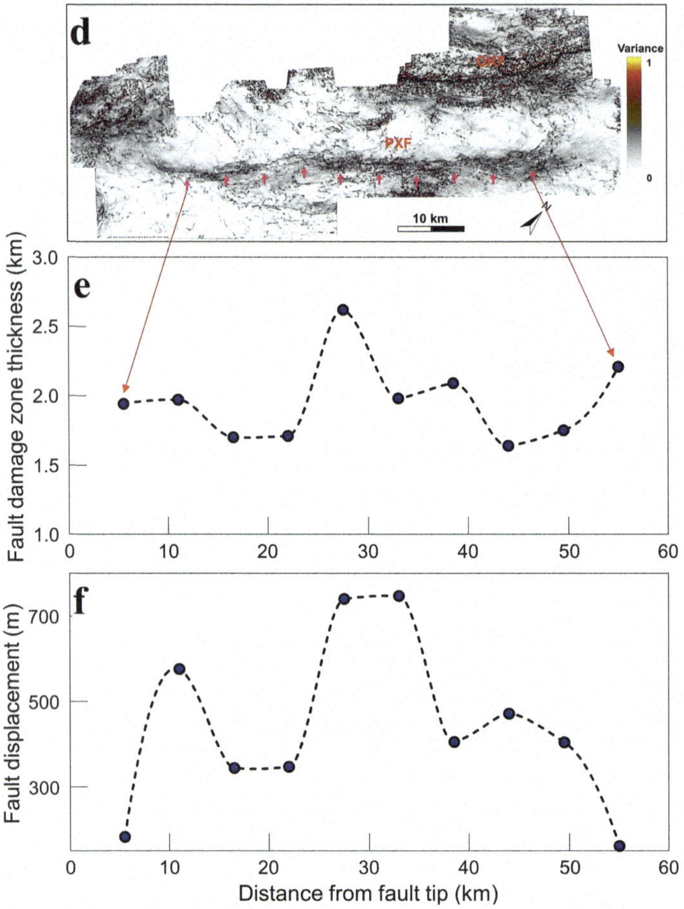

Fig. 2.7 Damage zone structure of the Pengxian fault on a map view. **d** Horizontal map view of the variance attribute (text) at 5.89 km depth (top of Middle Triassic detachment layer, Liao et al., 2023). The PXF is recognized as the broad zone of intense variance; the ten red arrows along the PXF are distributed at 5.5 km spacing from the SW tip of the fault. **e, f** Damage zone thickness (**e**) and displacement (**f**) measured at red arrows of 6d; dashed line-hand drawn lines connecting data points

fault-core thickness. Savage and Brodsky (2011), on the other hand, focused on the thickness of the fault damage as manifested by a damaged halo composed of small faults, fractures, and microfractures around the fault-core. Their compilation also revealed linear TD, but significantly larger with TD ~ 1 (hollow circles in Fig. 2.8b). They noted that this TD trend becomes much gentler for faults with displacements of ~2400 m which merges the fault-core observations of Scholz's compilation (dashed black line in Fig. 2.8b). These two compilations explicitly imply that for a given fault displacement, the fault-core thickness is about 1% of the damage-zone thickness.

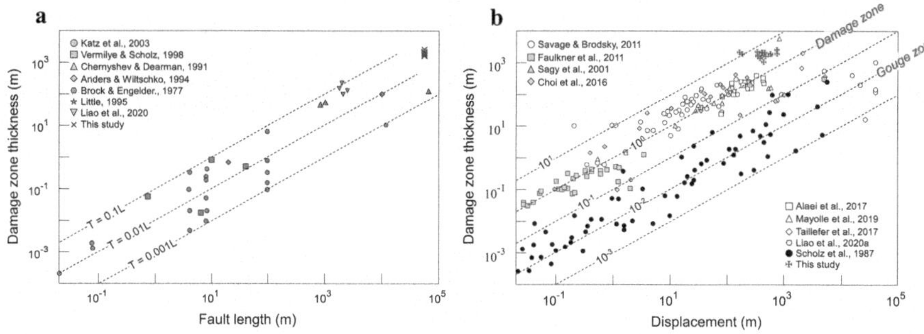

Fig. 2.8 **a** A compilation of data for damage zone thickness versus fault length. The black dashed lines indicate the simple proportionality between damage zone thickness and fault length. The ratios of damage zone thickness versus fault length are ranging from 10–3 to 10–1. *Data sources* (1) Katz et al. (2003); (2) Vermilye and Scholz (1998); (3) Chernyshev and Dearman (1991); (4) Anders and Wiltschko (1994); (5) Brock and Engelder (1977); (6) Little (1995), (7) Liao et al., (2020a, 2020b) and this study. **b** A compilation of data for damage zone (and/or gouge zone) thickness versus displacement. The black dashed lines indicate the simple proportionality between damage zone (and/or gouge zone) thickness and displacement. The ratios of damage zone and gouge zone thickness versus displacement range from 10–1 to 101 and 10–3 to 10–1, respectively. The constants of proportionality for damage zone and gouge zone thickness are commonly about 100 and 10–2, respectively. *Data sources* (1) Savage and Brodsky (2011); (2) Faulkner et al. (2011); (3) Sagy et al. (2001); (4) Choi et al. (2016); (5) Alaei and Torabi (2017); (6) Mayolle et al. (2019); (7) Taillefer et al. (2017); (8) Liao et al., (2020a, 2020b); (9) Scholz (1987) and this study

The analysis of Pengxian fault reveals total damage thicknesses that range from ~1 km to nearly 3 km, for corresponding segment displacements of 100 m to almost 800 m (Figs. 2.5, 2.6a, Table 2.1), and the 15 data points of PXF are plots in the range of TD = 100–101 (red crosses in Fig. 2.8b). The PXF data display the largest thicknesses (1–3 km) in the compiled data, and they form an extension of the damage-zones in Savage and Brodsky (2011) compilation (hollow circles). To explain this large thickness, we examined the seismic attributes sections (Fig. 2.4) and the variance detailed plots (Fig. 2.5). These figures reveal a few distinct zones of very intense variance values inside the 1–3 km thick zone (marked by red arrows in Fig. 2.5a). We interpret these distinct zones as fault-cores, namely, zones composed of anastomosing slip surfaces, and gouge of intense variance (Liao et al., 2017). These fault-cores are surrounded by thick damage zones, and the large total thickness for the PXF zones reflects composite fault zones with a few fault cores and associated damage zones (Fig. 2.9) (Faulkner et al., 2003; Powers & Jordan, 2010). In summary, the relatively large thickness of PXF reflects its composite nature and complex internal structure.

2.5.2 *Fault Zone Evolution Stages*

The quasi-linear relationship between damage-zone thickness and fault displacement (Fig. 2.8b) led to the concept that the damage-zone thickness increases with the

Table 2.1 Parameters of fault geometries

PXF														
Damage zone thickness (T), m	1940[a]	1970[a]	1700[a]	1710[a]	2620[a]	1980[a]	2090[a]	1640[a]	1750[a]	2210[a]	2260[b]	2190[b]	2150[b]	1420[b]	1080[b]
Displacement (D), m	182[a]	575[a]	344[a]	347[a]	740[a]	747[a]	405[a]	471[a]	404[a]	161[a]	335[b]	408[b]	455[b]	374[b]	435[b]

Note Superscript a—data for Fig. 2.8b, superscript b—data for Fig. 2.8

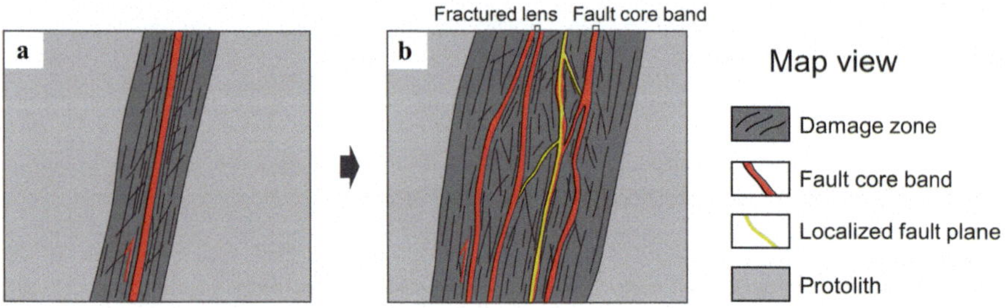

Fig. 2.9 Schematic view of a composite fault with a few fault-cores (red) and associated damage-zones (dark gray); (**a** an isolated single fault zone, modified from Faulkner et al., 2003) and (**b** a composite fault zone, modified from Powers & Jordan, 2010)

host fault displacements (Faulkner et al., 2003, 2011; Knott et al., 1996; Savage & Brodsky, 2011). This relationship may be affected by processes such as failure at segment interactions (e.g., Faulkner et al., 2011), wear controlled by normal stress (e.g., Scholz, 1987), propagation of damage front into protolith (Liao et al., 2014; Lyakhovsky et al., 2014), and lithological effects. We envision that the multiplicity of these processes produces the systematic (yet scattered) "universal" relationships between thickness and displacement over a wide range of fault structures, lithologies, depth intervals, and tectonic settings (Fig. 2.8b).

The formation of a composite fault zone (Fig. 2.9) (Faulkner et al., 2011) is an effective mechanism of damage-zone thickening. A composite fault-zone develops by generating new, separate fault-cores and damage-zones with increasing displacement (Aydin & Johnson, 1978; Faulkner et al., 2003; Liao et al., 2020b). Liao et al. (2020b) discussed three possible settings for the formation of a new fault-core/damage-zone along an existing fault-zone: (i) Strengthening of the existing fault-core to become stronger than the host rock, for example, by healing and cementation during sticking periods (Muhuri et al., 2003), grain-size reduction (Aydin & Johnson, 1978) or pseudotachylite freezing is stronger (Di Toro & Pennacchioni, 2005); (ii) the activation of the mechanical contrast at the boundaries between damaged and undamaged rocks (e.g., Ben-Zion & Andrews, 1998); (iii) Fault structural complexity such as "Segmented, non-planar faults which form by inherent randomness of fracturing or temporal stress rotation, are stronger than a planar fault by their contorted geometry…"(Faulkner et al., 2010). Lu et al. (2019) analyzed the geometry and kinematics of Pengxian buried structures and stated: "…the initial uplift of the Longquan Shan (LQS) could be traced back to the Late Miocene (~11–15 Ma). The slip and shortening in the LQS anticline could place constraints on the geometry and kinematics of the Pengxian anticline and PXF (Fig. 2.2b)." The segmented and non-planar geometry of the Pengxian thrust (Figs. 2.3, 2.4) indicate that it is in a mature stage (c.f., Manighetti et al., 2007) with primary components of damage zone and fault core. The evolution of the Pengxian composite fault damage zones (Fig. 2.9) possibly belongs to the third mechanism. Additionally, the second mechanism may play an important role derived from a massive accumulation of seismic events in the Longmen

Shan range front (e.g., Burchfiel et al., 2008; Hubbard & Shaw, 2009 and references therein).

2.5.3 *Detachment Controlling Fault Pattern*

The ~1 km thick detachment zone between the Middle and Lower Triassic layers (Jin et al., 2010; Lu et al., 2014, 2019) is composed of evaporites (most anhydrite), which is expressed on seismic data as chaotic and low-amplitude reflections (Fig. 2.4b, e.g., Jackson et al., 1994; Yu et al., 2009). We found that the PXF terminates into the Triassic detachment and shifts by it with a maximum detached distance approaching ~8 km toward the east (Figs. 2.3, 2.4), consistent with Liao et al. (2023). Lu et al. (2019) stated that the slip and shortening in the LQS anticline could place constraints on the geometry and kinematics of PXF. The structural deformation of LQS is highly associated with the horizontal detachment in the Middle-Lower Triassic layer (Fig. 2.2b; Burchfiel et al., 1995; Hubbard et al., 2010). It complies with the evaporitic layers of Middle-Upper Triassic segmented the faults in the brittle layers above and below in rifting off-shore Norway (e.g., Marsh et al., 2010; Wilson et al., 2013). The segmentation leads to the heterogeneous distribution of the Triassic detachment thickness along the western Sichuan fold-and-thrust belt (Cai & Liu, 1997; Wang et al., 2020). As a fault evolves into a complex fault system, individual faults may interact and overlap, forming relay ramps (Childs et al., 2009).

The evolution of fault damage zones for the buried reverse faults could be critical for evaluating the fault conduit capacity in hydrocarbon reservoirs in the sedimentary basins. The permeability architecture of a mature fault zone consists of two main components: the damage zone, which is expected to be very permeable, and the fault core, which constitutes a low-permeability sector (Billi et al., 2003; Caine et al., 1996). The average thickness of the Pengxian buried fault zone is more than 1 km, and the bilateral damage zones abutting fault cores can provide a preferential conduit for subsurface fluid flow, which possibly controls the high gas production within Leikoupo dolomite. Additionally, the damage zone is also an essential factor for estimating future seismic hazards, particularly in this seismic-active region (e.g., Huang et al., 2016; Liu et al., 2003; Powers & Jordan, 2010 and references therein).

2.6 Conclusions

The recognition of fault zone subsurface architecture is a challenging task that we approached here by using 3-D seismic data and volume attributes. The present analysis of the Pengxian Fault, a ~55 km long thrust at the margins of the Longmen Shan range, led to the following conclusions:

1. The PXF is a segmented fault system with km-scale damage zones that is correlated with the local displacements of the fault. The PXF is in a mature stage of its evolution with composite slip surfaces evolved from sequentially from a single slip surface.
2. The thickness variations of damage zones of PXF fit a power-law function with fault displacements. This scaled function agrees with previous studies and may be applied to characterize fault evolution in the upper crust.
3. Our analysis demonstrates the effectiveness of using 3-D seismic attributes (variance and curvature) to characterize the structural features of km-scale faults in the subsurface.

References

Alaei, B., & Torabi, A. (2017). Seismic imaging of fault damaged zone and its scaling relation with displacement. *Interpretation, 5*(4), 83–93. https://doi.org/10.1190/INT-2016-0230.1

Anders, M. H., & Wiltschko, D. V. (1994). Microfracturing, paleostress and the growth of faults. *Journal of Structural Geology, 16*(6), 795–815. https://doi.org/10.1016/0191-8141(94)90146-5

Aydin, A., & Johnson, A. M. (1978). Development of faults as zones of deformation bands and as slip surfaces in sandstone. *Pure and Applied Geophysics, 116*(4–5), 931–942. https://doi.org/10.1007/bf00876547

Ben-Zion, Y., & Andrews, D. J. (1998). Properties and implications of dynamic rupture along a material interface. *Bulletin of the Seismological Society of America, 88*(4), 1085–1094. https://doi.org/10.1785/BSSA0880041085

Billi, A., Salvini, F., & Storti, F. (2003). The damage zone-fault core transition in carbonate rocks: Implications for fault growth, structure, and permeability. *Journal of Structural Geology, 25*(11), 1779–1794. https://doi.org/10.1016/S0191-8141(03)00037-3

Brock, W. G., & Engelder, J. T. (1977). Deformation associated with the movement of the Muddy Mountain overthrust in the Buffington window, southeastern Nevada. *GSA Bulletin, 88*(11), 1667–1677. https://doi.org/10.1130/0016-7606(1977)88%3c1667:DAWTMO%3e2.0.CO;2

Burchfiel, B. C., Chen, Z., Liu, Y., & Royden, L. H. (1995). Tectonics of the Longmen Shan and adjacent regions. *International Geology Review, 37*, 661–738. https://doi.org/10.1080/00206819509465424

Burchfiel, B. C., Royden, L. H., Van der Hilst, R. D., & Hager, B. H. (2008). A geological and geophysical context for the Wenchuan earthquake of 12 May 2008, Sichuan, PRC. *GSA Today, 18*(7), 4–11. https://doi.org/10.1130/GSATG18A.1

Cai, L., & Liu, H. (1997). Structural styles and characteristics of fold-thrust belts in Sichuan foreland Basin. *Petroleum Geology and Experiment, 19*(2), 115–120.

Caine, J. S., Evans, J. P., & Forster, C. B. (1996). Fault zone architecture and permeability structure. *Geology, 24*(11), 1025–1028. https://doi.org/10.1130/00917613(1996)024%3c1025:FZAAPS%3e2.3.CO;2

Chen, S., Wilson, C. J. L., & Worley, B. A. (1995). Tectonic transition from the Songpan-Garzê Fold Belt to the Sichuan Basin, south-western China. *Basin Research, 7*(3), 235–253. https://doi.org/10.1111/J.1365-2117.1995.TB00108.X

Chernyshev, S. N., & Dearman, W. R. (1991). *Rock Fractures*. Butterworth-Heinemann.

Chester, F. M., Evans, J. P., & Biegel, R. L. (1993). Internal structure and weakening mechanisms of the San Andreas Fault. *Journal of Geophysical Research: Solid Earth, 98*(B1), 771–786. https://doi.org/10.1029/92JB01866

Childs, C., Manzocchi, T., Walsh, J. J., Bonson, C. G., Nicol, A., & Schöpfer, M. P. J. (2009). A geometric model of fault zone and fault rock thickness variations. *Journal of Structural Geology, 31*(2), 117–127. https://doi.org/10.1016/j.jsg.2008.08.009

Choi, J. H., Edwards, P., Ko, K., & Kim, Y. S. (2016). Definition and classification of fault damage zones: A review and a new methodological approach. *Earth-Science Reviews, 152*, 70–87. https://doi.org/10.1016/j.earscirev.2015.11.006

Chopra, S., & Marfurt, K. J. (2007). Seismic attributes for prospect identification and reservoir characterization. SEG, Geophysical Developments Series 11.

Di Toro, G., & Pennacchioni, G. (2005). Fault plane processes and mesoscopic structure of a strong-type seismogenic fault in tonalites (Adamello batholith, Southern Alps). *Tectonophysics, 402*(1–4), 55–80. https://doi.org/10.1016/j.tecto.2004.12.036

Evans, J. P. (1990). Thickness-displacement relationships for deformation zones. *Journal of Structural Geology, 10*(4), 431–435. https://doi.org/10.1016/0191-8141(88)90020-X

Faulkner, D. R., Jackson, C. A. L., Lunn, R. J., Schlische, R. W., Shipton, Z. K., Wibberley, C. A. J., & Withjack, M. O. (2010). A review of recent developments concerning the structure, mechanics, and fluid flow properties of fault zones. *Journal of Structural Geology, 32*(11), 1557–1575. https://doi.org/10.1016/j.jsg.2010.06.009

Faulkner, D. R., Lewis, A. C., & Rutter, E. H. (2003). On the internal structure and mechanics of large strike-slip fault zones: Field observations of the Carboneras fault in southeastern Spain. *Tectonophysics, 367*(3–4), 235–251. https://doi.org/10.1016/S0040-1951(03)00134-3

Faulkner, D. R., Mitchell, T. M., Jensen, E., & Cembrano, J. (2011). Scaling of fault damage zones with displacement and the implications for fault growth processes. *Journal of Geophysical Research Solid Earth, 116*, 1–11. https://doi.org/10.1029/2010JB007

Hale, D. (2013). Methods to compute fault images, extract fault surfaces, and estimate fault throws from 3D seismic images. *Geophysics, 78*(2), O33–O43. https://doi.org/10.1190/geo2012-0331.1

Huang, Y., Ampuero, J.-P., & Helmberger, D. V. (2016). The potential for supershear earthquakes in damaged fault zones-theory and observations. *Earth and Planetary Science Letters, 433*, 109–115. https://doi.org/10.1016/j.epsl.2015.10.046

Hubbard, J., & Shaw, J. H. (2009). Uplift of the Longmen Shan and Tibetan Plateau, and the 2008 Wenchuan (M = 7.9) earthquake. *Nature, 458*, 194–197. https://doi.org/10.1038/nature07837

Hubbard, J., Shaw, J. H., & Klinger, Y. (2010). Structural setting of the 2008 Mw 7.9 Wenchuan, China, earthquake. *Bulletin of the Seismological Society of America, 100*(5B), 2713–2735. https://doi.org/10.1785/0120090341

Iacopini, D., & Butler, R. W. H. (2011). Imaging deformation in submarine thrust belts using seismic attributes. *Earth and Planetary Science Letters, 302*(3–4), 414–422. https://doi.org/10.1016/j.epsl.2010.12.041

Jackson, M. P. A., Vendeville, B. C., & Schultz-Ela, D. D. (1994). Salt-related structures in the Gulf of Mexico: A field guide for geophysicists. *The Leading Edge, 13*(8), 837–842. https://doi.org/10.1190/1.1437040

Jia, D., Wei, G., Chen, Z., Li, B., Zeng, Q., & Yang, G. (2006). Longmen Shan fold-thrust belt and its relation to the western Sichuan Basin in central China: New insights from hydrocarbon exploration. *AAPG Bulletin, 90*, 1425–1447. https://doi.org/10.1306/03230605076

Jin, W., Tang, L., Yang, K., Wan, G., & Lü, Z. (2010). Segmentation of the Longmen Mountains thrust belt, Western Sichuan Foreland Basin, SW China. *Tectonophysics, 485*, 107–121. https://doi.org/10.1016/j.tecto.2009.12.007

Katz, O., Reches, Z. E., & Baer, G. (2003). Faults and their associated host rock deformation: Part I. Structure of small faults in a quartz–syenite body, southern Israel. *Journal of Structural Geology, 25*(10), 1675–1689. https://doi.org/10.1016/S0191-8141(03)00011-7

Kim, Y. S., Peacock, D. C. P., & Sanderson, D. J. (2004). Fault damage zones. *Journal of Structural Geology, 26*(3), 503–517. https://doi.org/10.1016/j.jsg.2003.08.002

Knott, S. D., Beach, A., Brockbank, P. J., Brown, J. L., & Mccallum, J. E. (1996). Spatial and large strike-slip fault zones: Field observations of the Carboneras fault in southeastern laws:

Preliminary results. *Journal of Geophysical Research: Solid Earth, 98*(B12), 21951–21961. https://doi.org/10.1029/93jb01008
Li, Z., Liu, S., Chen, H., Deng, B., Hou, M., Wu, W., & Cao, J. (2012). Spatial variation in Meso-Cenozoic exhumation history of the Longmen Shan thrust belt (eastern Tibetan Plateau) and the adjacent western Sichuan basin: Constraints from fission track thermochronology. *Journal of Asian Earth Sciences, 47*, 185–203. https://doi.org/10.1016/j.jseaes.2011.10.016
Li, Z., Liu-Zeng, J., Jia, D., Sun, C., Wang, W., Yuan, Z., & Liu, B. (2016). Quaternary activity of the range front thrust system in the Longmen Shan Piedmont, China, revealed by seismic imaging and growth strata. *Tectonics, 35*(12), 2807–2827. https://doi.org/10.1002/2015TC004093
Liao, Z., Chang, J. C., & Reches, Z. (2014). Fault strength evolution during high velocity friction experiments with slip-pulse and constant-velocity loading. *Earth and Planetary Science Letters, 406*, 93–101. https://doi.org/10.1016/j.epsl.2014.09.010
Liao, Z., Hu, L., Huang, X., Carpenter, B. M., Marfurt, K. J., Vasileva, S., & Zhou, Y. (2020a). Characterizing damage zones of normal faults using seismic variance in the Wangxuzhuang oilfield, China. *Interpretation, 8*(4), SP53–SP60.
Liao, Z., Li, W., Zou, H., Hao, F., Marfurt, K. J., & Reches, Z. (2020b). Composite damage zones in the subsurface. *Geophysical Journal International, 222*(1), 225–230. https://doi.org/10.1093/gji/ggaa158
Liao, Z., Liu, H., Carpenter, B. M., Marfurt, K. J., & Reches, Z. (2019). Analysis of fault damage zones using three-dimensional seismic coherence in the Anadarko Basin, Oklahoma. *AAPG Bulletin, 103*(8), 1771–1785. https://doi.org/10.1306/1219181413417207
Liao, Z., Liu, H., Jiang, Z., Marfurt, K. J., & Reches, Z. (2017). Fault damage zone at subsurface: A case study using 3D seismic attributes and a clay model analog for the Anadarko Basin, Oklahoma. *Interpretation, 5*(2), 141–150. https://doi.org/10.1190/INT-2016-0033.1
Liao, Z., Zhang, L., Zou, H., Hao, F., & Reches, Z. (2023). Damage stratigraphy as a controlling mechanism of earthquakes distribution and fault patterns. *Tectonophysics, 846*(5), 229661.
Little, T. A. (1995). Brittle deformation adjacent to the Awatere strike-slip fault in New Zealand: Faulting patterns, scaling relationships, and displacement partitioning. *GSA Bulletin, 107*(11), 1255–1271. https://doi.org/10.1130/0016-7606(1995)107%3c1255:BDATTA%3e2.3.CO;2
Liu, J., Sieh, K., & Hauksson, E. (2003). A structural interpretation of the aftershock "cloud" of the 1992 Mw 7.3 landers earthquake. *Bulletin of the Seismological Society of America, 93*(3), 1333–1344.
Liu, S., Li, Z., Cao, J., Liu, S., Deng, B., & Wang, G. (2009). 4-D textural and structural characteristics of Longmen intracontinental composite orogenic belt, southwest China. *Chinese Journal of Geology, 44*(4), 1151–1180.
Liu, S., Luo, Z., Dai, S., Dennis, A., & Wilson, C. J. L. (1996). The uplift of the Longmenshan thrust belt and subsidence of the West Sichuan Foreland Basin. *Acta Geologica Sinica, 9*, 16–26. https://doi.org/10.1111/j.1755-6724.1996.mp9001002.x
Lu, R., He, D., Suppe, J., Wu, J., Liu, B., & Chen, Y. (2014). Structural model of the central Longmen Shan thrusts using seismic reflection profiles: Implications for the sediments and deformations since the Mesozoic. *Tectonophysics, 630*, 43–53. https://doi.org/10.1016/j.tecto.2014.05.003
Lu, R., He, D., Xu, X., Tan, X., Li, Y., Cai, M., & Wang, Z. (2019). Geometry and kinematics of buried structures in the piedmont of the central Longmen Shan: Implication for the growth of the Eastern Tibetan Plateau. *Journal of the Geological Society, 176*(2), 323–333. https://doi.org/10.1144/jgs2018-015
Lyakhovsky, V., Sagy, A., Boneh, Y., & Reches, Z. (2014). Fault wear by damage evolution during steady-state slip. *Pure and Applied Geophysics, 171*, 3143–3157. https://doi.org/10.1007/s00024-014-0787-x
Manighetti, I., Campillo, M., Bouley, S., & Cotton, F. (2007). Earthquake scaling, fault segmentation, and structural maturity. *Earth and Planetary Science Letters, 253*, 429–438. https://doi.org/10.1016/j.epsl.2006.11.004

Manighetti, I., King, G. C. P., Gaudemer, Y., Scholz, C. H., & Doubre, C. (2001). Slip accumulation and lateral propagation of active normal faults in Afar. *Journal of Geophysical Research: Solid Earth, 106*(B7), 13667–13696. https://doi.org/10.1029/2000JB900471

Marsh, N., Imber, J., Holdsworth, R. E., Brockbank, P., & Ringrose, P. (2010). The structural evolution of the Halten Terrace, offshore Mid-Norway: Extensional fault growth and strain localisation in a multi-layer brittle-ductile system. *Basin Research, 22*(2), 195–214. https://doi.org/10.1111/j.1365-2117.2009.00404.x

Mayolle, S., Soliva, R., Caniven, Y., Wibberley, C., Ballas, G., Milesi, G., & Dominguez, S. (2019). Scaling of fault damage zones in carbonate rocks. *Journal of Structural Geology, 124*, 35–50. https://doi.org/10.1016/j.jsg.2019.03

Mitchell, T. M., & Faulkner, D. R. (2009). The nature and origin of off-fault damage surrounding strike-slip fault zones with a wide range of displacements: A field study from the Atacama fault system, northern Chile. *Journal of Structural Geology, 31*(8), 802–816. https://doi.org/10.1016/j.jsg.2009.05.002

Mitchell, T. M., & Faulkner, D. R. (2012). Towards quantifying the matrix permeability of fault damage zones in low porosity rocks. *Earth and Planetary Science Letters, 339–340*, 24–31. https://doi.org/10.1016/j.epsl.2012.05.014

Muhuri, S. K., Dewers, T. A., Scott, T. E., & Reches, Z. (2003). Interseismic fault strengthening and earthquake-slip instability: Friction or cohesion? *Geology, 31*(10), 881–884. https://doi.org/10.1130/g19601.1

Peacock, D. C. P. (2002). Propagation, interaction and linkage in normal fault systems. *Earth-Science Reviews, 58*(1–2), 121–142. https://doi.org/10.1016/S0012-8252(01)00085-X

Powers, P. M., & Jordan, T. H. (2010). Distribution of seismicity across strike-slip faults in California. *Journal of Geophysical Research Solid Earth, 115*(B5), 3040–3065. https://doi.org/10.1029/2008JB006234

Reches, Z., & Eidelman, A. (1995). Drag along faults. *Tectonophysics, 247*(1–4), 145–156. https://doi.org/10.1016/0040-1951(94)00170-E

Sagy, A., Reches, Z., & Roman, I. (2001). Dynamic fracturing: Field and experimental observations. *Journal of Structural Geology, 23*(8), 1223–1239. https://doi.org/10.1016/S0191-8141(00)00190-5

Savage, H. M., & Brodsky, E. E. (2011). Collateral damage: Evolution with displacement of fracture distribution and secondary fault strands in fault damage zones. *Journal of Geophysical Research Solid Earth, 116*, 7665.

Savage, H. M., Keranen, K. M., Schaff, D. P., & Dieck, C. (2017). Possible precursory signals in damage zone foreshocks. *Geophysical Research Letters, 44*(11), 5411–5417. https://doi.org/10.1002/2017GL073226

Scholz, C. H. (1987). Wear and gouge formation in brittle faulting. *Geology, 15*(6), 493–495. https://doi.org/10.1130/0091-7613(1987)15%3c493:WAGFIB%3e2.0.CO;2

Shipton, Z. K., & Cowie, P. A. (2001). Damage zone and slip-surface evolution over μm to km scales in high-porosity Navajo sandstone, Utah. *Journal of Structural Geology, 23*, 1825–1844. https://doi.org/10.1016/S0191-8141(01)00035-9

Sibson, R. H. (1990). Conditions for fault-valve behavior. *Geological Society of London Special Publication, 54*(1), 15–28. https://doi.org/10.1144/GSL.SP.1990.054.01.02

Taillefer, A., Soliva, R., Guillou-Frottier, L., Le Goff, E., Martin, G., & Seranne, M. (2017). Fault-related controls on upward hydrothermal flow: An integrated geological study of the Têt fault system, eastern Pyrénées (France). *Geofluids, 12*, 1–19. https://doi.org/10.1155/2017/8190109

Thomas, G.B. (1972). Calculus and Analytical Geometry. Addison-Wesley Co.

Vermilye, J. M., & Scholz, C. H. (1998). The process zone: A microstructural view of fault growth. *Journal of Geophysical Research Solid Earth, 103*(B6), 12223–12237. https://doi.org/10.1029/98JB00957

Wang, M., Wang, F., Jiang, D., Yan, B., Chen, Z., & Song, G. (2020). Interactions between thin- and thick-skinned tectonics at the western Sichuan Basin. *China. Tectonophysics, 12*, 796. https://doi.org/10.1016/j.tecto.2020.228628

Weng, H., Yang, H., Zhang, Z., & Chen, X. (2016). Earthquake rupture extents and coseismic slips promoted by damaged fault zones. *Journal of Geophysical Research: Solid Earth, 121*(6), 4446–4457. https://doi.org/10.1002/2015JB012713

Wibberley, C. A. J., Yielding, G., & Di Toro, G. (2008). Recent advances in the understanding of fault zone internal structure; a review. In: Wibberley, C. A. J., Kurz, W., Imber, J., Holdsworth, R. E., & Collettini, C. (Eds.), *Structure of fault zones: Implications for mechanical and fluid-flow properties*. Geological Society of London Special Publication, pp. 5–33. https://doi.org/10.1144/SP299.2

Wilson, J. E., Chester, J. S., & Chester, F. M. (2003). Microfracture analysis of fault growth and wear processes, Punchbowl Fault, San Andreas system. *California. Journal of Structural Geology, 25*(11), 1855–1873. https://doi.org/10.1016/S0191-8141(03)00036-1

Wilson, P., Elliott, G. M., Gawthorpe, R. L., Jackson, C.A.-L., Michelsen, L., & Sharp, I. R. (2013). Geometry and segmentation of an evaporite-detached normal fault array: 3D seismic analysis of the southern Bremstein Fault Complex, offshore mid-Norway. *Journal of Structural Geology, 51*, 74–91. https://doi.org/10.1016/j.jsg.2013.03.005

Yan, D. P., Zhou, M. F., Li, S. B., & Wei, G. Q. (2011). Structural and geochronological constraints on the Mesozoic-Cenozoic tectonic evolution of the Longmen Shan thrust belt, eastern Tibetan Plateau. *Tectonics, 30*, TC6005.

Yu, Y., Zhou, X., Tang, L., Peng, W., Lu, D., & Li, W. (2009). Salt structures in the Laizhouwan depression, offshore Bohai Bay basin, eastern China: New insights from 3D seismic data. *Marine and Petroleum Geology, 26*(8), 1600–1607. https://doi.org/10.1016/j.marpetgeo.2009.01.007

Chapter 3
Strike-Slip Fault Damage Zones in the Anadarko Basin, Oklahoma, USA

Abstract In this chapter, we expand these analyses to the subsurface by investigating the damage-zone structure of a 32-km (~105 ft) long right-lateral strike-slip fault in Oklahoma. We used the 3D seismic attribute of coherence to first define its regional and background levels, and then evaluated the damage-zone dimensions at multiple sites. We found damage-zone thickness of ~1615 m 1600 m (5300 ft) at a segment that is dominated by subsidiary faults, and it is slightly thicker at a segment with a pull-apart basin. The damage-zone intensity decays exponentially with distance from the fault core, in agreement with field observations and distribution of seismic events. The coherence map displays a strong asymmetry of the damage-zone between the two sides of the 3D fault, which is related to the subsidiary structures of the fault-zone. We also suggest that the damage zone contains a high density of subseismic fractures associated with multiscale faulting and secondary folding that may correspond to highly permeable features within the Woodford Shale.

3.1 Introduction

3.1.1 Fault-Zone Structure

Field analyses of fault-zones has revealed three primary components: fault core, damage-zone, and protolith (Fig. 3.1) (Caine et al., 1996; Kim et al., 2004; Sagy et al., 2001; Savage & Brodsky, 2011). The fault core is a discrete, quasi-tabular shear zone, comprised of gouge layers that accommodates most fault displacement. If the fault is composed of strands with several anastomosing segments, its core could be up to meters thick (e.g., Faulkner et al., 2010; Savage & Brodsky, 2011). The fault core could be a sealing zone with thick clay bodies (e.g. Billi et al., 2003), a permeable conduit (e.g. Caine et al., 1996), or both, depending on the fault's state in its seismic cycle (e.g. Sibson, 1990). The damage-zone is constituted of fractured, brecciated, and pulverized rocks derived from the protolith and are generally confined to a zone on the scale of a kilometer between the fault core and intact protolith (Busetti et al., 2012; Rempe et al., 2013; Sibson, 2003). The fracture sets within

Z. Liao, *Applied Seismic Geostructures*, https://doi.org/10.1007/978-981-97-7079-3_3

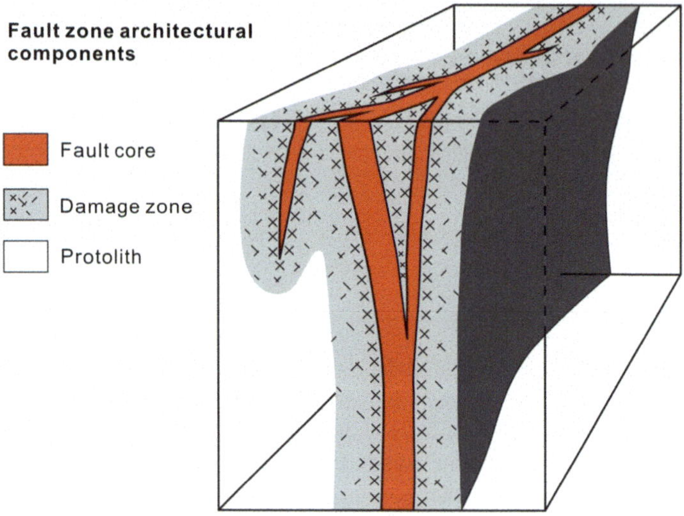

Fig. 3.1 Schematic diagram showing the fault zone architectural components for a strike slip fault (after Caine et al., 1996). Red indicates fault core, grey for damage-zones, and the protolith is removed

the damage-zone often provide a high permeability conduit for fluid flow (Billi et al., 2003). The fault core and damage-zone may vary along strike, owing to fault-related diagenesis, segmentation and evolution (Laubach et al., 2014). The structural complexity within the damage-zones, and particularly the distribution and openness of its fracture networks, can significantly affect the migration, accumulation and leakage of subsurface fluids (e.g., Caine et al., 1996; Ellis et al., 2012; Faulkner et al., 2010) and earthquake rupture characteristics (e.g., Weng et al., 2016).

Characterization of the structure of a subsurface fault zone, without borehole data, can be done only indirectly because fracture networks are invisible to seismic data. Without borehole data, the structure of a subsurface fault zone can only be indirectly characterized since fracture networks cannot be observed through seismic data. The properties of subsurface fault patterns, including geometry and internal architecture, can be determined, for example, by using seismic attributes (Chopra & Marfurt, 2007). Application of seismic mapping to a submarine fold thrust system can detect structural deformation by recognition of reduced signal through volumes (Iacopini & Butler, 2011; Iacopini et al., 2012). The concept of a seismic distortion zone enhanced the understanding of the associated damage of a thrust system at seismic scales. This seismic characterization method is further used for fault structure and its surrounding deformation that is defined as a seismic disturbance zone (Iacopini et al., 2016). Even though the fracture networks are invisible at seismic scale, their cumulative effects could be detected as distortion of the signal (Chopra & Marfurt, 2010; Li et al., 2015). Numerical simulations of synthetic fault models and associated seismic responses shows the potential to characterize the damage zones using seismic attributes and seismic tomography, as presented by Botter et al., (2016, 2017). Their workflow provides information on fault structure at different seismic resolutions, through the seismic images determined by the discrete element modeling.

The present study utilizes 3D seismic attributes for the analysis of damage and splays of a large, subsurface strike-slip fault in Oklahoma. It is demonstrated that the utilized seismic approach can reveal the dimensions and shapes of damage-zones with indications of the deformation intensity. We further show that the detected subsurface damage-zone display similar scaling relation to well documented field observations.

3.1.2 Damage-Zone Dimension

It is commonly observed that the intensity of fracturing and deformation within a fault damage-zone decays with distance from the fault core toward the protolith (e.g. Caine et al., 1996; Katz et al., 2003; Rempe et al., 2013; Sagy et al., 2001; Savage & Brodsky, 2011). Sagy et al. (2001) analyzed a system of joints within dolomite layers close to a large normal fault of the Dead Sea basin. The joint density, reported by the ratio of Fracture-Spacing-Ratio (FSR) = layer thickness/joint spacing, decreased significantly from FSR = 28 close to the fault core to a background value of 2–3 at ~100 m (328 330 ft) away from the core (Fig. 3.2a). Wilson et al. (2003) analyzed the brittle deformation around the Punchbowl fault in California. They found that the density of subsidiary fractures in the sandstone decreases from about ~90 fractures/m at the sub-fault core, to a regional background of ~20 fractures/m at about 10 m (33 ft) distance (Fig. 3.2b). In general, the observed damage decay can be fit by a power function or an exponential function (Fig. 3.2), for example, the fracture density (D) decays as fault-normal distance (x):

$$D = ae^{-bx}$$

where a and b are constants that reflect physical properties related to the layer thickness or brittleness of the rock (Cowie et al., 1995).

Determination of the dimension of subsurface damage-zones is challenging. Peng et al. (2003) used synthetic wave modeling to determine a thickness of ~100 m (328 330 ft) for a shallow fault in Landers, California. They found that the fault-zone has ~50% decreases in seismic velocity compared to the surrounding protolith. Powers and Jordan (2010) analyzed the variation of seismicity rate around right-lateral strike-slip faults in California (Fig. 3.2c). In the fault core, the number of seismic events is ~120 per km (~3300 ft) normal to the fault, and this seismicity rate decayed to 20/km (~66,000 ft) at a distance of 10 km (~323,8000 ft) by a power-law relationship with distance from the fault core. Their estimates of thicknesses of the damage-zones ranged from 120 m (~394 400 ft) to 440 m (~1443 1400 ft) along Elsinore-Temecula segments of the southern San Andreas fault system. Valoroso et al., (2014) used high-resolution earthquake locations to evaluate the damage zone of the L'Aquila normal fault, Italy. They found damage zone thicknesses ranging 0.5 km (~1640 600 ft) to

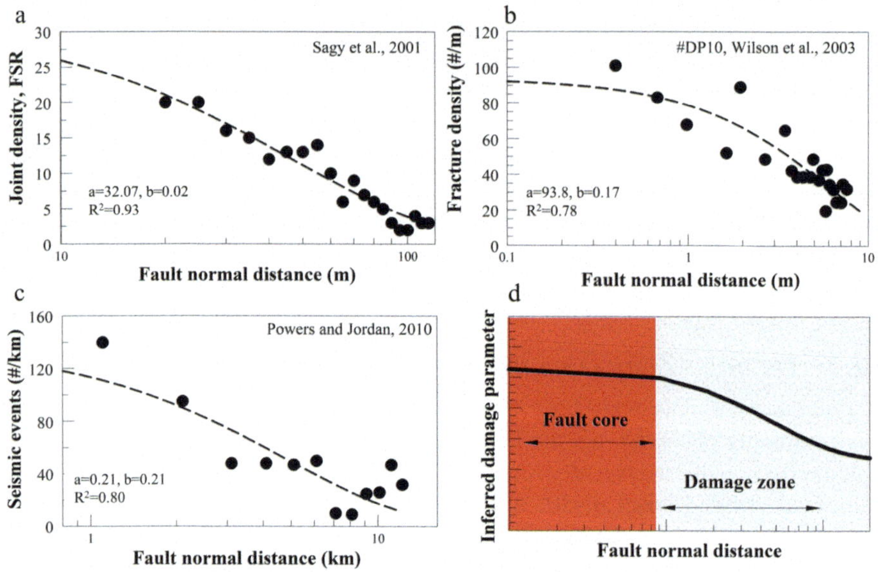

Fig. 3.2 Density of fault damage as a function of fault normal distance from the fault center: **a** Joint density versus fault normal distance by Sagy et al. (2001); **b** Fracture density versus fault normal distance (data number #DP10) by Wilson et al. (2003); **c** Seismic events versus fault normal distance by Powers and Jordan (2010); **d** and schematic diagram shows the decay relationship between inferred damage parameter versus fault normal distance

1.5 km (~4,925,000 ft) with damage intensity decaying at an exponential rate with distance from the fault core, which is in general agreement with field observations. Additional information can be derived from borehole logs. For example, the drilling across the San-Andreas fault near Parkfield, revealed a 200 m (~656 660 ft) thick damage-zone based on reduced seismic velocities (e.g. Zoback et al., 2011).

3.2 Approach and Observations

We investigate the damage-zone of the El-Reno fault (ERF), a 32 km (~105 ft) long, right-lateral, strike-slip fault in the Anadarko Basin, Oklahoma. The analysis utilizes the 3D seismic attribute "coherence" which is defined as the energy of the coherent part of seismic traces divided by the average acoustic energy of the input seismic traces (Chopra & Marfurt, 2007, 2010). This attribute is commonly used to identify lateral discontinuities, under the premise that its low values indicate discontinuities in layers, for example usage to detect faults and damage zones in the subsurface (e.g., Botter et al., 2017; Chopra & Marfurt, 2007; Iacopini et al., 2016; Liao et al., 2013). Here, we focus on utilizing coherence for characterization of seismic scale damage-zone in 3D to demonstrate the practical effectiveness of this attribute for damage-zone analysis for an onshore case of a large fault in an oil province.

The study area is in central-west Oklahoma (inset Fig. 3.3) where the Devonian Woodford Shale was deposited in the Anadarko-Arkoma-Ardmore Basins (Cardott, 2008; Paxton et al., 2006) during a global sea-level transgression (Johnson, 1988; Lambert, 1993). The Woodford Shale is an important petroleum source rock in the United States midcontinent, characterized as a laminated unit with alternating brittle and ductile layers (Slatt et al., 2012). The quartz- and calcite-rich brittle layers are fractured by layer-perpendicular open fractures (Bernal et al., 2012). Gale et al. (2014) observed widely distributed small fractures with heights <3 cm in thin chert layers of the Woodford.

The seismic data analyzed here were collected in 2012, and it includes nine narrow azimuth surveys that were reprocessed and prestack time migrated using a single datum and the same bin size (33.5 by 33.5 m or 110 by 110 ft). The frequency ranged from 10 to 60 Hz yielding the increased impedance as positive amplitude. The coherence volume calculations followed the procedure of Marfurt and Rich (2010).

The general features of the study area are displayed by a time structure map co-rendered with a map-based extraction from the seismic coherence volume at the Woodford Shale level (Fig. 3.3b). The dark zone (within the red box) indicates a north–south fault in the eastern part of the area that is the El-Reno fault (ERF).

3.3 Seismic Attribute Analysis

The 3D seismic reflection data we used covering the study area comprise nine narrow azimuth surveys acquired in central Oklahoma between 1994 and 2011 to image the Red Fork sandstones of Middle Pennsylvanian age. In 2012, the nine surveys were reprocessed together forto prestack time migration using a single datum and the same a common bin size (33.5 by 33.5 m, or 110 by 110 ft). At the top of the deeper Woodford Shale, that is deeper than the original seismic acquisition target, the frequencies range between 10 and 60 Hz, and the fold coverage varies from is 15 to 60.

We computed volumetric seismic attributes—coherence, curvature and dip-azimuth on the 3D seismic volume following Marfurt and Rich (2010). Next, we extracted the attribute values along the TWT map of the top of the Woodford Shale to reveal the deformation associated to the ERF.

Chopra and Marfurt (2007) define 'coherence' as the energy of the coherent part of seismic traces divided by the average acoustic energy of the input seismic traces. The energy-ratio coherence algorithm we applied is similar to the one described by Gersztenkorn and Marfurt (1999) and it is sensitive to lateral changes in waveform, but not in amplitude.

We extracted the attribute values along the TWT map of the top of the Woodford Shale to reveal the deformation associated to the ERF. The time TWT structural

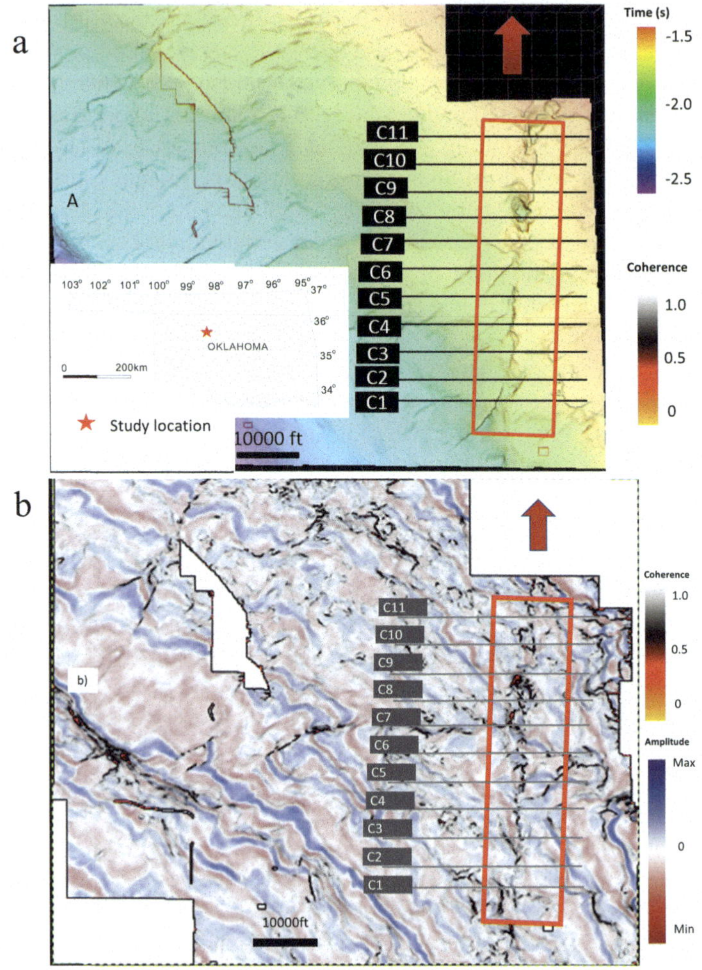

Fig. 3.3 **a** Two-way travel time (TWT) map of the top of the Woodford Shale indicating its large-scale structure of gently dipping (<2°) to the southwest. The time structural is co-rendered with the 3D determined coherence of a horizontal surface at this depth. The dark lineaments reveal structural elements, including the north–south El-Reno fault zone within the red rectangle. Index (C1, C11) show lines of sampling numbered S to N, with 1520 m (~5000 ft) in space. General location of the study area in the Anadarko Basin, Oklahoma (red rectangle star in lower left). **b** Seismic amplitude map co-rendered with the coherence along the top of the Woodford Shale

map of the Woodford Shale in the El-Reno fault (ERF)zone area, co-rendered with the 3D seismic coherence (Fig. 3.4b), reveals the following features: (1) a major north–south trending major structure with interrupted by several discontinuities that we refer to as the ERF, and locally characterized by red patches of low coherence; and (2) northeast-southwest trending black-red lineaments of (medium coherence) striking between at 15° and 60° relatively to the ERF; and (3) red patches (low coherence) along the ERF. We use the fault pattern we observed in The comparison with the results of the clay experiment (Fig. 3.4a) led us to interpret these the seismic features observations as follows (Fig. 3.4c): (1) the lineaments striking at 15–30° are the major Riedel (R) faults shears (black lines); (2) the lineaments striking at 60° are splay faults (green lines); and (3) the red-colored patches along the ERF are P faults shears (red lines).

Next, we examined the patterns seen on the horizon slice through the co-rendered dip-magnitude and dip-azimuth attribute volumes along the top of the Woodford Shale (Fig. 3.5). The dip magnitude attribute measures the magnitude of the local

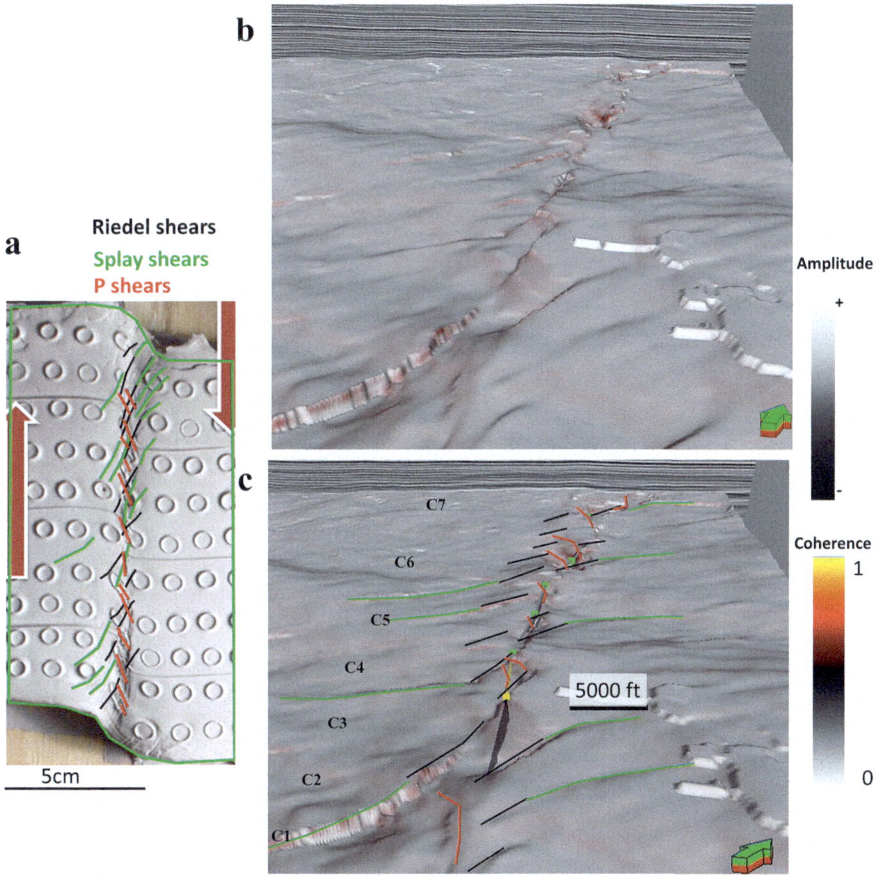

Fig. 3.4 **a** A 3D view of coherence co-rendered with seismic amplitude along the top of the Woodford Shale, and **b** its interpretation. The black lines indicate potential Riedel (R) faults, the green lines splay faults and red lines P shears. Index (C1–C7) show line location of sampling numbered in the after figure. Index (C1–C7) indicate corridors for coherence extraction perpendicular to the fault displayed in Fig. 3.6. **c** A photograph from the lab experiment taken after xx24 min. The fault interpretation on the model conforms to the structural interpretation of the top of the Woodford Shale. The connected points are interpreted fault plane

dip, whereas the dip azimuth that indicate the direction of maximum orientation of the maximum local layer inclination perpendicular to the strike (Chopra & Marfurt, 2007). The map of the ERF area displays three characteristic features: (1) dip layers dipping to 180° and 360°, marked by patches of yellow, green, and blue; (2) northeast-southwest trending lineaments in yellowred-orange (dipping towards southeast), cyan (dipping towards northwest), and darker blue (dipping towards southeast); (3) northwest-southeast oriented bands along the ERF in (pink-red (dipping towards east), and wide bands across the study area in green-yellow (dipping towards southwest). We interpret these features as: (1) different bedding dip directions, with yellow representing southwest-dipping beds and blue representing northeast-dipping beds; (2) the northeast-southwest trending lineaments fit the predicted orientation of R shears with respect to the main ERF trend; (3) the pink-red areas are ERF-related secondary anticlinal folds (ridges).

Following Chopra and Marfurt (2007), we also use horizon slices through the co-rendered the most-positive (K1) and the most-negative (K2) principal curvature volumes to identify folds and flexures related to the ERF and the associated Riedel structure (Fig. 3.6). Curvatures are presented as 2nd derivative of the surface. We interpreted the local most positive structures (red) as anticlinal foldsridges and the most negative structures (blue) as synclinesvalleys.

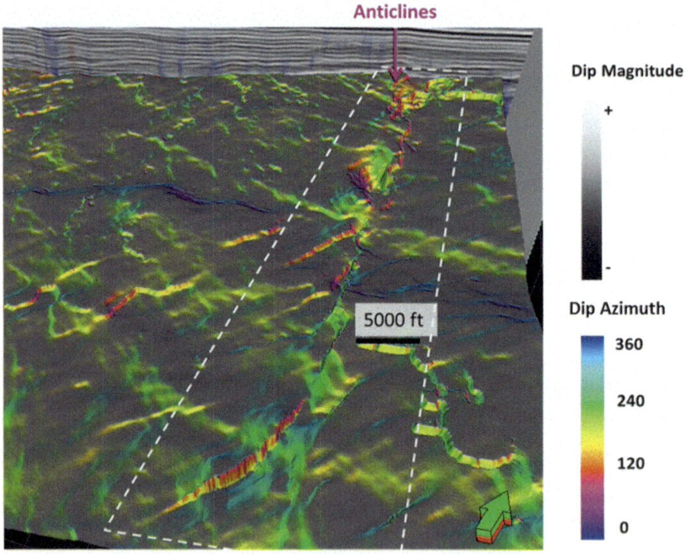

Fig. 3.5 Dip-magnitude map of the top of the Woodford Shale co-rendered with dip-azimuth. Area of pink color within reveals potential fracture zones with anticlinal compressed folds. Black/grey color show low dip magnitude area on the map. The bright part of the area presents higher dip magnitude. Based on the color palette of dip azimuth, relatively high magnitude dips dipping towards: North (0° or 360°) are blue; South (180°) are yellow; Green indicates high magnitude dips dipping towards SW (around 240°); Cyan indicates high magnitude dips dipping towards NW (around 300°)

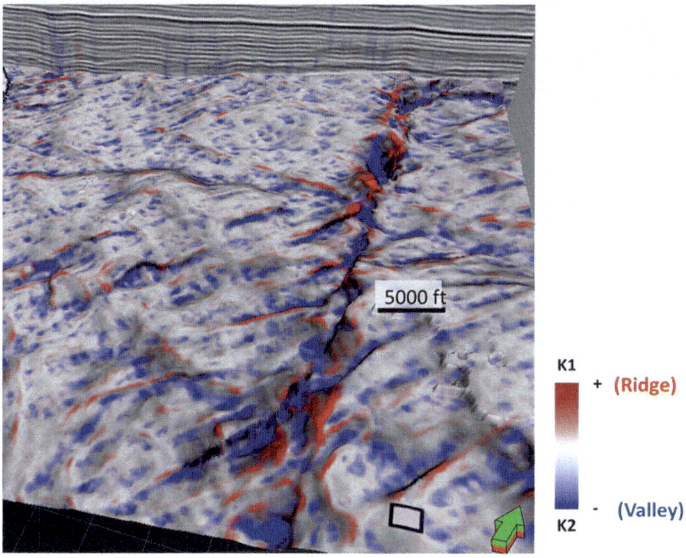

Fig. 3.6 Curvatures of most-positive (K1) and most-negative (K2) on the top of Woodford Shale. K1 and K2 are normal curvatures measured in planes perpendicular to each other on the surface, representing the maximum and minimum curvature. Areas with strong curvature (*brighter red indicate ridges and blue valleys*) correspond to strong flexures, folds, and possibly a high density of fractures

3.3.1 Damage Zones of the Strike-Slip Fault

Above section indicated that the ERF is a right-lateral strike-slip fault based on two distinguishing features: (1) it is a vertical fault with several sub-parallel vertical segments (Liao et al., 2013), which is typical feature of strike-slip faults (Christie-Blick & Biddle, 1985; Harding, 1985); and (2) the relatively small vertical throw (~80 m or 262 260 ft) is in contrast to the large fault length of 32 km (~105 ft), and vertical extent of at least 900 m (~2950 3000 ft) (Fig. 3.7).

The structure of the ERF at its intersection with the top of the Woodford Shale is displayed by the coherence map (Fig. 3.3). The structure includes a system of folds and flexures that are most intense within a zone around the primary fault zone (Liao et al., 2017). We interpret this structure as the damage-zone of ERF, and evaluate its thickness in 11 horizontal, fault-normal sections of coherence. These sections are spaced at ~1500 m (4921 5000 ft) intervals along the ERF (marked 'C' in Fig. 3.3). The seismic amplitude and coherence section samples are presented in Fig. 3.7, and the coherence profiles are displayed in Fig. 3.8.

It has been shown that between the Rayleigh limit and distinctive seismic response scale, seismic attributes could be interpreted to track structure details by an image processing procedure (Iacopini et al., 2016; Liao et al., 2013). Figure 3.7 displays vertical sections of amplitude (Fig. 3.7a, b) and coherence (Fig. 3.7c, d) along line C2

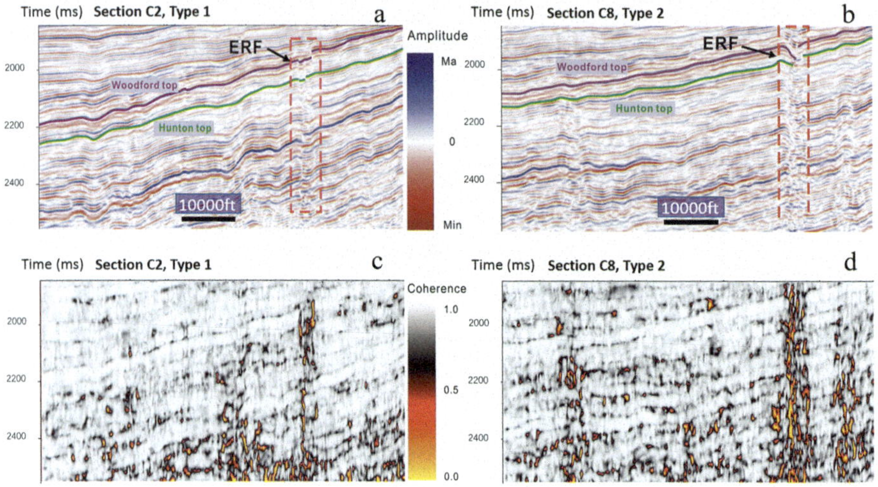

Fig. 3.7 Seismic amplitude and coherence across section maps normal to the El-Reno fault in the study area. (**a**) and (**b**) are amplitude maps of section C2 and C8 respectively (showed in Fig. 3.9), red rectangles indicate the fault area corresponding to the area of low coherence value in (**c**) and (**d**)

and C8 (defined in Fig. 3.3) that are perpendicular to ERF. These sections reveal a few discrete vertical zones, with the ERF (red, dashed box) as the most prominent zone. The amplitude signals are strongly disturbed around the vertical fault zones, which is enhanced by the low coherence maps. The vertical fault zones are comprised of several vertical segments that become wider with depth. These seismic disturbance zones are analyzed here as the seismic damage zone of two structural types (type 1 in Fig. 3.7a, c, and type 2 in Fig. 3.7b, d) along the strike-slip fault. The internal character of these structural types is discussed below.

The profiles display three general zones of coherence intensity (Fig. 3.8; note inverted scale of the coherence): (1) Zones of high coherence, >0.9, observed away from the ERF; (2) Zones of intermediate coherence, 0.8–0.9, within the ERF zone (gray in Fig. 3.8); and (3) A zone of low coherence, 0.4–0.8, within the ERF (pink in Fig. 3.8). The coherence levels in 3D-seismic analysis indicate the intensity of structural disturbance and discontinuities (Chopra & Marfurt, 2010). As fracturing and faulting disturb the continuity of geologic features, we regard the three coherence zones of Fig. 3.8 as indicating three levels of damage intensity. The high coherence zone is the protolith zone away from the fault, the intermediate level zone is the damage-zone, and the low coherence level zone is the fault core that is most intensely damaged, which is defined here as the 'seismic fault core'. We apply this interpreted zonation in the synthesis below.

We further explored the validity of the above interpretation of damage-zones by plotting the RMS of the seismic amplitude along the same profiles of Fig. 3.8. We regard the amplitude RMS as a proxy for the reduction of the seismic intensity due to damage and found that the amplitude RMS at the Woodford Formation horizon

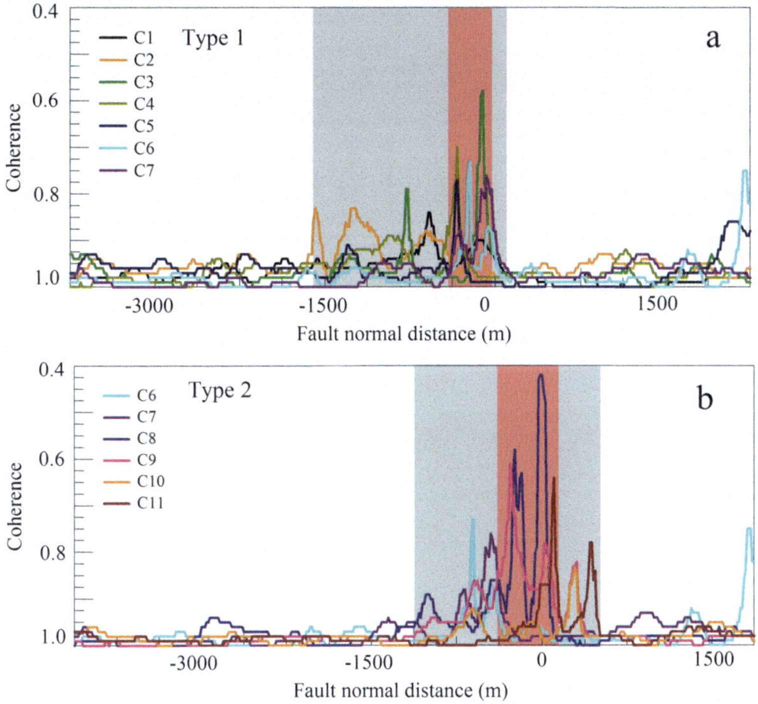

Fig. 3.8 Profiles of the coherence values across El-Reno fault at the Woodford Shale level. Profiles locations in Fig. 3.3. Note the inverse scale of the coherence. Zones of coherence values below background coherence are interpreted as damage-zones. Coherence reduction intensity shown in colors: Pink-intense; grey-gentle; white- background. **a** Damage-zones along the type 1 segment of El-Reno fault characterized by Riedel shear sub-faults (C1–C7 sections in Fig. 3.3b). **b** Damage-zones along the type 2 segment of El-Reno fault characterized by a pull-apart basin (C6–C11 sections in Fig. 3.3b)

corresponds well to the coherency plots of Fig. 3.8. Yet, while the coherency sections revealed both fault core (pink) and damage zone (grey), the amplitude RMS plots did not display the core (Fig. 3.9).

3.4 Synthesis

We noted that the width of the damage-zone (gray in Fig. 3.8) is asymmetric with respect to the seismic core-zone (pink in Fig. 3.8), and based on this asymmetry, we separated the coherence profiles into two types. Type 1, which includes profiles C1–C7 (Fig. 3.8a), displays a strong asymmetry in which the damage-zone is ~1100 m (3609 3600 ft) wide in the western block of ERF and only ~75 m (246 250 ft) wide in the eastern block. A core-zone (pink), of ~400 m (1312 ft) width, separates the

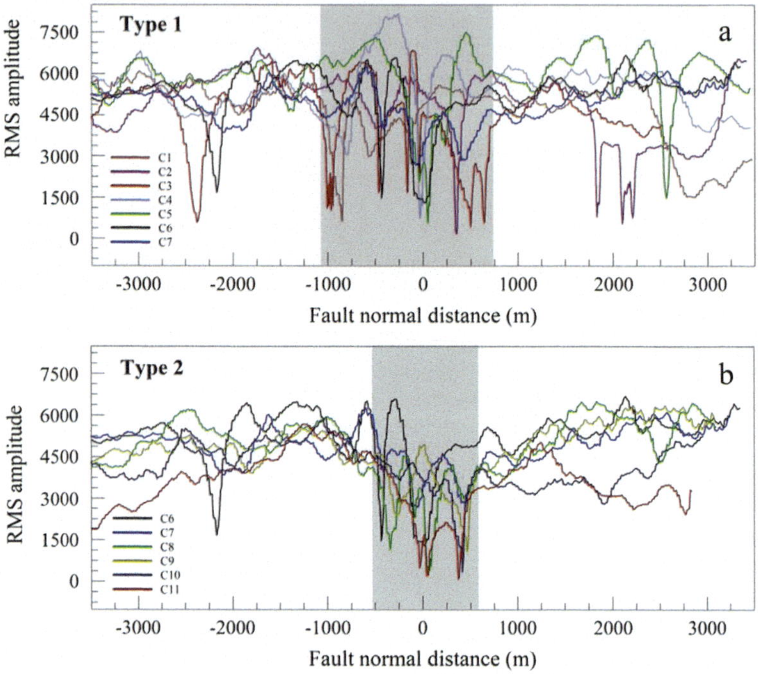

Fig. 3.9 Profiles of the RMS amplitude values across El-Reno fault at the Woodford Shale level. Profiles locations in Fig. 3.3. Zones of abnormal low values are interpreted as damage-zones. **a** Damage-zones along the type 1 segment of El-Reno fault (C1–C7 sections in Fig. 3.3b). **b** Damage-zones along the type 2 segment of El-Reno fault (C6–C11 sections in Fig. 3.3b)

two blocks. Type 2, which includes profiles C6–C11 (Fig. 3.8b), has a ~1645 1600 m (5397 5400 ft) thick damage-zone (coherence < 0.9), that includes a central core-zone of ~500 m (1640 600 ft) width. This type displays a gentler asymmetry with a western damage-zone of ~760 m (2493 2500 ft) width, and an eastern damage-zone of ~380 m (1246 1200 ft) width. These types correspond to different structural styles that were recognized by Liao et al. (2017). Type 1 corresponds to ERF segments with multiple, subsidiary Riedel faults trending at 15°–30° with respect to the main trend (R in Fig. 3.3, or refer to Liao et al., 2017). Type 2, on the other hand, is associated with segments of ERF with pull-apart basin (e.g., area between profiles C8–C9 in Fig. 3.3).

The dimensions and shapes of the identified coherence zones (Fig. 3.8), which we interpret as damage-zones, can be compared to equivalent features of exposed damage-zones. First, both types displayed asymmetry of damage-zone width with respect to the seismic fault-core (Fig. 3.8), and similar asymmetry has been observed in field cases and derived in theoretical models. Dor et al. (2006) found a systematic asymmetry of damage and pulverization distribution along multiple fault segments of the southern part of the San Andreas system. The pulverized rocks along these faults were typically associated with fault segments that separate rock bodies of different elastic properties. This association suggests that the asymmetric damage is related to preferential rupture propagation during earthquakes (Ampuero & Mao, 2016; Ben-Zion & Shi, 2005; Xu et al., 2012). It was modeled in rupture simulations of bi-material faults that this preferred propagation direction would lead to strong strain asymmetry between the two sides of the fault (Ampuero & Mao, 2016; Cochard & Rice, 2000; Dalguer & Day, 2009; Shi & Yehuda, 2006).

To examine depth variations of the damage-zones, we plotted a sequence of coherence profiles at 50 ms time intervals that are similar to the single depth sections of Fig. 3.8. The damage zones, marked grey in Fig. 3.10 for type 1 (a) and type 2 (b), are wider (>1000 m or 3280 ft) within the central part of the fault (e.g., intervals 1950–2000 ms in Fig. 3.10a), and are thinner upward and downward (e.g., −1800 or −2150 ms, Fig. 3.10a). Similar width variations can be observed for a type 2 segment (Fig. 3.10b). This reduction of damage-zone width from fault center towards its margins, fits the well-documented observation that the largest displacement along a fault which is, in general, within its central region (Cowie & Scholz, 1992; Walsh & Watterson, 1987). However, the change in damage zone width from shallow to deep could be possibly influenced by differences in the connectivity of the various strands in subsurface, which is not to be discussed in this paper.

In addition, we compare the coherence zones' geometry with the distribution of damage observed in field studies. Figure 3.11a shows the normalized density of fractures as a function of normalized distance from the fault zone for the aforementioned three examples using seismic data (Powers & Jordan, 2010) and outcrop data (Sagy et al., 2001; Wilson et al., 2003). The curves of normalized density fit well the above exponential model with slightly different coefficients, a and b, that reflect the fault lithology and geometry. Figure 3.11b shows the normalized coherence of the two fault blocks in type 1 (average values of C1–C7), displaying an exponential decay of the coherence as a function of increasing distance from the core. Within a wider damage-zone, the deformation consists of subsidiary faults indicated by two pulses of coherence values (Fig. 3.11c). Similar patterns of coherence (average of C6–C11) are observed in type 2 (Fig. 3.11d, e). Figure 3.11e illustrate the extreme coherence anomaly of the eastern fault block of the pull-apart basin. These two types indicate that the thickness of a damage-zone covers a distance of two orders of magnitude.

3.5 Discussion

The damage-zone of a fault can develop by various mechanisms. For example, earthquake propagation along a fault radiates seismic waves that could damage the surrounding blocks (e.g. Andrews, 1994; Dor et al., 2006). The intensity of this damage was analyzed and simulated based on the stress distribution during rupture and fault properties (Ben-Zion & Ampuero, 2009; Xu et al., 2012), and the analyses showed that the damage-zone thickness depends primarily on fault depth, pre-earthquake stresses, and the intensity of stress drop during rupture (Ampuero & Mao, 2017). For example, a 15 km (~5 × 104 ft) deep strike-slip fault is expected

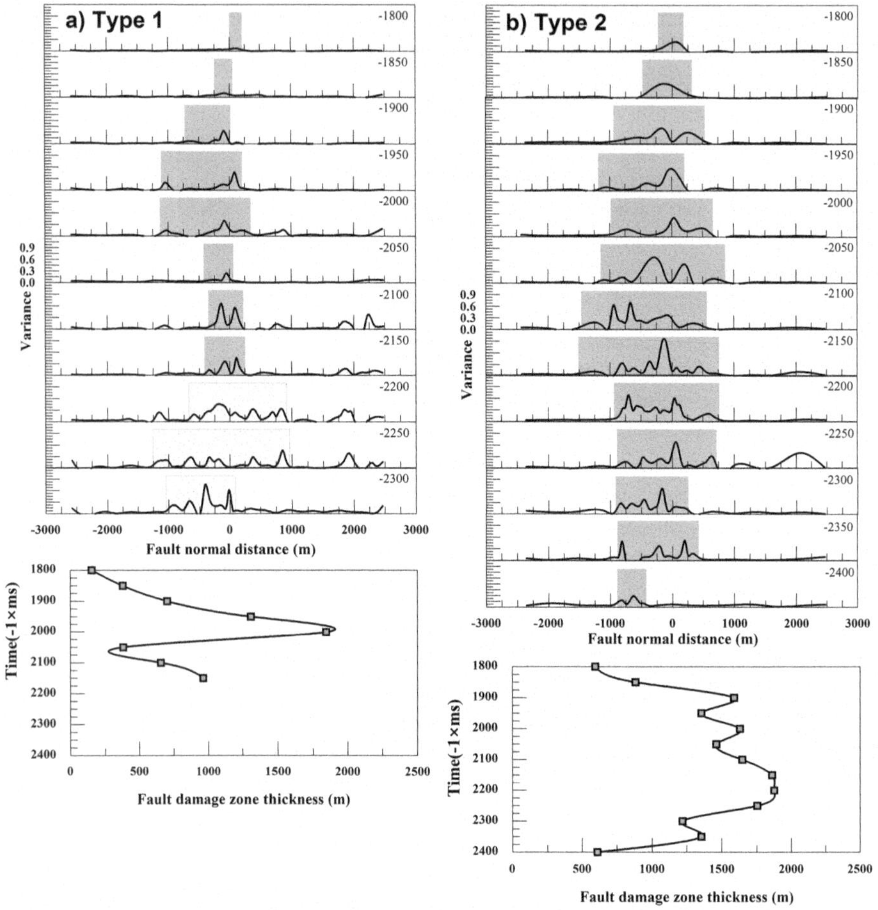

Fig. 3.10 Coherence damage-zones variations with depth intervals from −1800 to −2500 ms. **a** Type 1 segment with Riedel shear sub-faults). **b** Type 2 segment with pull-apart basin. The zone of low coherence values below background coherence indicates the damage-zone (colored in grey)

to generate a ~400 m (1312 1300 ft) thick damage zone for typical crustal conditions (Ampuero & Mao, 2017). The vertical extent of the present El-Reno fault is about 900 m (~2950 3000 ft) and thus the expected damage-zone due to earthquake rupture is less than 100 m. As our analysis revealed a much thicker damage-zone (~1600 m or 5248 5000 ft), we propose that most of the observed damage is associated with the following evolution of fault growth. First, the early stages of fault evolution is characterized by development of multiple fractures and small faults that precede the localized slip in the core-zone, due partly to the merger and coalescence of these smaller structures (e.g., Heesakkers et al., 2011a, 2011b; Reches & Lockner, 1994). A large strike slip fault, like the present El-Reno fault, evolves over extended time, and may develop a complex damage distribution that generates a wide zone.

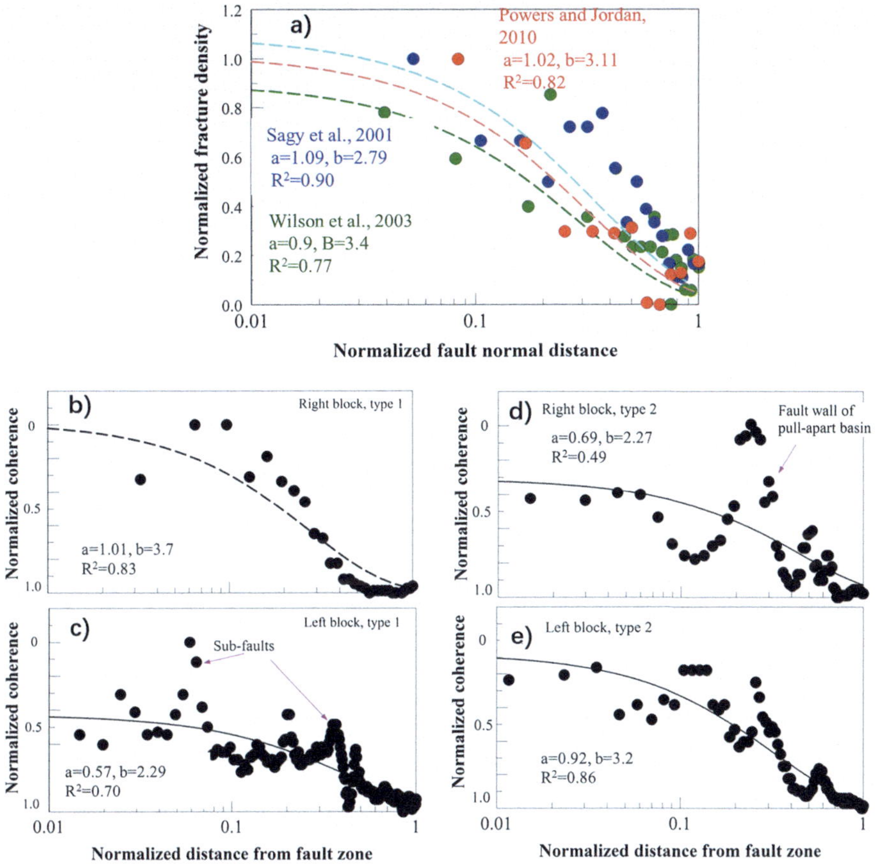

Fig. 3.11 **a** Normalized damage density (fractures or seismic events) as a function of normalized distance from the fault zone. Three examples from references (Powers & Jordan, 2010; Sagy et al., 2001; Wilson et al., 2003). All data are well fit by the model $D = ae^{-bx}$ $y = a{*}exp(-b{*}x)$ where coefficients a and b are determined by different fault lithology and geometry. **b** Normalized coherence (average of C1–C7) as a function of normalized distance from the fault zone for right block and **c** left block of type 1 segment of El-Reno fault with Riedel shear sub-faults (shown by the two arrows). **d** Normalized coherence (average of C6–C11) as a function of normalized distance from the fault zone for right block with fault wall of type 2 segment of El-Reno fault with pull-apart basin and **e** left block in ERF

Experimental works have shown that strike slip faults typically initiate as a wide, simple-shear zone with multiple secondary structures (Riedel shears, P shears), that eventually merge into a complex fault-zone (Liao et al., 2013; Naylor et al., 1986; Reches, 1988). This process forms a wide damage-zone that continues to deform internally due to the non-planar, intersecting relation of the coalesced secondary faults. Such evolution may lead to a rough fault core (Sagy et al., 2001), and the slip along such a rough fault generates a heterogeneous stress field comparable to the scale of the roughness (Dieterich & Smith, 2009; Powers & Jordan, 2010). Figure 3.12

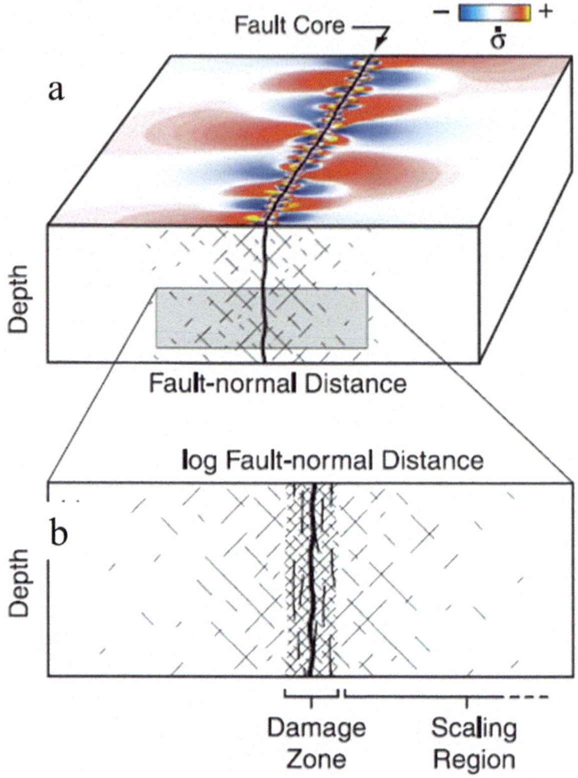

Fig. 3.12 Schematic presentation of the fault model with **a** heterogeneous stress field over a scaling region (Dieterich & Smith, 2009; Powers & Jordan, 2010) (**a**), and the associated damage-zones (Chester et al., 2005) (**b**)

displays a model calculation of the stress distribution at the proximity of a rough strike-slip fault (Chester et al., 2005). This stress field leads to further damage by branching of multiple secondary faults and general fracturing, particularly in the more tensile area (Reches, 1988), as well as multiple short folds and flexures. The ERF, studied here, is likely to be at a mature stage of its development, and we argue that the above processes prevailed during its activity forming the damage-zones with reduced seismic coherence.

3.6 Conclusions

The present analysis of the damage-zone of El-Reno fault in Oklahoma by using seismic attributes led to the following conclusions:

1. The analysis shows the effectiveness of using the 3D seismic attribute of coherence for characterization of the structural features of large fault-zones.

2. The thickness variations of the damage-zone of the El-Reno segments fit an exponential decay with distance from the fault core. This scaled decay function agrees with field observations over different scales, and may be applied to characterize damage zone dimensions in the subsurface.
3. It is suggested that the pattern and scale of damage-zone thickness is controlled by the secondary structures that develop during fault evolution.

References

Ampuero, J. P., & Mao, X. (2016). Scaling between fault length, damaged zone thickness and width of secondary fault fans derived from fracture mechanics. In *EGU general assembly conference abstracts* (pp. EPSC2016–10593).https://doi.org/10.1306/ad462538-16f7-11d7-8645000102c1865d

Ampuero, J. P., & Mao, X. (2017). Upper limit on damage zone thickness controlled by seismogenic depth. In *Fault zone dynamic processes: Evolution of fault properties during seismic rupture* (pp. 243–253). https://doi.org/10.1002/9781119156895.ch13

Andrews, D. J. (1994). Fault geometry and earthquake mechanics. *Annals of Geophysics, 37*, 6. https://doi.org/10.4401/ag-4136

Ben-Zion, Y., & Ampuero, J. P. (2009). Seismic radiation from regions sustaining material damage. *Geophysical Journal International, 178*(3), 1351–1356. https://doi.org/10.1111/j.1365-246x.2009.04285.x

Ben-Zion, Y., & Shi, Z. (2005). Dynamic rupture on a material interface with spontaneous generation of plastic strain in the bulk. *Earth and Planetary Science Letters, 236*, 486–496. https://doi.org/10.1016/j.epsl.2005.03.025

Bernal, A. S., Mayorga, L. C., Prada, A. G., & Slatt, R. M. (2012). Geological characterization of the Woodford Shale, McAlester Cemetery Quarry. *Oklahoma. Shale Shaker, 63*(3), 202–212.

Billi, A., Salvini, F., & Storti, F. (2003). The damage zone-fault core transition in carbonate rocks: Implications for fault growth, structure, and permeability. *Journal of Structural Geology, 25*, 1779–1794. https://doi.org/10.1016/S0191-8141(03)00037-3

Botter, C., Cardozo, N., Hardy, S., Lecomte, I., Paton, G., & Escalona, A. (2016). Seismic characterization of fault damage in 3D using mechanical and seismic modeling. *Marine and Petroleum Geology, 77*, 973–990. https://doi.org/10.1016/j.marpetgeo.2016.08.002

Botter, C., Cardozo, N., Qu, D., Tveranger, J., & Kolyukhin, D. (2017). Seismic characterization of fault facies models. *Interpretation, 5*, 1–53. https://doi.org/10.1190/INT-2016-0226.1

Busetti, S., Mish, K., Hennings, P., & Reches, Z. (2012). Damage and plastic deformation of reservoir rocks: Part 2. Propagation of a hydraulic fracture. *AAPG Bulletin, 96*, 1711–1732.

Caine, J. S., Evans, J. P., & Forster, C. B. (1996). Fault zone architecture and permeability structure. *Geology, 24*, 1025–1028. https://doi.org/10.1306/02011211011

Cardott, B. (2008). Overview of Woodford gas-shale play of Oklahoma, US. AAPG Annual Convention and Exhibition.

Chester, J. S., Chester, F. M., & Kronenberg, A. K. (2005). Fracture surface energy of the Punchbowl fault, San Andreas system. *Nature, 437*, 133–136. https://doi.org/10.1038/nature03942

Chopra, S., & Marfurt, K. J. (2007). Seismic attributes for prospect identification and reservoir characterization (pp. 11). SEG publisher.

Chopra, S., & Marfurt, K. J. (2010). Interpreting fractures through 3-D seismic discontinuity attributes and their visualization. *CSEG Recorder, 34*(8), 5946.

Christie-Blick, N., & Biddle, K. T. (1985). Deformation and basin formation along strike-slip faults. In K. T. Biddle, & N. Christie-Blick (Eds.), Strike-slip deformation, basin formation, and sedimentation (pp. 1–34). SEPM Special Publications. https://doi.org/10.2110/pec.85.37.0001

Cochard, A., & Rice, J. R. (2000). Fault rupture between dissimilar materials: Ill-posedness, regularization, and slip-pulse response. *Journal of Geophysical Research Atmospheres, 105*(B11), 25891–25907. https://doi.org/10.1029/2000JB900230

Cowie, P. A., & Scholz, C. H. (1992). Displacement-length scaling relationship for faults: Data synthesis and discussion. *Journal of Structural Geology, 14*, 1149–1156. https://doi.org/10.1016/0191-8141(92)90066-6

Cowie, P. A., Sornette, D., & Vanneste, C. (1995). Multifractal scaling properties of a growing fault population. *Geophysical Journal International, 122*, 457–469. https://doi.org/10.1111/j.1365-246x.1995.tb07007.x

Dalguer, L. A., & Day, S. M. (2009). Asymmetric rupture of large aspect-ratio faults at bimaterial interface in 3D. *Geophysical Research Letters, 36*, 195–215. https://doi.org/10.1029/2009gl040303

Dieterich, J. H., & Smith, D. E. (2009). Nonplanar faults: Mechanics of slip and off-fault damage. *Pure and Applied Geophysics, 166*, 1799–1815. https://doi.org/10.1007/978-3-0346-0138-2_12

Dor, O., Rockwell, T. K., & Ben-Zion, Y. (2006). Geological observations of damage asymmetry in the structure of the San Jacinto, San Andreas, and Punchbowl faults in Southern California: A possible indicator for preferred rupture propagation direction. *Pure and Applied Geophysics, 163*, 301–349. https://doi.org/10.1007/s00024-005-0023-9

Ellis, M. A., Laubach, S. E., Eichhubl, P., Olson, J. E., & Hargrove, P. (2012). Fracture development and diagenesis of Torridon Group Applecross Formation, near An Teallach, NW Scotland millennia of brittle deformation resilience. *Journal of the Geological Society, London, 169*(3), 297–310. https://doi.org/10.1144/0016-76492011-086

Faulkner, D. R., Jackson, C. A. L., Lunn, R. J., Schlische, R. W., Shipton, Z. K., Wibberley, C. A. J., & Withjack, M. O. (2010). A review of recent developments concerning the structure, mechanics and fluid flow properties of fault zones. *Journal of Structural Geology, 32*, 1557–1575. https://doi.org/10.1016/j.jsg.2010.06.009

Gale, J. F., Laubach, S. E., Olson, J. E., Eichhubl, P., & Fall, A. (2014). Natural fractures in shale: A review and new observations. *AAPG Bulletin, 98*(11), 2165–2216. https://doi.org/10.1306/08121413151

Gersztenkorn, A., & Marfurt, K. J. (1999). Eigenstructure-based coherence computations as an aid to 3-D structural and stratigraphic mapping. *Geophysics, 64*(5), 1468–1479. https://doi.org/10.1190/1.1444651

Harding, T. P. (1985). Seismic characteristics and identification of negative flower structures, positive flower structures and positive structural inversion. *AAPG Bulletin, 69*(4), 1016–1058.

Heesakkers, V., Murphy, S., & Reches, Z. (2011a). Earthquake rupture at focal depth, Part I: Structure and rupture of the pretorius fault, TauTona Mine, South Africa. *Pure and Applied Geophysics, 168*, 2395–2425. https://doi.org/10.1007/s00024-011-0354-7

Heesakkers, V., Murphy, S., Lockner, D. A., & Reches, Z. (2011b). Earthquake rupture at focal depth, Part II: Mechanics of the 2004 M2.2 earthquake along the pretorius fault, TauTona Mine. *South Africa. Pure and Applied Geophysics, 168*, 2427–2449. https://doi.org/10.1007/s00024-011-0355-6

Iacopini, D., & Butler, R. W. H. (2011). Imaging deformation in submarine thrust belts using seismic attributes. *Earth and Planetary Science Letters, 302*, 414–422. https://doi.org/10.1016/j.epsl.2010.12.041

Iacopini, D., Butler, R. W. H., & Purves, S. (2012). Seismic imaging of thrust faults and structural damage visualization workflow for deepwater thrust belts. *First Break, 30*, 39–46. https://doi.org/10.3997/1365-2397.30.5.58681

Iacopini, D., Butler, R. W. H., Purves, S., McArdle, N., & De Freslon, N. (2016). Exploring the seismic expression of fault zones in 3D seismic volume. *Journal of Structural Geology, 89*, 54–73. https://doi.org/10.1016/j.jsg.2016.05.005

Johnson, K. S (1988) Geologic evolution of the Anadarko basin. In K. S. Johnson (Ed.), *Anadarko basin Symposium: Circular 90, Oklahoma Geological Survey* (pp. 3–12).

Katz, O., Reches, Z., & Baer, G. (2003). Faults and their associated host rock deformation: Part I. Structure of small faults in a quartz–syenite body, southern Israel. *Journal of Structural Geology, 25*, 1675–1689. https://doi.org/10.1016/s0191-8141(03)00011-7
Kim, Y. S., Peacock, D. C. P., & Sanderson, D. J. (2004). Fault damage zones. *Journal of Structural Geology, 26*, 503–517. https://doi.org/10.1016/j.jsg.2003.08.002
Lambert, M. W. (1993). Internal stratigraphy and organic facies of the Devonian-Mississippian Chattanooga (Woodford) Shale in Oklahoma and Kansas: Source rocks in a sequence stratigraphic framework. *AAPG Studies in Geology, 37*, 163–176. https://doi.org/10.1306/st37575c11
Laubach, S. E., Eichhubl, P., Hargrove, P., Ellis, M. A., & Hooker, J. N. (2014). Fault core and damage zone fracture attributes vary along strike owing to interaction of fracture growth, quartz accumulation, and differing sandstone composition. *Journal of Structural Geology, 68*, 207–226. https://doi.org/10.1016/j.jsg.2014.08.007
Li, F., Zhao, T., Lin, T., & Marfurt, K. J. (2015). Fracture characterization based on attenuation estimation from seismic reflection data using well-log-based localized spectral correction. *Unconventional Resources Technology Conference (URTeC)*. https://doi.org/10.2118/178529-ms
Liao, Z., Reches, Z., Marfurt, K. J., & Gupta, N. (2013). Characterizing a fault-zone and associated fractures using lab experiments and attribute-based seismic analysis: An example from Woodford Shale. In *Proceedings of the 83rd annual international meeting of the SEG expanded abstracts* (pp. 1416–1420). https://doi.org/10.1190/segam2013-0656.1
Liao, Z., Liu, H., Jiang, Z., Marfurt, K. J., & Reches, Z. (2017). Fault damage zone at subsurface: A case study using 3D seismic attributes and a clay model analogue for the Anadarko Basin, Oklahoma. *Interpretation, 5*, 143–150. https://doi.org/10.1190/int-2016-0033.1
Marfurt, K. J., & Rich, J. (2010) Beyond curvature: Volumetric estimation of reflector rotation and convergence. In *Proceedings of the 80th annual international meeting of the SEG, expanded abstracts* (pp. 1467–1472). https://doi.org/10.1190/1.3513118
Naylor, M. A., Mandl, G., & Supesteun, C. H. K. (1986). Fault geometries in basement-induced wrench faulting under different initial stress states. *Journal of Structural Geology, 8*, 737–752. https://doi.org/10.1016/0191-8141(86)90022-2
Paxton, S. T., Cruse, A. M., & Krystyniak, A. M. (2006). Detailed fingerprints of global sea-level change revealed in Upper Devonian/Mississippian Woodford Shale of south-central Oklahoma. AAPG Annual Meeting.
Peng, Z., Ben-Zion, Y., Michael, A. J., & Zhu, L. (2003). Quantitative analysis of seismic fault zone waves in the rupture zone of the 1992 Landers, California, earthquake: Evidence for a shallow trapping structure. *Geophysical Journal International, 155*, 1021–1041. https://doi.org/10.1111/j.1365-246x.2003.02109.x
Powers, P. M., & Jordan, T. H. (2010). Distribution of seismicity across strike-slip faults in California. *Journal of Geophysical Research, 115*(B05), 3040. https://doi.org/10.1029/2008jb006234
Reches, Z. (1988). Evolution of fault patterns in clay experiments. *Tectonophysics, 145*(1), 141–156. https://doi.org/10.1016/0040-1951(88)90322-8
Reches, Z., & Lockner, D. A. (1994). Nucleation and growth of faults in brittle rocks. *Journal of Geophysical Research Atmospheres, 99*, 18159–18173. https://doi.org/10.1029/94jb00115
Rempe, M., Mitchell, T., Renner, J., Nippress, S., Ben-Zion, Y., & Rockwell, T. (2013). Damage and seismic velocity structure of pulverized rocks near the San Andreas Fault. *Journal of Geophysical Research, 118*, 2813–2831. https://doi.org/10.1002/jgrb.50184
Sagy, A., Reches, Z., & Roman, I. (2001). Dynamic fracturing: Field and experimental observations. *Journal of Structural Geology, 23*, 1223–1239. https://doi.org/10.1016/s0191-8141(00)00190-5
Savage, H. M., & Brodsky, E. E. (2011). Collateral damage: Evolution with displacement of fracture distribution and secondary fault strands in fault damage zones. *Journal of Geophysical Research, 116*(B03405), 428–452. https://doi.org/10.1029/2010jb007665

Shi, Z., & Yehuda, B. (2006). Dynamic rupture on a bimaterial interface governed by slip-weakening friction. *Geophysical Journal International, 165*, 469–484. https://doi.org/10.1111/j.1365-246x.2006.02853.x

Sibson, R. H. (1990). Conditions for fault-valve behaviour. *Geological Society London Special Publications, 54*, 15–28. https://doi.org/10.1144/gsl.sp.1990.054.01.02

Sibson, R. H. (2003). Thickness of the seismic slip zone. *Bulletin of the Seismological Society of America, 93*(3), 1169–1178. https://doi.org/10.1785/0120020061

Slatt, R. M., Portas, R., Buckner, N., Abousleiman, Y., O'Borien, N. M., Sierra, R., et al. (2012). *Outcrop-behind outcrop (quarry): Multiscale characterization of the Woodford gas shale, Oklahoma* (pp. 382–402). AAPG Memoir.https://doi.org/10.1306/13321481m97441

Valoroso, L., Chiaraluce, L., & Collettini, C. (2014). Earthquakes and fault zone structure. *Geology, 42*, 343–346. https://doi.org/10.1130/g35071.1

Walsh, J. J., & Watterson, J. (1987). Distributions of cumulative displacement and seismic slip on a single normal fault surface. *Journal of Structural Geology, 9*, 1039–1046. https://doi.org/10.1016/0191-8141(87)90012-5

Weng, H., Yang, H., Zhang, Z., & Chen, X. (2016). Earthquake rupture extents and coseismic slips promoted by damaged fault zones. *Journal of Geophysical Research Solid Earth, 121*, 4446–4457. https://doi.org/10.1002/2015jb012713

Wilson, J. E., Chester, J. E., & Chester, F. M. (2003). Microfracture analysis of fault growth and wear processes, Punchbowl Fault, San Andreas system, California. *Journal of Structural Geology, 25*, 1855–1873. https://doi.org/10.1016/s0191-8141(03)00036-1

Xu, S., Ben-Zion, Y., & Ampuero, J. P. (2012). Properties of inelastic yielding zones generated by in-plane dynamic ruptures-II. Detailed parameter-space study. *Geophysical Journal International, 191*(3), 1343–1360. https://doi.org/10.1111/j.1365-246x.2012.05685.x

Zoback, M. D., Hickman, S., & Ellsworth, W. (2011). Scientific drilling into the San Andreas Fault Zone - An overview of SAFOD's first five years. *Scientific Drilling, 11*, 14–28. https://doi.org/10.5194/sd-11-14-2011

Chapter 4
Composite Damage Zones

Abstract The cumulative displacement of multiple slip events along faults may generate composite damage zones of increasing width and modification of hydraulic and mechanical properties. The structure of composite damage zones at the subsurface is analyzed here by using seismic attributes of a three-dimensional, seismic reflection survey. We investigate the geometry, internal architecture, and fracture distribution within tight sandstone reservoirs in northeast Sichuan, China. Based on the damage intensity, we identified six composite damage zones with thickness approaching 1 km along faults ranging 3–15 km in length, and up to 600 m of cumulative slip. The recognition of multiple fault cores and associated damage zones inside the composite damage zones revealed three classes of composite damage zones that are defined as banded-shape, box-shape, and dome-shape. The observed mechanical contrasts and distortion of fault cores indicate an evolution path for the growth of composite damage zones. The analyses could extend to complex, brittle crustal faulting, e.g., for earthquake rupture characteristics and hydraulic behavior.

4.1 Introduction

Faulting is typically preceded by fragmentation and slip along microfractures that eventually coalesce to form a macroscopic fault zone (Reches and Lockner, 1994; Scholz, 2002; Savage and Brodsky, 2011). The fragmented and fractured rock body form a wide damage-zone with a central core of the localized slip as recognized in multiple experiments, field exposures and subsurface analyses (Fig. 4.1) (Billi et al., 2003; Faulkner et al., 2003; Katz et al., 2003; Wilson et al., 2003). The existence of the damage zones affects the strength of the local crust, and the migration, accumulation, and leakage of subsurface fluids (Caine et al., 1996; Ellis et al., 2012; Faulkner et al., 2010; Lyakhovsky & Ben-Zion, 2009; Savage & Brodsky, 2011). Renew deformation at the proximity of an existing fault-zone can occur at two general locations: (1) reactivation of fault with slip along the existing fault core, a process that leads to long-term fault weakening, or (2) slip away of the fault-core, within the damage-zone or at the contact with the intact host rock. This process widens the deformed zone to

Z. Liao, *Applied Seismic Geostructures*, https://doi.org/10.1007/978-981-97-7079-3_4

develop a complex structure (Fig. 4.1) that can be described as composite-damage-zone (CDZ) (Caine et al., 1996; Faulkner et al., 2003; Johansen et al., 2005). Analyses of exposed CDZ reveals multiple fault cores of localized slip with associated damage zones that can be hundreds of meters wide (e.g., Di Toro & Pennacchioni, 2005; Faulkner et al., 2003). Typically, the intensity of fracture damage is higher close to the fault core, and diminishes outward (e.g., Billi et al., 2003; Faulkner et al., 2010; Sagy et al., 2001; Savage & Brodsky, 2011; Wilson et al., 2003). The present study focuses on a seismic analysis of CDZ in the subsurface where rock fractures and pervasive damage are invisible (Chopra & Marfurt, 2007; Iacopini & Butler, 2011), and we determine the damage distribution by utilizing three-dimensional seismic data.

Damage zones can be detected in the subsurface by using synthetic wave modeling (Peng et al., 2003; Xu et al., 2012) or seismological events at large scale (1–10 km), e.g., the variation of seismicity rate (Powers & Jordan, 2010), and earthquake locations (Valoroso et al., 2014). We use seismic attributes that can provide good resolution (~50 m) for characterizing fault structure and its surrounding deformation at medium-large scale by inferring from seismic disturbance volume indirectly (Iacopini et al., 2012; Liao et al., 2019).

The analysis utilizes 3D seismic data from the Tongnanba anticline in the northeast of the Sichuan Basin, China, which is part of the Sichuan gas filed. The anticline is bounded by the Micangshan thrust system in the north, and Dabashan thrust fold belt in the east (Fig. 4.2), and Longmenshan fault and fold belt in the far west. The Tongnanba anticline underwent two major tectonic stages (Dai et al., 2013; Li et al.,

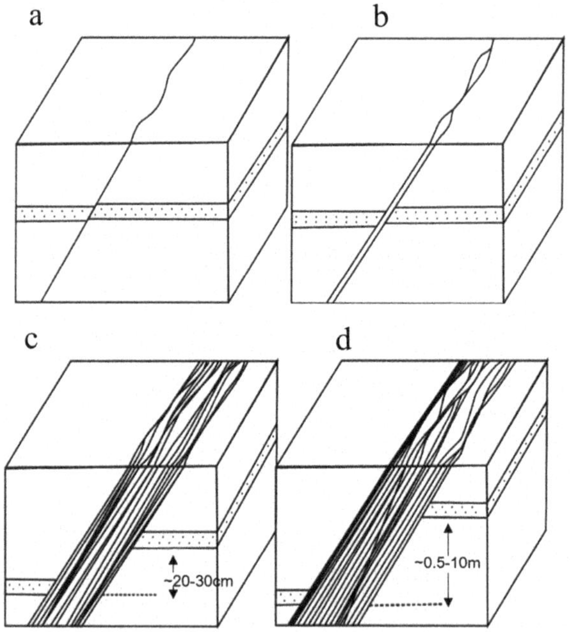

Fig. 4.1 Sequential development from **a** a single damage band to **b** two inosculating bands, **c** a zone of bands, and **d** multi-banded damage zones (after Aydin and Johnson, 1978)

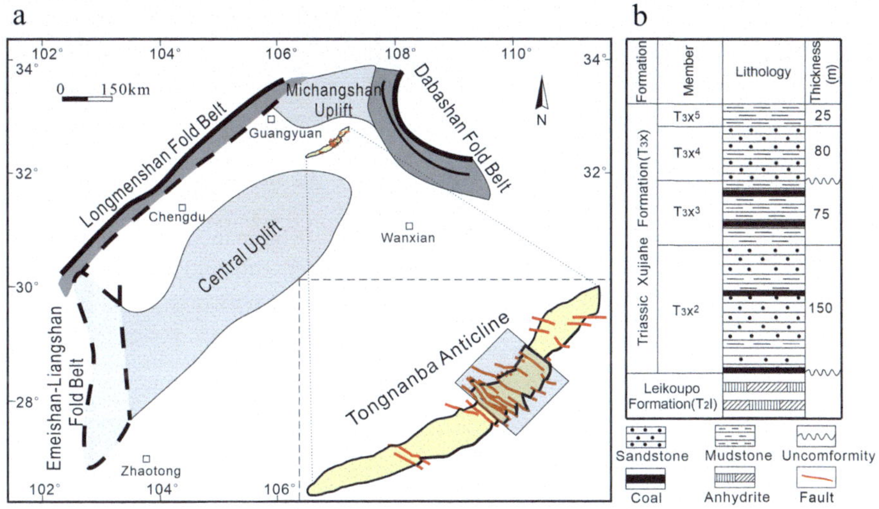

Fig. 4.2 **a** Study area of the Tongnanba anticline in northeast of Sichuan Basin (inset figure), China. Seismic data covers the shading area. The reverse faults are marked red. The yellow is the roll anticline striking northeast. **b** Stratigraphy of the Xujiahe Formation

2016). First, NW–SE compression of the middle-to-late Jurassic, of the Micangshan deformation that led to the formation of the 100-km-long 8-km-wide Tongnanba anticlinal fold (Fig. 4.2). Second, the NE-SW compression by the Dabashan thrust of the Himalayan orogenic movement that extensively deforms the Tongnanba anticline by a system of reverse faults trending NW–SE (Fig. 4.3). The present analysis focuses on the structure and evolution of the NW–SE reverse faults and their associated damage zones. These faults intersect and deform the principal gas-bearing tight sandstone, and it is expected that the high, natural production capacity of this member is controlled by extensive fault deformation and damage.

4.2 Seismic Attributes

The structure of the reverse faults is analyzed by using a 3D seismic reflection dataset of an area of about 225 km2, which was provided by Sinopec China. This dataset covers parts of the Xujiahe sandstones of Upper Triassic age, with the major reverse faults and Tongnanba anticline, northeast Sichuan Basin (Fig. 4.3a). The dataset was collected at frequency ranged from 10 to 60 Hz yielding the increased impedance as positive amplitude, with Kirchhoff pre-stack time migration. To identify the subsurface structure and characterize the damage zones, we computed a set of four volumetric seismic attributes (Fig. 4.3): reflection amplituide, dip-azimuth, curvature, and seismic variance (Marfurt & Rich, 2010; Zonghu et al., 2018).

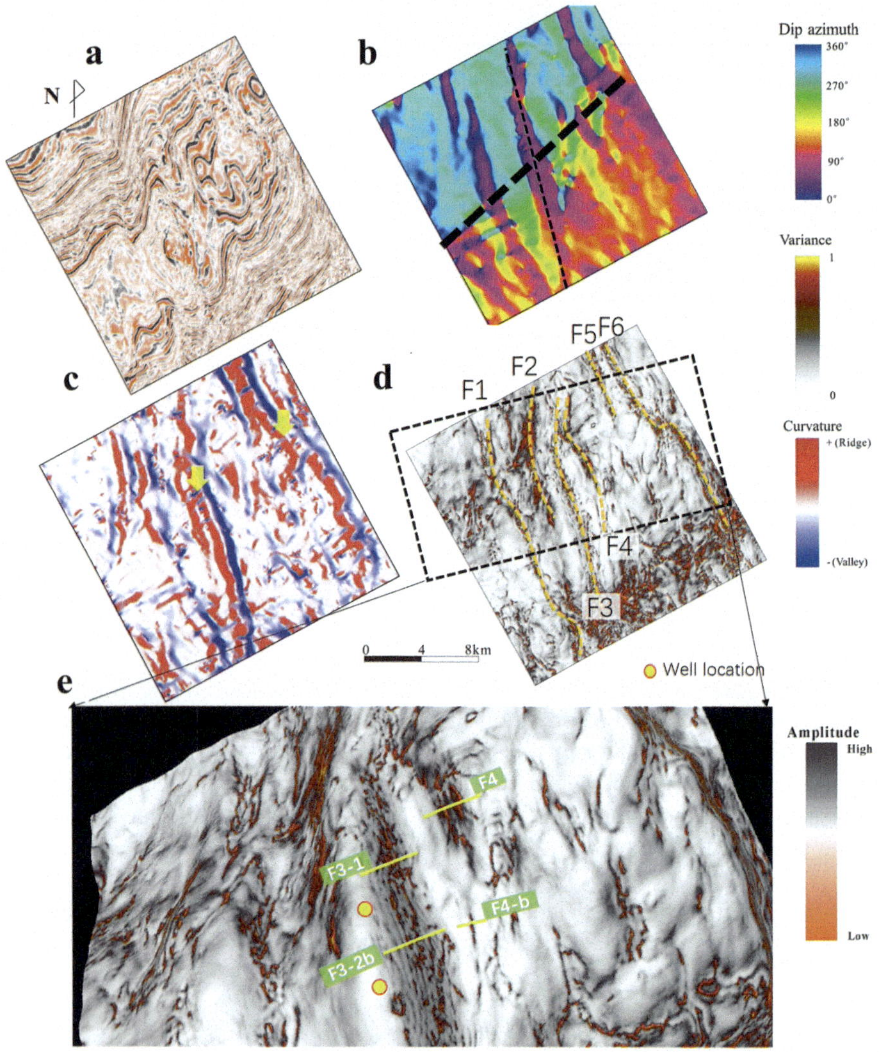

Fig. 4.3 **a** Seismic amplitude map and seismic attributes calculated for tight sandstone in 2nd member of Xujiahe Formation in the study area, including **b** dip-azimuth, **c** curvatures, **d** variance and zoom-in figure (**e**). Note the yellow dash lines are interpreted as faults (in **d**). Locations of wells and fault profiles for Fig. 4.3 are marked

The 'dip-azimuth' attribute, which has the standard geological definition, is mapped in Fig. 4.3b for a volume reflector of the Xujiahe Formation in the time-migrated seismic profiles (Marfurt et al., 1998; Marfurt, 2006). This map displays two superposed structural features (Fig. 4.3b): First, the Tongnanba anticline, with general axis marked by a thick, dashed line, has one limb (green–blue) dipping toward 335°-345°, and the other limb (red-yellow) dipping to the southeast, 90°-120°. Second, three large zones (pink), with trend marked by a thin, dotted line, of local dipping reflector toward 50°-80°, that cut across the Tongnanba anticline. To explore these zones, we computed (Fig. 4.3c) the most-positive (K1) and most-negative (K2) principal curvature volumes. These curvatures are the 2nd derivatives with most positive curvature (red) for anticlinal flexures and negative curvature (blue) for synclinal flexures, and together they reveal flexures related to the reverse faults (Chopra & Marfurt, 2007). The curvature map (Fig. 4.3c) reveals abundant folding information: (1) two major fold zones (green arrows) striking 335° to 345°; (2) each fold zone is composed of a red anticlines in west and blue synclines in east; (3) many small northeast-trending lineaments could be recognized on top of the two major folds.

Finally, to identify the damage zones, we mapped the variance attribute (Fig. 4.2d) that is defined as the energy of the coherent part of seismic reflector divided by the average acoustic energy of the input seismic traces (Chopra & Marfurt, 2007). The energy-ratio coherence algorithm that is applied here, has high coherence (no fractures), which is equivalent to low variance. The 3D variance of the mapped Tongnanba area is computed along reflector dip, and the result reveals (Fig. 4.3d) that the Tongnaba anticline is interrupted by three long high variance lineaments accompanied by three shorter lineaments that strike 335° to 345°, parallel to the Dabashan thrust fold belt, about 60 km northeast of the mapped area (Fig. 4.2). We refer to this system as the northwest-trending reverse faults.

Based on the above three seismic attributes (Fig. 4.3), we recognize six major reverse faults striking between 335° and 345° that are marked by yellow dash lines lineament and numbered as F1 to F6 (Fig. 4.3d). The curvature and dip-azimuth attribute indicate that F3 and F6 are associated with intense deformation and wide secondary anticlinal and synclinal flexures. The other faults display relatively smaller deformation intensity.

4.3 Composite Damage Zones

Fracture networks and small faults are typically invisible by seismic data. To detected the subsurface damage zones, we used the 'seismic variance signal' that is a 3D attribute which integrates the disturbance to seismic waves due to the cumulative effects of small faults and fractures (Chopra and Marfurt, 2007; Li et al., 2016; Liao et al., 2019). This attribute was also defined as the 'seismic disturbance zone' (Iacopini et al., 2012). The detected seismic disturbance zones detected in the present analysis are regarded as CDZs associated with the major reverse faults described above. The seismic variance levels in 3D-seismic analysis indicate the intensity of structural disturbance and discontinuities, which we interpret as three levels of damage intensity. The high variance zone is fault core that is most intensely damaged, the intermediate level zone is the damage zone, and the low variance zone is protolith zone away from the fault.

Liao et al. (2019) showed that the variance zonation in the subsurface fits well damage zones observations in field exposures the previous researches (Mitchell & Faulkner, 2009; Sagy et al., 2001; Savage & Brodsky, 2011; Wilson et al., 2003). Superposition of multiple faults is expected to generate CDZs (Chester et al., 2004; Savage & Brodsky, 2011).

We prepared 18 vertical profiles of the seismic variance that are arranged in four sections across faults F3 and F4 (yellow lines in Fig. 4.3e). The profiles in each section display a similar distribution of the seismic variance that we generalized as the characteristic damage zones shown by dashed red curves in Fig. 4.4d, f, h. These sections reveal four styles of damage zones.

Pulse section (F4-b in Fig. 4.4a) is a typical damage zone of a single fault (Zonghu et al., 2018) with a central zone of high variance, >0.4, (pink in Fig. 4.4a), and two zones of intermediate variance, 0.1–0.4, that decay exponentially to the background variance, 0–0.1 (Fig. 4.4b). A schematic section is shown in Fig. 4.4c.

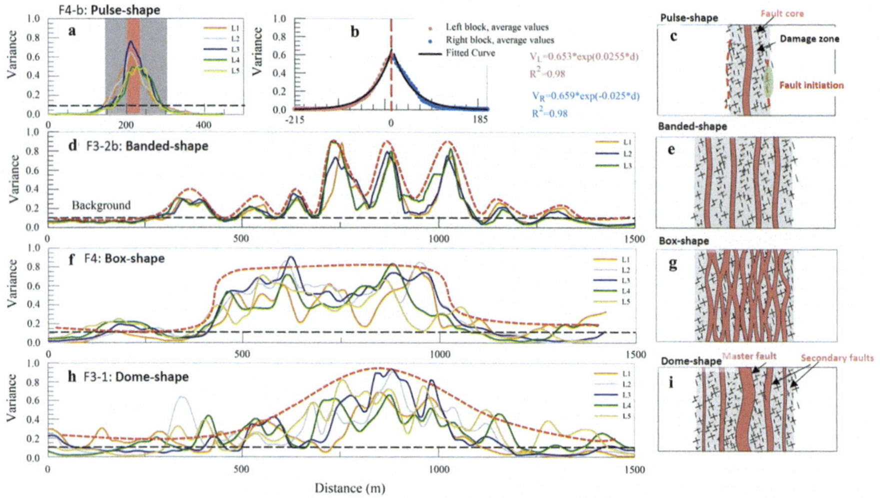

Fig. 4.4 Profiles of the seismic variance values across reverse faults at the Xujiahe tight sandstone level and internal structures of composite damage zones. Profiles locations in Fig. 4.3e. Zones of variance values above background variance are interpreted as damage zones. **a** Single core damage zone along Fault 4-b. Variance intensity shown in colors: Pink- intense; light pink -less intense; grey- gentle; light grey-close to background; white- background; **b** Synthesis of the pulse-shape individual damage zone; **c** Individual fault and pulse-shape damage zone; New faults develop on the contacting area, marked by dashed red lines; **d** banded damage zones along Fault 3-2b; **e** banded-shape damage zones with equal-size faults parallel/sub-parallel to each other; **f** Box damage zones along Fault 4; **g** Box-shape damage zones with intensive and similar size faults inosculated; **h** Dome damage zones along Fault 3–1; **i** Dome-shape damage zones with a favored fault in the center and a sequence of sub-faults on two sides. Note all three classes of composite damage zones are ideal schemes in dash red lines

Banded section of ~900 m width (F3-2b in Fig. 4.4d) with three large faults cores (high variance) that are separated by corresponding damage zones; note three smaller cores on the left. A schematic section is shown in Fig. 4.4e.

Box section of ~500 m width (F4 in Fig. 4.4f) with multiple anastomosing fault cores and damage zones without distinct variance pulses of individual cores. A schematic section is shown in Fig. 4.4g.

Dome section of ~1,200 m width (F3-1 in Fig. 4.4h) with multiple cores and damage zones that gradually decay in variance intensity from the central core. A schematic section is shown in Fig. 4.4i.

The analyzed sections of Fig. 4.4 are located close to two gas-producing wells (yellow dots in Fig. 4.3e). The drilling core of Well M101 show the microfractures of tight sandstone buried in 3230–3245-m deep within the 2nd Xujiahe formation. These samples are at the outmost area of the dome damage zones of F3-1 of F3 where the seismic variance is 0.2–0.3. Recent research of the core and its thin section (Dai et al., 2013; Li et al., 2016) indicates a diagenesis history of the sandstone tightening including compaction, dissolution, quartz overgrowth, and carbonate cementations. While the estimated pore space is ~1% in the thin sections, they also display many microfractures with dominating intragranular tensile fractures and a few transgranular fractures (Mizoguchi & Ueta, 2013). These open microfracture are expected to be part of an interconnected fracture network. We envision that the damage intensity and large width of the dome section of F3 contributed to the large volume of fractured-space for natural gas of Well M101 (dots in Fig. 4.3e).

4.4 Discussion: Evolution of Composite Damage Zones

Increasing displacement along a fault system may generate new faults with separate cores and damage zones that form a composite damage zone, as described above and elsewhere (Aydin and Johnson, 1978; Faulkner et al., 2003). The formation of a CDZ is somewhat puzzling because it is expected that an existing fault-zone is the weakest regional feature, and continuous deformation should localize along it. Indeed, this style prevails along many faults with slip localization within zones of $\leq$ 1 m thickness along faults of tens of km, e.g., the San-Andreas fault (Chester & Logan, 1986), or the Glarus thrust (Badertscher & Burkhard, 2000). This localization indicates that the fault maintains low static strength relative to the host blocks (Lockner et al., 2011). CDZs, on the other hand, will form when the existing fault is stronger than the surrounding rock mass, and three main settings could lead to this situation:

CDZ can develop if the slip along a fault produces fault-core that is stronger than the host rock. Aydin and Johnson (1978) investigated the evolution of fault zones in porous sandstone layers. They recognized a hierarchy of three structural levels of shear structures: deformation bands of a few mm slip and ~1 mm thickness, zones of deformation bands that are composed of tens, sub-parallel deformation bands, and slip surfaces of localized slip on the order of meters (Fig. 4.1a). Aydin and Johnson (1978) attributed this evolution to the strain hardening of the deformation bands due to the reduction of the sandstone grain-size. Another example was presented in Di Toro and Pennacchioni (2005) study of the Gole Larghe fault in the Adamello granitoid intrusion, the Alps. The fault-zone is ~550 m wide that is consists of ~200 main segments composed of pseudotachylites and cataclasites, which carry displacements up to a few meters each, and are separate by damage host rocks. The continuous shear along Gole Larghe fault-zone formed hundreds of new segments in a CDZ (Di Toro & Pennacchioni, 2005), because the shear strength of the glassy segments was higher than the strength of the joints in the damaged host rock.

CDZ can form due to healing and strengthening of a fault-zone during periods of no-slip as demonstrated experimentally (e.g., Muhuri et al., 2003). Heesakkers et al. (2011) map the rupture zone of a M2.2 earthquake that reactivated an ancient fault network in Tautona Mine, South Africa. They found that the slip was systematically localized along the contacts between the ancient fault-core composed of healed cataclasite, and the damaged quartzitic host rocks. Their analysis indicated that this localization is controlled by the contrast of mechanical properties between the cataclasite and the host rocks (Heesakkers et al., 2011). Multiple occurrences of this process could lead to a widening of the damage-zone and formation of a CDZ.

A third mechanism is related to structural complexity. Segment linkage of non-planar faults (Caine et al., 1996; Faulkner et al., 2010; Laubach et al., 2014), that by virtue of their contorted geometry become stronger than a planar fault (Faulkner et al., 2010). Distorted fault segments will then interact with adjacent ones, and incorporate wall rocks into the CDZs with various structures. Also, the structural complexity can lead to co-existence of distributed shear in phyllosilicate, strain-hardening fault cores and localized shear in carbonate (Fig. 4.1b) (Faulkner et al., 2003).

The 3D seismic attribute method provides an effective tool to quantify the relative damage intensity in the subsurface, and led to the recognition of three CDZs types termed as the banded, box, and dome CDZs (Fig. 4.4e, g, i). We evaluate the evolution of these CDZ by using the variance intensity as an indicator of the internal damage (Fig. 4.4a). It is suggested that the banded-CDZ (Fig. 4.4d, e) formed by the superposition of several sub-parallel faults of an approximately similar amount of slip that is spaced at equal distance from each other. This evolution style fits the above mechanisms in which ancient fault cores are stronger than the surrounding rock bodies either during faulting (mechanism A) or by healing during quiescence periods (mechanism B). The box-CDZ (Fig. 4.4f, g) evolved by coalescence of multiple, closely-spaced faults that linked separated fault segments that formed in earlier slip phases, in agreement with mechanism C (above). We envision that dome-CDZ (Fig. 4.4h, i) started as banded-CDZ with multiple subparallel, strong faults, and during an advanced stage, one of the fault-cores, in the center or margins, became permanently weaker to accommodate later deformation by localized slip. This evolution fits the structural model of Aydin and Johnson (1978) (Fig. 4.1a) in which after the development of multiple, closely spaced deformation-bands, each with ~1 mm slip, a weak slip-surface with meters of slip develops at the margin of the damage-zone.

4.5 Conclusions

Our data indicate that seismic attributes could be used to characterize fault structure and related damage zones in the subsurface. The analysis of the tight sandstone reservoir in northeast Sichuan Basin revealed CDZs composed of multiple cores and damage zones with a total width up to 1 km. Based on the damage distribution, we recognized three types of CDZ: the banded-shape with a few, clearly separated fault cores; the box-shape with a few fault-cores that coalesce with each other; and dome-shape with one dominant fault-core and a few secondary ones. We infer that these CDZs evolved due to the post-slip strengthening of the existing fault-cores, and geometric complexity of segmented fault-zones. The present utilization of the 3D seismic attributes appears as a reliable tool to evaluate the internal structure of CDZs in the subsurface that provides an effective tool for reservoir analysis.

References

Aydin, A., & Johnson, A. M. (1978). Development of faults as zones of deformation bands and as slip surfaces in sandstone. *Pure and Applied Geophysics, 116*(4–5), 931–942. https://doi.org/10.1007/BF00876568

Badertscher, N. P., & Burkhard, M. (2000). Brittle-ductile deformation in the Glarus thrust Lochseiten (LK) calc-mylonite. *Terra Nova, 12*(6), 281–288. https://doi.org/10.1046/j.1365-3121.2000.00308.x

Ben-Zion, Y. (2006). Comment on "Material contrast does not predict earthquake rupture propagation direction" by RA Harris and SM Day. *Geophysical Research Letters, 33*(13), 25652. https://doi.org/10.1029/2005GL025652

Billi, A., Salvini, F., & Storti, F. (2003). The damage zone-fault core transition in carbonate rocks: Implications for fault growth, structure and permeability. *Journal of Structural Geology, 25*, 1779–1794. https://doi.org/10.1016/S0191-8141(02)00206-0

Botter, C., Cardozo, N., Hardy, S., Lecomte, I., Paton, G., & Escalona, A. (2016). Seismic characterization of fault damage in 3D using mechanical and seismic modelling. *Marine and Petroleum Geology, 77*, 973–990. https://doi.org/10.1016/j.marpetgeo.2016.07.007

Botter, C., Cardozo, N., Qu, D., Tveranger, J., & Kolyukhin, D. (2017). Seismic characterization of fault facies models. *Interpretation, 5*(4), 9–26. https://doi.org/10.1190/INT-2016-0163.1

Busetti, S., Mish, K., Hennings, P., & Reches, Z. E. (2012). Damage and plastic deformation of reservoir rocks: Part 2. Propagation of a hydraulic fracture. *AAPG Bulletin, 96*(9), 1711–1732. https://doi.org/10.1306/03201211129

Caine, J. S., Evans, J. P., & Forster, C. B. (1996). Fault zone architecture and permeability structure. *Geology, 24*(11), 1025–1028. https://doi.org/10.1130/0091-7613(1996)024%3c1025:FZAAPS%3e2.3.CO;2

Cavailhes, T., Soliva, R., Labaume, P., Wibberley, C., Sizun, J. P., Gout, C., et al. (2013). Phyllosilicates formation in faults rocks: Implications for dormant fault-sealing potential and fault strength in the upper crust. *Geophysical Research Letters, 40*(16), 4272–4278. https://doi.org/10.1002/grl.50821

Chester, F. M., Chester, J. S., Kirschner, D. L., Schulz, S. E., & Evans, J. P. (2004). Structure of large-displacement, strike-slip fault zones in the brittle continental crust. *Rheology and Deformation in the Lithosphere at Continental Margins, 1*, 223–260. https://doi.org/10.1017/CBO9780511529349.009

Chester, F. M., & Logan, J. M. (1986). Implications for mechanical properties of brittle faults from observations of the Punchbowl fault zone, California. *Pure and Applied Geophysics, 124*(1–2), 79–106. https://doi.org/10.1007/BF00876546

Chopra, S., & Marfurt, K. J. (2007). Seismic attributes for prospect identification and reservoir characterization. *Society of Exploration Geophysicists and European Association of Geoscientists and Engineers, 12*, 2170. https://doi.org/10.1190/1.9781560802170

Dai, J., Fengrong, L., & Yunyan, N. (2013). Discussions on the gas source of the Triassic Xujiahe Formation tight sandstone gas reservoirs in Yuanba and Tongnanba, Sichuan Basin: an answer to Yin Feng et al. *Petroleum Exploration and Development, 40*(2), 268–275. https://doi.org/10.1016/S1876-3804(13)60034-3

Di Toro, G., & Pennacchioni, G. (2005). Fault plane processes and mesoscopic structure of a strong-type seismogenic fault in tonalites (Adamello batholith, Southern Alps). *Tectonophysics, 402*(1–4), 55–80. https://doi.org/10.1016/j.tecto.2005.03.010

Ellis, M. A., Laubach, S. E., Eichhubl, P., Olson, J. E., & Hargrove, P. (2012). Fracture development and diagenesis of Torridon Group Applecross Formation, near An Teallach, NW Scotland: Millennia of brittle deformation resilience? *Journal of the Geological Society, 169*(3), 297–310. https://doi.org/10.1144/jgs2011-129

Faulkner, D. R., Jackson, C. A. L., Lunn, R. J., Schlische, R. W., Shipton, Z. K., Wibberley, C. A. J., & Withjack, M. O. (2010). A review of recent developments concerning the structure, mechanics and fluid flow properties of fault zones. *Journal of Structural Geology, 32*(11), 1557–1575. https://doi.org/10.1016/j.jsg.2010.06.009

Faulkner, D. R., Lewis, A. C., & Rutter, E. H. (2003). On the internal structure and mechanics of large strike-slip fault zones: Field observations of the Carboneras fault in southeastern Spain. *Tectonophysics, 367*, 235–251. https://doi.org/10.1016/S0040-1951(03)00174-2

Gersztenkorn, A., & Marfurt, K. J. (1999). Eigenstructure-based coherence computations as an aid to 3-D structural and stratigraphic mapping. *Geophysics, 64*(5), 1468–1479. https://doi.org/10.1190/1.1444651

Heesakkers, V., Murphy, S., Lockner, D. A., & Reches, Z. (2011). Earthquake rupture at focal depth, Part II: Mechanics of the 2004 M2.2 earthquake along the Pretorius fault, TauTona Mine, South Africa. *Pure and Applied Geophysics, 168*(12), 2427–2449. https://doi.org/10.1007/s00024-011-0298-0

Iacopini, D., & Butler, R. W. (2011). Imaging deformation in submarine thrust belts using seismic attributes. *Earth and Planetary Science Letters, 302*(3–4), 414–422. https://doi.org/10.1016/j.epsl.2010.12.003

Iacopini, D., Butler, R. W. H., & Purves, S. (2012). Seismic imaging of thrust faults and structural damage: A visualization workflow for deepwater thrust belts. *First Break, 30*(5), 77–84. https://doi.org/10.3997/1365-2397.2012007

Johansen, T. E. S., Fossen, H., & Kluge, R. (2005). The impact of syn-faulting porosity reduction on damage zone architecture in porous sandstone: An outcrop example from the Moab Fault, Utah. *Journal of Structural Geology, 27*(8), 1469–1485. https://doi.org/10.1016/j.jsg.2005.04.001

Katz, O., Reches, Z. E., & Baer, G. (2003). Faults and their associated host rock deformation: Part I. Structure of small faults in a quartz–syenite body, southern Israel. *Journal of Structural Geology, 25*(10), 1675–1689. https://doi.org/10.1016/S0191-8141(03)00027-3

Laubach, S. E., Eichhubl, P., Hargrove, P., Ellis, M. A., & Hooker, J. N. (2014). Fault core and damage zone fracture attributes vary along strike owing to interaction of fracture growth, quartz accumulation, and differing sandstone composition. *Journal of Structural Geology, 68*, 207–226. https://doi.org/10.1016/j.jsg.2014.07.018

Li, J., Hu, D., Zou, H., Shang, X., Ren, H., & Wang, L. (2016). Coupling relationship between reservoir diagenesis and gas accumulation in Xujiahe Formation of Yuanba–Tongnanba area, Sichuan Basin, China. *Journal of Natural Gas Geoscience, 1*(5), 335–352. https://doi.org/10.1016/j.jnggs.2016.11.001

Liao, Z., Liu, H., Carpenter, B. M., Marfurt, K. J., & Reches, Z. E. (2019). Analysis of fault damage zones using three-dimensional seismic coherence in the Anadarko Basin, Oklahoma. *AAPG Bulletin, 103*(8), 1771–1785. https://doi.org/10.1306/1219181413417207

Lockner, D. A., Morrow, C., Moore, D., & Hickman, S. (2011). Low strength of deep San Andreas fault gouge from SAFOD core. *Nature, 472*, 82–85. https://doi.org/10.1038/nature09927

Lyakhovsky, V., & Ben-Zion, Y. (2009). Evolving geometrical and material properties of fault zones in a damage rheology model. *Geochemistry, Geophysics, Geosystems, 10*, Q11011. https://doi.org/10.1029/2009GC002543

Marfurt, K. J., & Rich, J (2010) Beyond curvature: Volumetric estimates of reflector rotation and convergence. In *SEG technical program expanded abstracts* (pp. 1467–1472)

Marfurt, K. J. (2006). Robust estimates of 3D reflector dip and azimuth. *Geophysics, 71*(4), 29–40. https://doi.org/10.1190/1.2194923

Marfurt, K. J., Kirlin, R. L., Farmer, S. L., & Bahorich, M. S. (1998). 3-D seismic attributes using a semblance-based coherency algorithm. *Geophysics, 63*(4), 1150–1165. https://doi.org/10.1190/1.1444415

Mitchell, T. M., & Faulkner, D. R. (2009). The nature and origin of off-fault damage surrounding strike-slip fault zones with a wide range of displacements: A field study from the Atacama fault system, northern Chile. *Journal of Structural Geology, 31*(8), 802–816. https://doi.org/10.1016/j.jsg.2009.05.002

Mizoguchi, K., & Ueta, K. (2013). Microfractures within the fault damage zone record the history of fault activity. *Geophysical Research Letters, 40*(10), 2023–2027. https://doi.org/10.1002/grl.50469

Muhuri, S. K., Dewers, T. A., Scott, T. E., Jr., & Reches, Z. (2003). Interseismic fault strengthening and earthquake-slip instability: Friction or cohesion? *Geology, 31*, 881–884. https://doi.org/10.1130/G19601.1

Peng, Z., Ben-Zion, Y., Michael, A. J., & Zhu, L. (2003). Quantitative analysis of seismic fault zone waves in the rupture zone of the 1992 Landers, California, earthquake: Evidence for a shallow trapping structure. *Geophysical Journal International, 155*(3), 1021–1041. https://doi.org/10.1046/j.1365-246X.2003.02060.x

Powers, P. M., & Jordan, T. H. (2010). Distribution of seismicity across strike-slip faults in California. *Journal of Geophysical Research: Solid Earth, 115*(B05305), 6361. https://doi.org/10.1029/2009JB006361

Reches, Z. E., & Lockner, D. A. (1994). Nucleation and growth of faults in brittle rocks. *Journal of Geophysical Research: Solid Earth, 99*(B9), 18159–18173. https://doi.org/10.1029/94JB00115

Sagy, A., Reches, Z. E., & Roman, I. (2001). Dynamic fracturing: Field and experimental observations. *Journal of Structural Geology, 23*(8), 1223–1239. https://doi.org/10.1016/S0191-8141(01)00019-8

Savage, H. M., & Brodsky, E. E. (2011). Collateral damage: Evolution with displacement of fracture distribution and secondary fault strands in fault damage zones. *Journal of Geophysical Research: Solid Earth, 116*(B3), 428–452. https://doi.org/10.1029/2010JB007907

Scholz, C.H. (2002). *The mechanics of earthquakes and faulting*. Cambridge University Press.

Staples, E. R. (2011). Subsurface and experimental analyses of fractures and curvature. *Doctoral dissertation, University of Oklahoma*

Valoroso, L., Chiaraluce, L., & Collettini, C. (2014). Earthquakes and fault zone structure. *Geology, 42*(4), 343–346. https://doi.org/10.1130/G35291.1

Wibberley, C. A., & Shimamoto, T. (2005). Earthquake slip weakening and asperities explained by thermal pressurization. *Nature, 436*, 689–692. https://doi.org/10.1038/nature03901

Wibberley, C. A., Yielding, G., & Di Toro, G. (2008). Recent advances in the understanding of fault zone internal structure: A review. *Geological Society, London, Special Publications, 299*(1), 5–33. https://doi.org/10.1144/SP299.2

Wilson, J. E., Chester, J. S., & Chester, F. M. (2003). Microfracture analysis of fault growth and wear processes, Punchbowl Fault, San Andreas system, California. *Journal of Structural Geology, 25*(11), 1855–1873. https://doi.org/10.1016/S0191-8141(03)00044-6

Xu, S., Ben-Zion, Y., & Ampuero, J. P. (2012). Properties of inelastic yielding zones generated by in-plane dynamic ruptures—II. Detailed parameter-space study. *Geophysical Journal International, 191*(3), 1343–1360. https://doi.org/10.1111/j.1365-246X.2012.05688.x

Part II
Structure Systems

Chapter 5
Multiscale Fracture Within Fault Damage Zones

Abstract In this chapter, we investigate the fractures and fault damage zones in a tight sandstone coal gas-bearing Xujiahe Formation in northeast Sichuan Basin, China, based on comprehensive study of seismic data, well log, core as well as field observations. We demonstrate the structure and distribution of damage zones are demonstrated from basin-scale down to micro-scale. The core samples show mostly opening-mode microfractures that can be subcritically generated under low tectonic stress fields since the Late Triassic-Middle Jurassic. These opening-mode microfractures in such ultra-low permeability sandstone are likely to provide a storage volume for coal gas to accumulate. Macro-fractures from image logging and outcrop display well-developed joint networks within the sandstone. The basin-wide distribution of such macro-fractures implies potential conduits for gas migration during basin uplift in the Cretaceous. The damage zones are formed and controlled by a system of reverse faults, with thicknesses ranging 100–1500 m. The multi-scale analysis of fractures implies that, for the tight sandstone within the Xujiahe Formation, the fractures and damage zones are likely to control the gas migration and accumulation in the upper crust.

5.1 Introduction

Faults are commonly described as zones with a fault core and a surrounding damage zone, which differs structurally, mechanically, and petrophysically from the undeformed host rock (protolith) (Chester and Logan, 1986; Caine et al., 1996; Katz et al., 2003; Sagy et al., 2001; Savage and Brodsky, 2010). The damage zone is determined by the structures that formed as a result of the faulting process, which may present complex nature of its fracture networks, owing to fault-related diagenesis, segmentation, and evolution (Laubach et al., 2014). With limited data, it is challenging to evaluate the structure of damage zones in the subsurface that significantly affects the migration, accumulation, and leakage of subsurface fluids (e.g., Busetti et al., 2012; Caine et al., 1996; Ellis et al., 2012; Faulkner et al., 2010).

Z. Liao, *Applied Seismic Geostructures*, https://doi.org/10.1007/978-981-97-7079-3_5

Fault damage zones that contain fractures may seal the petroleum trap in some areas, but leak in others (Rawling et al., 2001). The heterogeneous structures of fault damage zones can strongly influence hydrologic properties in hydrocarbon basin, including: (1) Fault core thickness and gouge grain size. Eichhubl et al. (2005) presented clay (or gouge) smear within fault core and increasingly localized deformation of the clay component at a sandstone/shale bed interface. Such clay smear is commonly used to describe a collection of fault gouge acting as a capillary seal preventing cross-fault flow (Eichhubl et al., 2005; Vrolijk et al., 2016). Seals could fail due to high-permeability "hydraulic holes" (Caine & Minor, 2009); (2) Fracture network distribution. Fault propagation leads to the creation of a cloud of microfractures as a result of stress concentration at the fault tip. These fractures are distributed and connected within the damage zones, providing primary fluid conduits in the subsurface (Caine et al., 1996; Faulkner et al., 2010); (3) The lithification of protolith rocks. The lithification could be used to predict the juxtapositions for a given fault geometry (Knipe, 1997) and evaluate the sealing potential of the fault (Peacock et al., 2000). Brittle lithified sediments usually enhance horizontal flow in fault damage zones. However, faulting within poorly lithified sediments has less potential to act as vertical and horizontal flow conduits (Rawling et al., 2001). The difference is due to whether there is an absence of macroscopic fractures associated with fault damage zones.

Fractures surrounding faults are usually investigated by outcrop analyses on the surface (Mitchell & Faulkner, 2009; Sagy et al., 2001; Savage and Brodsky, 2009; Wilson et al., 2003). The fractures tend to locate near the fault core, and the fracture density decreases outward from the fault core. The damage zone extends outward from the fault core to a point where structure density approximately equals background density. Such observational outcrop studies are difficult for subsurface fracture characterizations as direct subsurface observation is costly and only in areas around the wellbore. Peng et al. (2010) and Xu et al. (2012) employed synthetic wave and real seismic wave events to characterize subsurface damage zones indirectly. The variation of seismicity rate is applied for damage zones in Elsinore-Temecula segment of the southern San Andreas fault system (Powers, 2010), and high-resolution earthquake locations are used to evaluate the damage zone in the L'Aquila normal fault, Italy (Valoroso et al., 2014). They found damage zone with damage intensity decaying from the fault core outward, agreeing with field observations. However, the poor seismic resolution limited the characterization to only—km scale. With the advance of seismic survey in petroleum industry, seismic attributes with good resolution is used to characterize fault and its surrounding deformation by inferring from seismic disturbance volume indirectly (Chopra & Marfurt, 2007; Iacopini, 2011; Iacopini et al., 2012; Liao et al., 2019; Qi et al., 2014). However, the properties of subsurface damage zones, including fractures, superposition architecture, and regional distribution, have been poorly studied and discussed.

The growing need for fracture evaluation extends from mineralized veins to shale play and tight gas reservoirs. In this study, we focus on fractures in tight sandstone within the Xujiahe Formation in Sichuan Basin, China. In general, a tight sandstone reservoir means a reservoir with permeability less than 0.1 mD and porosity less than 10%. In the Xujiahe Formation, the matrix porosity is less than 5%, and the permeability is less than 0.05 mD. Tight gas usually cannot be produced naturally without drilling intervention, e.g., hydrofracturing and horizontal drilling (Zou et al., 2012). In this case, however, high production from several pioneering wells in the tight sandstone Xujiahe Formation exhibited greater hydrocarbon potential than expected, suggesting that the permeability is controlled by the presence of fractures rather than low-permeability sandstone matrix. The essential characteristics of the tight gas reservoir in Sichuan Basin include but not limited to: (1) proximity relation of coal sources and reservoir rocks (Wang et al., 2013; Qin et al., 2018); (2) overpressure in tight sandstone reservoir; (3) unique migration and accumulation mechanisms coupling fractures and overpressure (Hao et al., 2015; Tingay et al., 2009). All these features are related to fractures, and therefore, the Xujiahe Formation provides an ideal case for studying the fractured tight gas reservoir. This study displays the damage zones of a subsurface sandstone tight gas reservoir in northeast Sichuan, China at multiple scales: from km scale using seismic attributes towards outcrop and image logging observations and down to microfracture characterizations from thin sections. The results indicate that the fractures within damage zones have the potential to act as reservoir storage space and horizontal-flow conduits over a regional scale. We further illustrate the relation between damage zones and overpressure, and the main factors controlling gas migration and accumulation. The result could provide insights for the next exploration deployment in tight sandstones.

5.2 Geological Setting

The Tongnanba anticline is one of the superimposed basins located northeast of the Sichuan Basin, China (Fig. 5.1a). Its evolution is controlled by the Micangshan thrust in the northwest, Dabashan thrust-fold belt in the northeast, and the Longmenshan fault belt far in the west. The study focuses on the middle segment of Tongnanba fold within the 2nd Member of the Upper Triassic Xujiahe Formation (T3x). It is an anticlinal fold striking NE–SW, structurally consistent with the two segments not covered by the data in the study: one segment in the northeast and the other one in the southwest. The axis of the Tongnanba anticline is tilted from high in the southwest to low in the northeast, with both limbs dipping gently. It is a large anticline with about 12.2 km in width and 80 km in length, among which a sequence of northwest-striking reverse faults developed with sub-folds (red lines in Fig. 5.1a). By analyzing the evolution of the superposition system of folds and faults, two stages of significant tectonic movements could be recognized (Dai et al., 2013; Li et al., 2016). In the first stage, the Tongnanba anticline formed by northwest-southeast compression in the Middle-Late Jurassic Micangshan deformation. In a later stage, Tongnanba anticline is further ruptured by the northeast–southwest compression from Dabashan thrust in the Himalayan orogeny, resulting in the formation of sub-folds and reverse faults. Details of the associated deformation are discussed later.

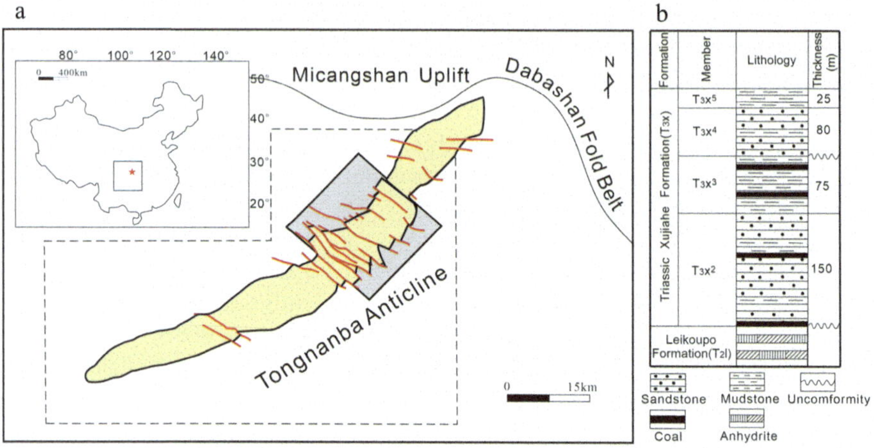

Fig. 5.1 Study area within Xujiahe Formation (**a**) and strata column in the middle of the Tongnanba fold (**b**) in northeast Sichuan, China. Red lines are generalized faults, and not all faults are shown

There are three major gas-bearing formations in the study area (Qin et al., 2018), including the Upper Permian Changxing Formation (P2c), the Lower Triassic Feixianguan Formation (T1f), and the Upper Triassic Xujiahe Formation (T3x) (this study). Our focus here is the Xujiahe Formation, which is a set of tight continental sandstone interbedded with gas-rich coal seams and silty mudstone that could be further divided into five bottom-top members (Fig. 5.1b): (1) the 1st Member is a 40-m thick silty mudstone. A thick layer of anhydrite from the Leikoupo Formation lays below the 1st Member; (2) the 2nd Member is comprised of 200-m thick sandstone under overpressure and two coal seams in the lower part, which is the focus of this study. The coal seams are rich in gas and are source rocks for commercial explorations. (3) the 3rd Member is a 100-m thick silty mudstone which acts as a direct seal for the tight sandstone of the 2nd Member. (4) the 4th Member is a 100-m thick sandstone, and (5) the 5th Member is a 50-m thick silty mudstone. The 2nd Member of tight sandstone of the Xujiahe Formation is the principal gas-producing layer while the 4th Member is also of great potential but will not be discussed here. As the 2nd Member tight sandstone has ultra-low porosity (<5%) and ultra-low permeability (<0.1 mD), the high production wells in Tongnanba lead to significant interests in studying the factors controlling the gas migration and accumulations (Li et al., 2016; Yue et al., 2018).

5.3 Microfractures in Tight Sandstone

Previous work reveals that microfractures strongly influence rock strength, elastic wave velocities, and permeability (Anders et al., 2013; Kranz, 1983). Here we present thin section images of the samples taken from tight sandstone of the 2nd Xujiahe Formation, with abundant microfractures (Fig. 5.2). The present tight sandstone has three facies, a pure quartz sandstone with substantial quartz overgrowth as evidenced by the low grain boundary contrast (Fig. 5.2a–c), a sandstone with chlorite coating (thick dark lines bounding the grain boundaries in Fig. 5.2d–f), and a clastic sandstone with muddy fillings (Fig. 5.2g–i). No appreciable void spaces can be visually observed in these images, indicating very low porosity in the sandstone matrix. The sand grain texture has a strong influence on the microfractures, as most of the microfractures tend to follow grain boundaries (Fig. 5.2a, b, d, f, h). Besides the predominant intergranular fractures, transgranular fractures also exist, cutting across hosting grains and straightening the fracture traces (Fig. 5.2c, g, i). Considering the multi-stage tectonic histories in the area (Dai et al., 2013; Li et al., 2016), these microfractures are likely developed subcritically under relatively small driving stresses (Chen et al., 2017, 2019; Olson et al., 2009) such as episodes of overpressure caused by gas expulsion during the Jurassic, local tensile stress by Tongnanba folding during the Jurassic and the Early Cretaceous, and unloading by basin uplift during the Later Cretaceous.

The tight nature of the sandstone is further manifested in the petrophysical measurements from a total of 58 core samples, laid out as compared to other tight sandstone gas reservoirs around the world (Fig. 5.3). The porosity is generally less than 2% for the quartz sandstone and the sandstone with chlorite-coat, and 2–5% for the sandstone mixed with muddy grains, giving an average porosity of 1% for the Tongnanba sandstone reservoir (Fig. 5.3a), which is significantly lower than others (Fig. 5.3a). Not only with low porosity, but the permeability of the present sandstone is also low as compared to other tight sandstone gas reservoirs, with the permeability distributed mainly between 0.01 and 0.05 mD (Fig. 5.3b). Water saturation (e.g., 60% in this case) is higher than others. The wider scatter of permeability range than the less scattered porosity range (Fig. 5.3a, b) suggests that fractures provide significant permeability than porosity in the tight sandstone.

5.4 Macro-Fractures in Fault Damage Zones

At the outcrop scale, fractures further develop into joints (Fig. 5.4a, b). These fractures extend vertically across the bedding layers, forming multiple sets of fractures cross-linked to a fracture network as commonly seen in brittle formations of carbonates and shales (Gale et al., 2007; Guerriero et al., 2010). The spacing between fractures can be as narrow as a few centimeters (Fig. 5.4a) but can also be as wide as—meters (Fig. 5.4b). The high angle fracture sets observed in logging image (Fig. 5.4d,

Fig. 5.2 Thin section imaging of microfractures from three wells in the tight sandstone reservoir under plane-polarized light. **a–c** Quartz-sandstone with multiple opening-mode microfractures (*arrows*) from well M101. Substantial quartz overgrowth is evidenced by the low grain boundary contrast. **d–f** Microfractures in the sandstone with chlorite-film (*dark intergranular marks*) coating the grains from well M201. **g–i** Microfractures (*blue lines*) in a quartz-rich clastic sandstone from well M3. Muddy fillings (*dark areas*) reside the intergranular spacing

e) is consistent with outcrop observations. The vertically extending fractures mostly end into intact mudstone/siltstone, together with the horizontally extending fractures, significantly enhancing fluid flow properties within the layer. The formation of joint network indicates that the fractures were mainly developed subcritically (Pollard and Aydin, 1988; Olson et al., 2009), which further suggests that the coupled mechanical (rock strength, elastic properties) and chemical (stress corrosion, dissolution–precipitation) effects contribute to the fracture system development (Atkinson, 1984; Chen et al., 2017, 2019).

5.5 Damage Zones Distribution by Seismic Attributes

The basin is majorly comprised of a large anticlinal Tongnanba fold (Fig. 5.5a) that is crosscut by a series of reverse faults into isolated sub-folds (Fig. 5.2a). As observed in microfractures from core samples and fracture networks from outcrops and image logging, the distribution of fracture bearing damage zones is critical

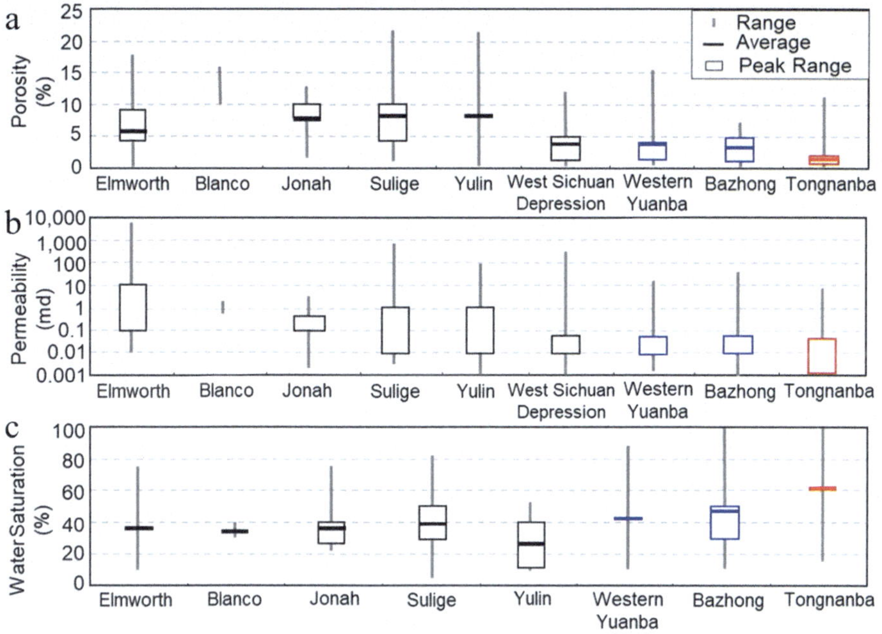

Fig. 5.3 Comparison of the rock properties of tight sandstone in the Xujiahe Formation to the other tight sandstone in the world, including porosity (**a**), permeability (**b**), and water saturation (**c**). Red is data in Tongnanba, blue is data in Sichuan Basin, and black is referred to data (Song et al., 2018; Zou et al., 2012)

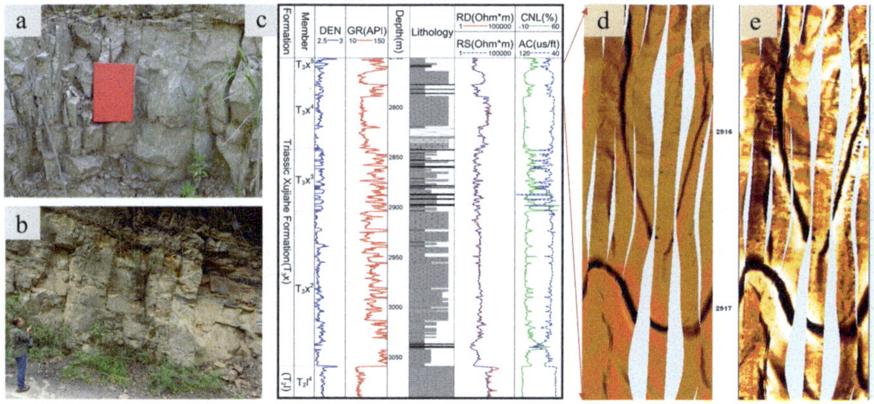

Fig. 5.4 Fractures at **a, b** Outcrop (20 km north of the study area) of the 2nd Member of Xujiahe Formation and **c** stratigraphy with logging data and **d, e** imaging log (well M103, 2916 m in depth)

for understanding the migration conduits on a basin scale. It is expected that the area is densely fractured due to the superposition of multiple sequences of tectonic movements and associated stress localizations around the anticline structure (Liao et al., 2019; Reches, 1988; Staples, 2011). We use seismic attributes calculated from the 3-dimensional seismic reflection data to image the seismic deformations within the Xujiahe Formation (Botter et al., 2016, 2018; Iacopini et al., 2012; Liao et al., 2019; Marfurt and Rich, 1949).

We compute variance and enhance the variance image based on the ant-tracking technique. The coherence attribute is defined as the energy of the coherent part of seismic traces divided by the average acoustic energy of the input seismic traces (Chopra & Marfurt, 2007). We applied statistical variance, a reversed version of the coherence described by Gersztenkorn and Marfurt (2012), with the high variance values equivalent to faults. Figure 5.5b shows the variance image computed from Petrel, which is a patented algorithm. The ant-tracking attribute is an algorithm

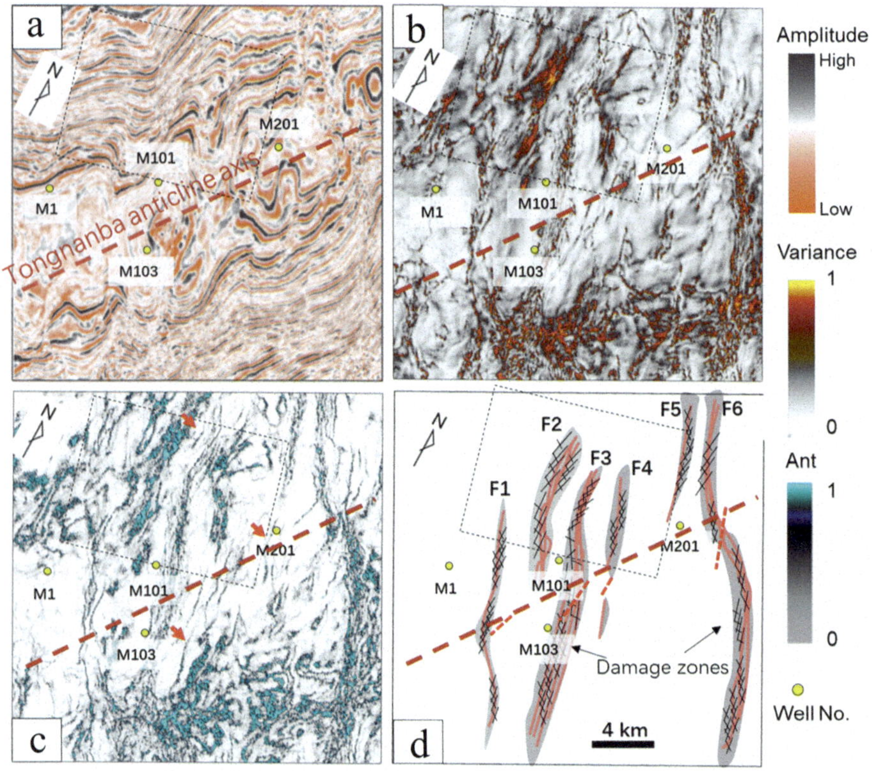

Fig. 5.5 **a** Time slice map of seismic amplitude and seismic attributes of coherences calculated for tight sandstone in the 2nd Member of Xujiahe Formation in the study area, including **b** variance and **c** ant-tracking. **d** The interpretation of damage zones on the crest of the Tongnanba fold. Note the red dash line in **a** is Tongnanba fold axis, red arrows in **c** indicate fault lineaments. Locations of wells marked. The area in the black dash box is to present later

adapted from the ant-colony systems to capture trends in noisy data of the other attributes in three dimensions. This approach enhances the structure of discontinuities and filters non-structural features (Jansen, 2005). It is a statistical edge enhancement tool, allowing the interpreter to better characterize faults in the seismic data. It is important to note that ant-tracking may produce artificial faults, e.g., connecting inosculating fault segments, or misinterpretation of footprints (Jansen, 2005).

The variance and ant-tracking attributes are computed along the top of the Xujiahe Formation, and our observations include the following two parts: (1) two large areas with high variance values distributed in upper-left and lower-right parts (Fig. 5.5b), which are transition areas of limbs and envelope layers. (2) Distinguishable strips of high variance trending north–south (red in Fig. 5.5b). The structures of the strips are further illustrated in the ant-tracking attribute map, where additional small faults are indicated by red arrows (Fig. 5.5c). These strips are interpreted as fault damage zones formed due to compression from Dabashan thrust in the northeast (Fig. 5.2a) during the Himalayan orogeny. These strips strike between 335° and 345°. We picked up six major damage zones of reverse faults (Fig. 5.5d) divided into two groups based on fault length. (1) The long faults that cross the Tongnanba fold, including F1, F3, F6 with fault lengths ranging from 12 to 20 km. These long faults consist of two segments slightly separated by axis-normal sub-faults on Tongnanba fold hinge, where dense fractures are inferred owing to the superposition of folding flexures and sub-faults. (2) The short faults that mostly appear on the north limb of the Tongnanba fold, including F2, F4, and F5 with lengths around 8–10 km.

Furthermore, in order to investigate the internal structure of the damage zones, we crop a cube of the dataset from the Tongnanba fold (black dashed box in Fig. 5.5b, c) containing the sandstone Xujiahe Formation. Figure 5.6a presents a 3D seismic amplitude data co-rendered with variance attribute (color bar differed from Fig. 5.5b for clarity). The amplitude generally shows continuous reflectors and define the sub-anticlines of the Xujiahe Formation in grey. The yellow–red color clusters (F1, F2, F3 in Fig. 5.6a) are high seismic variance anomalies that can be readily interpreted as fracture networks within fault damage zones (Chopra & Marfurt, 2007; Li et al., 2016; Liao et al., 2019). It sometimes is defined as a seismic disturbance zone (Iacopini et al., 2012) as a result of the cumulative effects of fault and fractures. The enhanced discontinuity (Fig. 5.6b) by amplitude data co-rendered with ant-tracking presents the potential slip surfaces of the reverse faults. It indicates that the faults may be segmented or inosculated vertically. Figure 5.6c presents the distribution patterns of damage zones, including multilayers of anhydrite, sandstone, and mudstone/siltstone. The thickness of these damage zones ranges from 100 to 1500 m (Fig. 5.6c). The two wells, M101 and M103, are located precisely by the side of a most substantial reverse fault, F3 (Figs. 5.5b, d and 5.6a). It is likely that the damage zones (Figs. 5.5d, 5.6c) inferred from the disturbed seismic signals represent a volume of densely developed fractures as observed in the field (Figs. 5.2, 5.4c and d), impacting the migration and accumulations of gas.

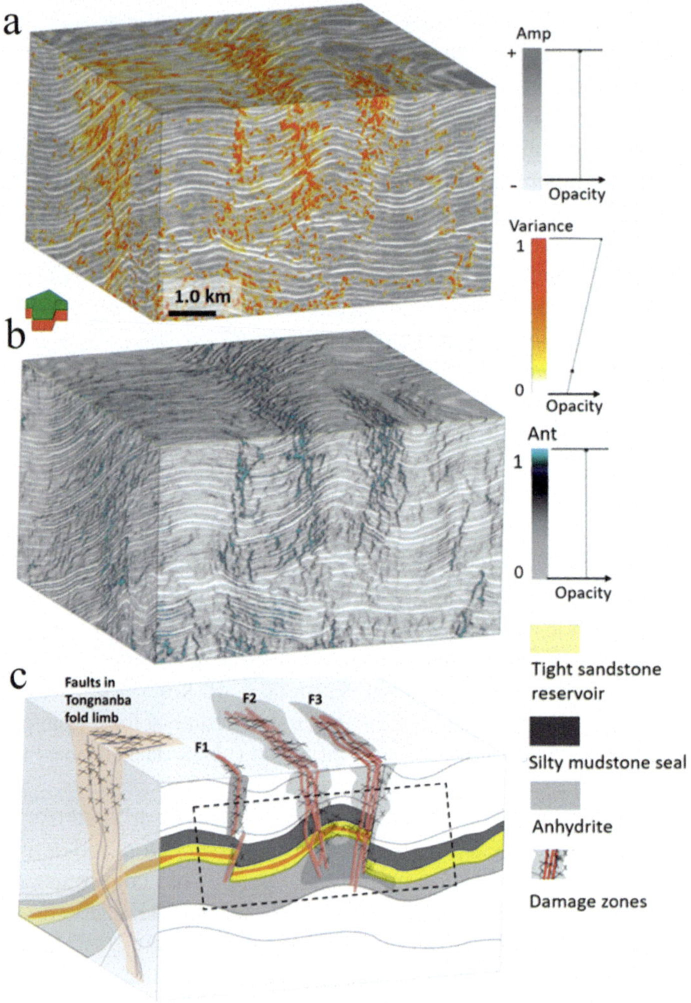

Fig. 5.6 3D seismic volume visualizations for the damage zones of the fold-thrust structure in Tongnanba fold. **a** Seismic cube of amplitude co-rendered with variance. The red/yellow scale gradational color has been adopted to highlight major discontinuities defined as damage zones. **b** Seismic cube of amplitude co-rendered with ant-tracking calculated from the amplitude. **c** Structure interpretation model of the study area with fault damage zones and major strata of the Xujiahe sandstone/mudstone as well as the Leikoupo anhydrite. Note the damage zones are composed of red lines (as fault core) surrounded with a grey area (as damage zones). The damage zones on the left of the figure in light orange color are fault-zones in Tongnanba fold limb that is not discussed here. The black dashed box is the study area for Fig. 5.7

5.6 Implication and Discussions

5.6.1 Fractures, Damage Zones and Gas Migration

Fracture is a significant geological factor controlling shale plays (Decker et al., 1992; Hill and Nelso, 2000). Higher gas productivity is often linked to fracture zones while low or no gas production to the poorly deformed area. A densely developed fracture network will significantly increase gas production in a low-porosity reservoir (Hill and Nelso, 2000). In the present case, commercial gas production is obtained in the middle part of the Tongnanba fold crest where fractures densely developed. The fractures (Figs. 5.2, 5.4) form subcritically during multiple stages of tectonic movements and assisted by the episodes of overpressure, which we argue supply large storage volume for free gas and desorption gas in an originally tight and less permeable sandstone (Curtis, 2002). In addition, fractures within damage zones further provide major migration channels (Figs. 5.5d, 5.6c), including three distinct stages based on geochemical and geostructural data (Li et al., 2016; Song et al., 2018).

(1) Late Triassic-Middle Jurassic (208–160MA): the early stage of gas accumulation in the lithologic reservoir. Gas generation occurs from the source rocks of the Xujiahe Formation in Tongnanba area at the end of the Late Triassic Xujiahe Formation, but at this time, the Tongnanba anticline has not yet formed, and the horizontal stratum is stable with few faults. A considerable amount of gas is expulsed (Li et al., 2016; Song et al., 2018; Wang et al., 2013) into the surrounding sandstone of the 2nd Member through subcritical microfractures in the Middle Jurassic. It is lithologic-controlled reservoirs without much interaction of fractures or faults at this stage.
(2) Late Jurassic-Early Cretaceous (160–100 Ma): intermediate stage of gas accumulation in the tectonic-lithologic reservoir. Under the influence of the movement of the Yanshan Mountains, the thrusting and extrusion of the Dabashan structural belt caused the initial formation of the Tongnanba anticline and formed a large number of fold-controlled tensile fractures. A significant amount of fractures formed and the gas generated by the Xujiahe Formation coal seams starts to migrate and accumulate in the high structural area (sub-horizontal red arrows in Fig. 5.7). The lithologic-controlled reservoir is transformed into a tectonic-lithologic gas reservoir.
(3) Late Cretaceous-present (~100 Ma): late stage of gas re-distribution and reservoir adjustment. Since the late Cretaceous, due to the influence of the Yanshan Movement and the Xishan Movement, the northeastern Sichuan region has been dramatically uplifted, and the source rocks of the Xujiahe Formation have ceased to produce hydrocarbons. The Tongnanba anticline is further uplifted and deformed with a large number of faults and fractures through the Permian and Triassic. The unloading and compression turn damage zones to mature stage during the reservoir transformation, providing conduits for gas flowing upward. The favored area, e.g., the crest of the anticlinal fold (Well M101), is

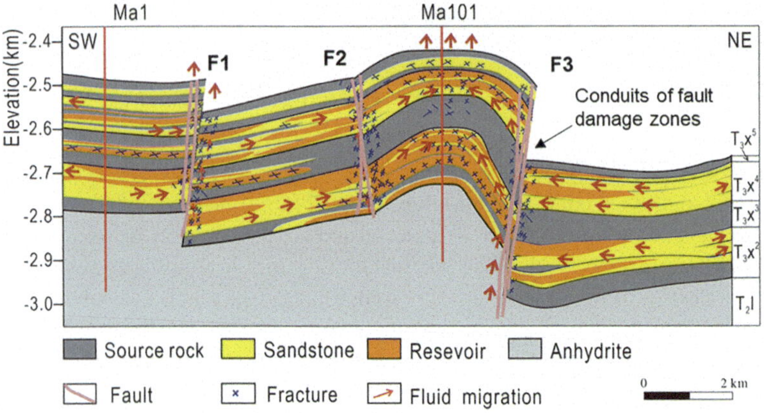

Fig. 5.7 Model of conduits for gas migration and accumulation in Tongnanba fold. Location of well Ma101 and Ma1 are shown

charged by surrounding gas and enriched by mixing the potential marine gas migrated through damage zones (red arrows in F3, Fig. 5.7). The gas is accumulated in the anticlinal top and damage zones with densely fractures in tight sandstone, forming the distribution characteristics of today's gas reservoirs. A certain amount of gas has been leaked upward to the 4th Member through the damage zones, e.g., F3.

5.6.2 Coupling Between Fractures and Overpressure

The tight sandstone in the Tongnanba area has a high degree of consolidation (Fig. 5.2) with poor reservoir petrophysical properties (Fig. 5.3) (low-porosity and low-permeability) and substantial structural heterogeneity (Figs. 5.5, 5.6, 5.7). Therefore, faults and fractures act as effective channels for the migration, accumulation, and dissipation of gas. In the Tongnanba area, the drilling positions (e.g., wells Ma 101 and Ma 103; Fig. 5.5) are in the Tongnanba anticline core, which is at a structural high point and is conducive to the accumulation of gas. We argue that the folding flexures provide the conductivity in the crest of Tongnanba fold aided by the fractures in the damage zones of F2 and F3. The in-situ pressure coefficients (~1.5, Fig. 5.8) indicate that the reservoir of the Xujiahe Formation is under overpressure. The overpressure in the lower formations, e.g., the Jurassic J1, J2 formations in Fig. 5.8, indicates excellent sealing by the mudstone interlayers in this area. In the current reservoir of the Xujiahe Formation, due to the complexity of geological conditions, the overpressure is resulted from a combination of mechanisms, including diagenesis and gas generation, compaction, and tectonic compression (Dai et al., 2013; Li et al., 2016).

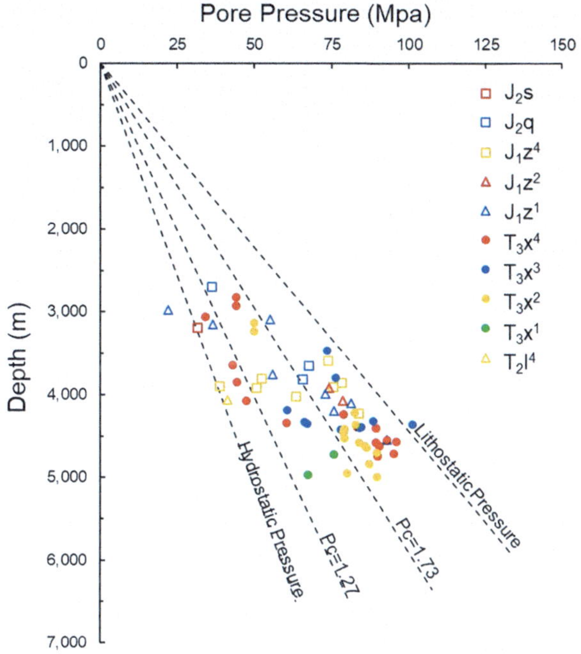

Fig. 5.8 Measured pore pressure versus depth in Tongnanba area, Sichuan Basin (adapted from Song et al., 2018). Red dots are well data from Xujiahe Formation, Triassic (T) while the others are from Jurassic formations (J)

The coupling effect between fracture and overpressure is mainly reflected in three aspects. (1) The overpressure in the reservoir increases the potential for the development and preserve of opening mode microfractures. This increases the storage volume and permeability significantly (e.g., Cavailhes et al., 2013; Faulkner et al., 2010; Lockner et al., 2011; Wibberley and Shimamoto, 2005); (2) Rupture of the formation or opening of fracture network leads to release of overpressure. Thus, overpressure is challenging to sustain during the long geological time (Hao et al., 2015; Tingay et al., 2009). Hydraulic fractures were found to reduce the overpressure in the Yuanba area, south of the study area (Li et al., 2016); (3) The fracture networks provide conduits for redistribution of overpressure compartments (Hao et al., 2015; Song et al., 2018). The overpressured gas migrates through fracture network to the higher structures, forming a high-pressure and high-yield gas reservoir. Industrial gas flows (e.g., well M101) were obtained in the crest of Tongnanba fold surrounded by damage zones of F2 and F3, which is a favored structurally high zone of accumulated gas with high fracture density. The wells Ma102, M3 near the fault damage zones also obtained industrial gas flow, while the well M2 slightly away from the fault zone did not receive industrial gas flow. This confirms that the enrichment and high yield of gas need to be based on the development of fractures. The associated core samples from the Xujiahe Formation display unfilled and active fractures (Figs. 5.2, 5.4), which we argue provide large fracture-based volume for gas storage within the low-porosity tight sandstone, particularly under overpressure conditions.

5.7 Conclusions

In this study, we display that seismic attributes could be used to characterize the structure and distribution of fault damage zones in the subsurface, taking tight sandstone in northeast Sichuan Basin as an example. We have demonstrated characteristics of the microfractures and macro-fractures within these damage zones, which we argue provide ample potential volumetric storage space for gas and primary migration conduits. This approach and data extend the reliable evaluation of fault damage zones on a regional scale that significantly controls the basin-wide migration and distribution of gas. Finally, the coupling effect between fractures and overpressure in tight sandstone provide insights for understanding the accumulation mechanisms of tight gas, which could also extend to CO_2 sequestration.

References

Anders, M. H., Schneider, J. R., Scholz, C. H., & Losh, S. (2013). Mode i microfracturing and fluid flow in damage zones: The key to distinguishing faults from slides. *Journal of Structural Geology, 48*, 113–125. https://doi.org/10.1016/j.jsg.2012.11.010

Atkinson, B. K. (1984). Subcritical crack growth in geological materials. *Journal of Geophysical Research Atmospheres, 89*(B6), 4077–4114. https://doi.org/10.1029/JB089iB06p04077

Botter, C., Cardozo, N., Hardy, S., Lecomte, I., Paton, G., & Escalona, A. (2016). Seismic characterisation of fault damage in 3D using mechanical and seismic modelling. *Marine and Petroleum Geology, 77*, 973–990. https://doi.org/10.1016/j.marpetgeo.2016.08.002

Botter, C., Cardozo, N., Qu, D., Tveranger, J., & Kholuykhin, D. (2018). Seismic characterization of fault facies models. In *EGU general assembly 2018*. https://doi.org/10.1190/INT-2016-0226.1

Busetti, S., Mish, K., Hennings, P., & Reches, Z. (2012). Damage and plastic deformation of reservoir rocks: Part 2. Propagation of a hydraulic fracture. *AAPG Bulletin, 96*(9), 1687–1709. https://doi.org/10.1306/02011211010

Caine, J. S., Evans, J. P., & Forster, C. B. (1996). Fault zone architecture and permeability structure. *Geological Society of America, 32*, 1130.

Caine, J. S., & Minor, S. A. (2009). Structural and geochemical characteristics of faulted sediments and inferences on the role of water in deformation, rio grande rift, new mexico. *Geological Society of America Bulletin, 121*(9–10), 1325–1340. https://doi.org/10.1130/B26164.1

Cavailhes, T., Soliva, R., Labaume, P., Wibberley, C., Sizun, J. P., Gout, C., et al. (2013). Phyllosilicates formation in faults rocks: Implications for dormant fault-sealing potential and fault strength in the upper crust. *Geophysical Research Letters, 40*(16), 4272–4278. https://doi.org/10.1002/grl.50829

Chen, X., Eichhubl, P., & Olson, J. E. (2017). Effect of water on critical and subcritical fracture properties of woodford shale. *Journal of Geophysical Research Solid Earth, 122*(4), 2736–2750. https://doi.org/10.1002/2016JB013708

Chen, X., Eichhubl, P., Olson, J. E., & Dewers, T. A. (2019). Effect of water on fracture mechanical properties of shales. *Journal of Geophysical Research: Solid Earth, 124*(3), 16479. https://doi.org/10.1029/2018JB016479

Chester, F. M., & Logan, J. M. (1986). Implications for mechanical properties of brittle faults from observations of the Punchbowl fault zone, California. *Pure and applied geophysics, 124*, 79–106. https://doi.org/10.1007/BF00875720

Chopra, S., & Marfurt, K. J. (2007). *Seismic attributes for prospect identification and reservoir characterization.*

Curtis, J. B. (2002). Fractured shale-gas systems. *AAPG Bulletin, 86*(11), 1921–1938. https://doi.org/10.1306/61EEDDBE-173E-11D7-8645000102C1865D

Dai, J., Liao, F., & Ni, Y. (2013). Discussion on the gas source of the triassic xujiahe formation tight sandstone gas reservoirs in yuanba and tongnanba, sichuan basin: an answer to Yin Feng et al. *Petroleum Exploration and Development, 40*(2), 268–275. https://doi.org/10.1016/S1876-3804(13)60033-6

Decker, S., Coates, J., & Wicks, M. P. (1992). Stratigraphy, gas occurrence, formation evaluation and fracture characterization of the antrim shale, Michigan basin. Topical report, January 1990March 1992. https://doi.org/10.1016/j.gsf.2011.10.001

Eichhubl, P., D'Onfro, P. S., Aydin, A., Waters, J., & Mccarty, D. K. (2005). Structure, petrophysics, and diagenesis of shale entrained along a normal fault at black diamond mines, California—implications for fault seal. *AAPG Bulletin, 89*(9), 1113–1137. https://doi.org/10.1306/04220504099

Ellis, M., Laubach, S., Eichhubl, E., Olson, A., et al. (2012). Fracture development and diagenesis of torridon group applecross formation, near an Teallach, NW Scotland: Millennia of brittle deformation resilience? https://doi.org/10.1144/0016-76492011-086

Faulkner, D. R., Jackson, C. A. L., Lunn, R. J., Schlische, R. W., Shipton, Z. K., Wibberley, C. A. J., et al. (2010). A review of recent developments concerning the structure, mechanics and fluid flow properties of fault zones. *Pergamon, 11*, 9. https://doi.org/10.1016/j.jsg.2010.06.009

Gale, J. F. W., Reed, R. M., & Holder, J. (2007). Natural fractures in the Barnett shale and their importance for hydraulic fracture treatments. *AAPG Bulletin, 4*, 91. https://doi.org/10.1306/11010606061

Gersztenkorn, A., & Marfurt, K. J. (2012). Eigenstructure-based coherence computations as an aid to 3-D structural and stratigraphic mapping. *Geophysics, 64*(5), 1468. https://doi.org/10.1190/1.1444651

Guerriero, V., Iannace, A., Mazzoli, S., Parente, M., Vitale, S., & Giorgioni, M. (2010). Quantifying uncertainties in multi-scale studies of fractured reservoir analogues: Implemented statistical analysis of scan line data from carbonate rocks. *Journal of Structural Geology, 32*(9), 1271–1278. https://doi.org/10.1016/j.jsg.2009.04.016

Hao, F., Zhu, W., Zou, H., & Li, P. (2015). Factors controlling petroleum accumulation and leakage in overpressured reservoirs. *AAPG Bulletin, 99*(5), 831–858. https://doi.org/10.1306/01021514145

Hill, D. G., & Nelso, C. R. (2000). Abstract: reservoir properties of the upper cretaceous lewis shale: a new natural gas play in the san juan basin. *AAPG Bulletin, 84*, 1865. https://doi.org/10.1306/A9673D60-1738-11D7-8645000102C1865D

Iacopini, D. B. R. W. H. (2011). Imaging deformation in submarine thrust belts using seismic attributes. In *Earth and planetary science letters: A letter journal devoted to the development in time of the earth and planetary system*. https://doi.org/10.1016/j.epsl.2010.12.041

Iacopini, D., Butler, R. W. H., & Purves, S. (2012). Seismic imaging of thrust faults and structural damage: A visualization workflow for deepwater thrust belts. *First Break, 30*(5), 58681. https://doi.org/10.3997/1365-2397.30.5.58681

Jansen, K. (2005). *Seismic investigation of wrench faulting and fracturing at Rulison Field, Colorado*.

Katz, O., Reches, Z. E., & Baer, G. (2003). Faults and their associated host rock deformation: Part I. Structure of small faults in a quartz–syenite body, southern Israel. *Journal of Structural Geology, 25*(10), 1675–1689. https://doi.org/10.1016/S0191-8141(03)00011-7

Knipe, R. J. (1997). Juxtaposition and seal diagrams to help analyze fault seals in hydrocarbon reservoirs. *AAPG Bulletin, 81*(2), 187–195. https://doi.org/10.1306/522B42DF-1727-11D7-8645000102C1865D

Kranz, R. L. (1983). Microcracks in rocks: A review. *Tectonophysics, 100*(1–3), 449–480. https://doi.org/10.1016/0040-1951(83)90198-1

Laubach, S. E., Eichhubl, P., Hargrove, P., Ellis, M. A., & Hooker, J. N. (2014). Fault core and damage zone fracture attributes vary along strike owing to interaction of fracture growth, quartz accumulation, and differing sandstone composition. *Journal of Structural Geology, 68*, 207–226. https://doi.org/10.1016/j.jsg.2014.08.007

Li, J., Hu, D., Zou, H., Shang, X., Ren, H., & Wang, L. (2016). Coupling relationship between reservoir diagenesis and gas accumulation in Xujiahe Formation of Yuanba–Tongnanba area, Sichuan Basin, China. *Journal of Natural Gas Geoscience, 1*(5), 335–352. https://doi.org/10.1016/j.jnggs.2016.11.003

Liao, Z., Liu, H., Carpenter, B. M., Marfurt, K. J., & Reches, Z. (2019). Analysis of fault damage zones using three-dimensional seismic coherence in the anadarko basin, oklahoma. *AAPG Bulletin, 103*(8), 1771–1785. https://doi.org/10.1306/1219181413417207

Lockner, D. A., Carolyn, M., Moore, S., Diane, C., et al. (2011). Low strength of deep san andreas fault gouge from safod core. *Nature, 32*, 9927. https://doi.org/10.1038/nature09927

Marfurt, K. J., & Rich, J. (1949). Beyond curvature: Volumetric estimates of reflector rotation and convergence. *SEG Technical Program Expanded Abstracts, 1*, 4453. https://doi.org/10.1190/1.3513118

Mitchell, T. M., & Faulkner, D. R. (2009). The nature and origin of off-fault damage surrounding strike-slip fault zones with a wide range of displacements: A field study from the atacama fault system, northern Chile. *Journal of Structural Geology, 31*(8), 802–816. https://doi.org/10.1016/j.jsg.2009.05.002

Mizoguchi, K., & Ueta, K. (2013). Microfractures within the fault damage zone record the history of fault activity. *Geophysical Research Letters, 40*(10), 2023–2027. https://doi.org/10.1002/grl.50469

Olson, J. E., Laubach, S. E., & Lander, R. H. (2009). Natural fracture characterization in tight gas sandstones: Integrating mechanics and diagenesis. *AAPG Bulletin, 93*(11), 1535–1549. https://doi.org/10.1306/08110909100

Peacock, D. C. P., Knipe, R. J., & Sanderson, D. J. (2000). Glossary of normal faults. *Journal of Structural Geology, 22*(3), 291–305. https://doi.org/10.1016/S0191-8141(00)80102-9

Peng, Z., Ben-Zion, Y., Michael, A. J., & Zhu, L. (2010). Quantitative analysis of seismic fault zone waves in the rupture zone of the 1992 landers, California, earthquake: Evidence for a shallow trapping structure. *Geophysical Journal International, 155*(3), 1021–1041. https://doi.org/10.1111/j.1365-246X.2003.02109.x

Pollard, D. D., & Aydin, A. (1988). Progress in understanding jointing over the past century. *Geological Society of America Bulletin, 100*(8), 1181–1204. https://doi.org/10.1130/0016-7606(1988)100%3c1181:PIUJOT%3e2.3.CO;2

Powers, P. M. (2010). Distribution of seismicity across strike-slip faults in California. *Journal of Geophysical Research: Solid Earth, 56*, 6234. https://doi.org/10.1029/2008JB006234

Qi, J., Zhang, B., Zhou, H., & Marfurt, K. (2014). Attribute expression of fault-controlled karst: Fort worth basin, Texas—a tutorial. *Interpretation, 2*(3), SF91–SF110. https://doi.org/10.1190/INT-2013-0188.1

Qin, S., Zhang, Y., Zhao, C., & Zhou, Z. (2018). Geochemical evidence for in situ accumulation of tight gas in the Xujiahe Formation coal measures in the central Sichuan Basin, China. *International Journal of Coal Geology, 196*, 173–184. https://doi.org/10.1016/j.coal.2018.07.009

Rawling, G. C., Goodwin, L. B., & Wilson, J. L. (2001). Internal architecture, permeability structure, and hydrologic significance of contrasting fault-zone types. *Geology, 29*(1), 43–46. https://doi.org/10.1130/0091-7613(2001)029%3c0043:IAPSAH%3e2.0.CO;2

Reches, Z. (1988). Evolution of fault patterns in clay experiments. *Tectonophysics, 145*, 141–156. https://doi.org/10.1016/0040-1951(88)90322-8

Sagy, A., Reches, Z., & Roman, I. (2001). Dynamic fracturing: Field and experimental observations. *Journal of Structural Geology, 23*, 1223–1239. https://doi.org/10.1016/S0191-8141(00)00190-5

Savage, H. M., & Brodsky, E. E. (2009). Collateral damage: How damage zone fracture density profiles reflect slip distribution. In *AGU Fall Meeting Abstracts*.

Savage, H. M., & Brodsky, E. E. (2010). Collateral damage: Evolution with displacement of fracture distribution and secondary fault strands in fault damage zones. *Journal of Geophysical Research Atmospheres, 116*(B3), B03405. https://doi.org/10.1029/2010jb007665

Song, Y., Zhang, L., Luo, D., Yang, S., Hao, F., & Zou, H. (2018). Characteristics of overpressure regime in compartmentalized over pressured system within tight sandstones, Xujiahe Formation, Yuanba-Tongnanba areas, Sichuan Basin. *Petroleum Geology and Experiment, 40*, 613–620.

Staples, E. R. (2011). *Subsurface and experimental analyses of fractures and curvature.*

Tingay, M. R. P., Hillis, R. R., Morley, C. K., King, R. C., Swarbrick, R. E., & Damit, A. R. (2009). Present-day stress and neotectonics of Brunei: Implications for petroleum exploration and production. *AAPG Bulletin, 93*(1), 75–100. https://doi.org/10.1306/08080808031

Valoroso, L., Chiaraluce, L., Collettini, S., & Cristiano, C. (2014). Earthquakes and fault zone structure. *Geology, 356*, 5071. https://doi.org/10.1130/G35071.1

Vrolijk, P. J., Urai, J. L., & Kettermann, M. (2016). Clay smear: Review of mechanisms and applications. *Journal of Structural Geology, 86*, 95–152. https://doi.org/10.1016/j.jsg.2015.09.006

Wibberley, C. A., & Shimamoto, T. (2005). Earthquake slip weakening and asperities explained by thermal pressurization. *Nature, 436*(7051), 689–692. https://doi.org/10.1038/nature03901

Wibberley, C. A. J., & Shimamoto, T. (2009). Earthquake slip weakening and asperities explained by thermal pressurization. *Nature, 436*(7051), 689–692. https://doi.org/10.1038/nature03901

Wilson, J. E., Chester, J. S., & Chester, F. M. (2003). Microfracture analysis of fault growth and wear processes, punchbowl fault, san andreas system, California. *Journal of Structural Geology, 25*(11), 1855–1873. https://doi.org/10.1016/S0191-8141(03)00036-1

Xu, S., Ben-Zion, Y., & Ampuero, J. P. (2012). Properties of inelastic yielding zones generated by in-plane dynamic ruptures—II. Detailed parameter-space study. *Geophysical Journal International, 191*(3), 1343–1360. https://doi.org/10.1111/j.1365-246X.2012.05685.x

Yue, D., Wu, S., Xu, Z., Xiong, L., Chen, D., Ji, Y., et al. (2018). Reservoir quality, natural fractures, and gas productivity of upper *Triassic xujiahe* tight gas sandstones in western sichuan basin, China. *Marine and Petroleum Geology, 926*, 7. https://doi.org/10.1016/j.marpetgeo.2017.10.007

Zecheng, W., Shipeng, H., Deyu, G., Wei, W., & Chong, Y. (2013). Geochemical characteristics of natural gases in the upper *Triassic xujiahe* formation in the southern Sichuan basin, SW China. *International Journal of Coal Geology, 120*, 15–23. https://doi.org/10.1016/j.coal.2013.09.002

Zou, C., Zhu, R., Liu, K., Su, L., Bai, B., Zhang, X., et al. (2012). Tight gas sandstone reservoirs in China: Characteristics and recognition criteria. *Journal of Petroleum Science and Engineering, 88*, 82–91. https://doi.org/10.1016/j.petrol.2012.02.001

Chapter 6
Lithologically-Controlled Fault Networks

Abstract The structure of subsurface fault networks, which has a significant effect on hydrocarbon migration, is inherently difficult to recognize and map. 3D seismic surveys provide an opportunity to map the fault system within a carbonate reservoir of Xinchang in the western Sichuan Basin, China. We calculated attributes from the seismic reflection data that covers an area of 1330 km^2, including variance, edge detection, dip-magnitude and dip azimuth. We focused on the Triassic Leikoupo Formation of dolostone, at ~6 km depth, adjacent to the Longmen Shan thrusting range. The mapped attributes display (1) a primary network of faults in an orthogonal pattern in the anticline, striking East–West and North–South, and (2) a secondary, conjugate fault set in the fold limb, striking Northeast and Northwest with a length range of 1–10 km. The latter set of conjugate faults are confined to a 300 m-thick layer of dolostone, diminishing into underlaying unit of anhydrite. We envision that the conjugate fault networks and associated fractures enhance gas charging and reflect a model of self-sourced migration and accumulation in the Leikoupo Formation. These results are significant for sweet-spot evaluation of carbonate reservoirs in the Sichuan Basin, and provide insights for understanding the migration of subsurface fluids.

6.1 Introduction

The presence of fault networks and associated fractures in carbonate and evaporate lithologies, controls the subsurface location and migration of fluid in hydrocarbon reservoirs (Ameen, 2003; Bhattacharya & Verma, 2019; Chilingarian et al., 1992; Lunn et al., 2008), as well as potential earthquakes that often nucleate or propagate through these sequences (Barchi, 2002; De Paola et al., 2008; Miller et al., 2004). Fault networks are basin-scale connected discontinuities with fractures or joints in the rock due to brittle deformation of the stratigraphy (Crider & Peacock, 2004; Flodin & Aydin, 2004; Kim et al., 2003; Segall & Pollard, 1983). Previous analyses of the geometries and interrelationships of the carbonate- and evaporite-hosted faults are mostly based on outcrop observations and field measurements (Caine et al., 1996;

Z. Liao, *Applied Seismic Geostructures*, https://doi.org/10.1007/978-981-97-7079-3_6

De Paola et al., 2008; Putz-Perrier & Sanderson, 2010). The air photo of fold-related fault/joints on Teton Anticline in Montana provides a particularly clear illustration of a fault network in the field (Fig. 6.1, Burberry et al., 2020; Stearns, 1967). A regular array of faults/joints within the rock is also reported as a brittle structure network within Cretaceous limestones in Big Bend National Park in Texas (Harmon, 2019, a web reference). The fault network, and associated fractures, are important for many reasons including as reservoirs and conduits for fluids moving through rock bodies (Caine et al., 1996; Chester & Chester, 1998; Chester et al., 1993). However, the structure of fault networks in the subsurface, that may behave as a conduit system to control the fluid flow, are inherently difficult to characterize and remain enigmatic due to lack of data.

In recent decades, seismic data acquisition has been applied to unlock hydrocarbon exploration potential (Bhattacharya & Verma, 2019; Eng & Tsuji, 2019; Nixon et al., 2014). More recently, seismic attributes computed from the seismic survey illustrate the potential to characterize fault structures in the subsurface at a depth range of 0–10 km (Gao, 2013; Liao et al., 2017, 2019; Wilson et al., 2015). Seismic attributes are based on neighborhood analysis of the seismic signal corresponding to the discontinuities, which is helpful in identifying subtle features that are difficult to interpret directly on the seismic cube (Chopra & Marfurt, 2007). Many attributes have been investigated to detect faults in the basins, including semblance (Marfurt et al., 1998), coherence (Marfurt et al., 1999; Prentice et al., 2006; Wu, 2017), curvature (Al-Dossary & Marfurt, 2006; Roberts, 2001), dip-magnitude and dip-azimuth (Chopra & Marfurt, 2007). Iacopini and Butler (2011) determined the structure and distribution of deformation in submarine thrust structures by using attributes of coherence, curvatures, and spectral decomposition. Liao et al. (2017, 2019) analyzed a km-scale

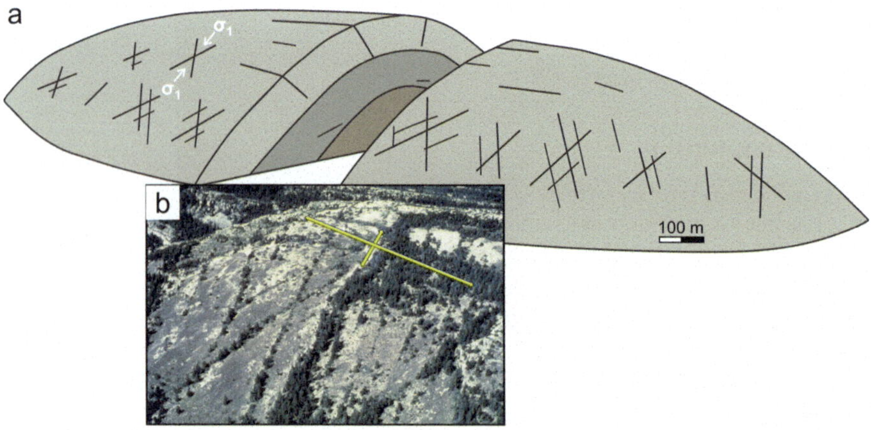

Fig. 6.1 **a** Conceptual fold model showing a set of conjugate shear fractures and their associated fractures (modified from Burberry et al., 2020; Stearns, 1967). **b** Air photo of tectonic fold-related faults/fractures on Teton Anticline, Sawtooth Mountains, W. Montana. Trees growing in the fractures are about 10 m tall. (Photo courtesy of Ron Nelson)

subsurface strike-slip fault in Oklahoma by using attributes of coherences, curvatures, dip-magnitude and dip-azimuth. The degree to which useful fault information can be extracted depends on the seismic resolution and the natural complexity of the structures.

In this study, seismic attributes were applied to investigate the fault networks in the Xinchang gas reservoir that is composed of carbonates and evaporates, in the western Sichuan Basin, China. The attributes of variance, edge detection, dip-magnitude and dip-azimuth are used to better reveal the spatial distribution of faults as a system. We then discuss the implications of the results and suggest the structural controls on the gas reservoir at the basin scale.

6.2 Geological Setting

The Sichuan Basin is one of the most natural gas-rich basins in China (Dai et al., 2013; He et al., 2019; Li et al., 2015; Wang et al., 2014). The western Sichuan depression is a northeast-trending foreland basin that formed during the late Triassic Indosinian Orogeny (Liu et al., 2010; Dominguez, 1992; Yong et al., 1995, 2003). The basin lies along the northwest Yangze Block, which has experienced two complex tectono-sedimentary stages of evolution during the Phanerozoic eon (Guo et al., 2014; Zhang et al., 2019): (a) a passive margin, with extension, accompanied with the sedimentation of marine carbonate and clastic sequences from the Ediacaran through middle Triassic, and (b) a foreland basin as a result of compression and uplift, with a fill of continental siliciclastic sediments from the Late Triassic to the Quaternary of less than 10-km thickness (Luo et al., 1996). The western depression is bounded by two seismically-active mountainous ranges, the Longmen Shan in the west and the Longquan Shan in the east. Our study area is located in the middle part of western Sichuan depression, adjacent to the multi-million-populated Chengdu city (Fig. 6.2a, b). Since the Middle Triassic, the study area is under the nappe compression by the Himalaya orogeny in the west (Dai et al., 2013; Li et al., 2012; Wang et al., 2014).

The identified strata from bottom to up include: the Precambrian crystalline basement, Cambrian Formation of shale, Permian Formation of Liangshan (P2l) mudstone and limestone, Qixia (P2q) limestone, Maokou (P2m) limestone, Longtan (P3l) limestone, shale and intercalated with coal seams, and Changxing (P3ch) limestone and shale, Triassic Formation of Feixianguan (T1f) shale, Jialingjiang (T1j) anhydrite, Leikoupo (T2l) dolostone, limestone and anhydrite, Ma'antang (T3m) mudstone and limestone, Xiaotangzi (T3xt) and Xujiahe (T3x) shale and sandstone, Jurassic and Quaternary Formation of sandstone, mud-shale and conglomerate, and the Leikoupo Formation is generally divided into four members based on lithological characteristics. We focus on the marine strata of carbonates and evaporates from Permian to Triassic periods: the organic-rich source rocks include Permian Longtan (P3l) and Maokuo (P2m) limestone, and Triassic Leikoupo Formation (T2l) dolostone; the main reservoir rock is dolostone in the Triassic Lei-4-3 submember (T2l43); the caprock is low-permeability terrigenous mudstone deposited in the Upper Triassic.

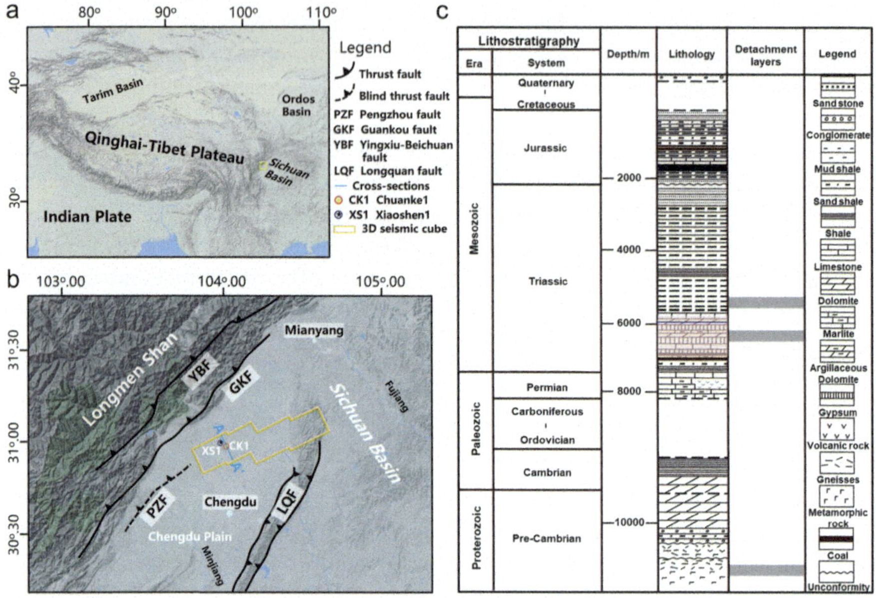

Fig. 6.2 General region of the study area (**a**) and the tectonic setting of the study area (**b**). Map **b** displays: (1) the main faults, Yingxiu-Beichuan (YB), Guankou (GK), Pengzhou (PZ), Longquan (LQ); (2) the study area; (3) seismic profile A–A′; (4) drilling well, CK1 and XS1. **c** The sedimentary stratigraphy of the Xingchang area (modified after Lu et al., 2019). Shaded red zone depicting the strata from T_1j to T_2l Formation; blue line depicting the layer of dolostone in Fig. 6.3

The target layer is the Triassic Leikoupo Formation dolostone, at an average depth of ~6000 m, which sits on a 500-m-thick ductile layer of anhydrite in the Lower Triassic Jialingjiang Formation (T1j; Sun et al., 2020) (Fig. 6.2c).

6.3 Methodology of Seismic Attributes

Detailed interpretation of seismic attributes along horizon slices, strata slices or seismic profiles revealed more changes in subsurface structural characterization than observable from the seismic amplitude data alone (Bahorich & Farmer, 1995; Sarkar et al., 2010; Weimer & Davis, 1996). In this study, in order to analyze the structure of the fault networks in the subsurface, we used a set of high-resolution 3D seismic data (Fig. 6.2b) collected during 2010–2012. It was processed using a method of post-stack time migration with a dominant frequency of 22 Hz. The seismic data were recorded with two-way travel time to depth, which covers an area of ~1330 km^2, at a depth range from ~2000 to ~4000 ms. We computed volumetric seismic attributes including dip-magnitude, dip-azimuth, variance, and edge detection.

The dip is the angle measured between the horizontal plane and the targeted layer, which is often expressed as dip magnitude (Chopra & Marfurt, 2007). Dip azimuth is the strike measured in the direction of maximum downward dip. The dip angle can be calculated from the apparent angle on the inline (longitudinal line) and crossline (transverse line). Assuming that the apparent angle in the direction of inline is θi and in the direction of crossline is θc, then θi and θc are related to the dip angle (θ) as follows:

$$\theta = (\theta i + \theta c)^{1/2} \tag{6.1}$$

θi and θc are related to the dip azimuth (φ) as follows:

$$\varphi = atan(\theta i, \theta c) \tag{6.2}$$

Variance is one of the seismic coherence attributes, which is a measure of discontinuities (D, Eq. 6.3) of waveforms and traces. High values indicate discontinuities due to strata deformation e.g., faults, or fracturing. Determinations of variance change are based on amplitude, frequency and phase, depending on the acoustic-impedance contrast (Chopra & Marfurt, 2007). It is defined as:

$$D^2 = \frac{\sum_i^n (x_i - \overline{x})^2}{n} \tag{6.3}$$

in which, x indicates the energy value of a sample point, $\overline{x}$ indicating the mean energy value of the sample points, n indicating the number of the samples.

The edge detection method of Luo et al. (SEG' 1996 Expanded Abstracts) measures the change of adjacent traces (Carter & Lines, 2001). With a signal A on the target trace and signal B on an adjacent trace, we can calculate the difference between the two traces by the following equation:

$$d = \|A - B\| / (\|A\| + \|B\|) \tag{6.4}$$

in which a high value indicates discontinuities of subsurface structure.

In addition to four geometrical seismic attributes, core samples from one gas drilling well, CK1, were collected, which has been drilled through Cenozoic, Cretaceous, Jurassic, Middle-Upper Triassic layers and terminated at the Lower Triassic Jialingjiang Formation. Thin sections of dolostone from Triassic Leikoupo Formation were analyzed.

6.4 Results

6.4.1 Fault Networks Within Dolostone by Seismic Attributes

The variance attribute is mapped in Fig. 6.3a for a volume reflector of the Leikoupo Formation in Middle Triassic. This map displays two superposed structural features (Fig. 6.3a): First, a fault network developed in the dolostone layer, with general faults marked by black arrows. This is a primary fault network that has two sets of faults in an orthogonal relationship, one set striking East–West, and the other set striking North–South. The lengths of North–South faults range from 5 km to tens of kms with variable spacing that averages ~3 km. Second, adjacent to the well CK1, there is a secondary set of conjugate faults. The conjugate faults are also fold zones in this region (Figs. 6.3a, 6.4a). These secondary faults strike Northeast–Southwest (45° in Figs. 6.3a, 6.4a, e, f) and Northwest-Southeast (Figs. 6.3a, 6.4a, e, f). There is a total of 16 sub-faults constrained by the fold limb, with an intersection angle of about 70° (Fig. 6.4e). The average spacing of the secondary fault network is estimated to be ~0.9-km.

Edge-detection maps (Figs. 6.3b, 6.4b) reveal abundant information: (1) several major fault damage zones (horizontal white arrows) striking East–West; (2) two North–South fault zones; (3) a damage zone composed of the small conjugate faults (inside the blue box) striking North-East. The high production well, CK1, is located on the anticline top next to the conjugate fault network that dominates the south limb of the anticline. These conjugate fault networks are also distinguishable by attributes of dip-magnitude and dip azimuth (Figs. 6.3c, d, 6.4c, d). On the other hand, the north fold-limb is also damaged as shown by a red-yellow stripe in the edge-detection map, however, without a seismically-detectable fault network in the attribute maps, e.g., variance map. The conjugate pattern of the secondary fault network detected in this study compares well with the field observations of joints in the Teton Anticline in Montana (Bergbauer & Pollard, 2004; Burberry et al., 2020; Stearns, 1967; Fig. 6.1).

To identify the fold axis of the fold that may control the conjugate fault network, the dip-azimuth attribute is calculated (Fig. 6.3d) to show the strike of the seismic reflector (Chopra & Marfurt, 2007). Our observations from the dip-azimuth map suggests a fold axis (black dashed line) that divides the fold into north limb (blue patches in Fig. 6.3d) and south limb (yellow patches in Fig. 6.3d).

6.4.2 Fault Networks and Folds

Complementary profiles of seismic amplitude and variance (Fig. 6.5) are selected along line AA′ (defined in Fig. 6.2b) that crosses both well CK1 and the conjugate fault network. These sections reveal the anticline within the Middle Triassic Leikoupo Formation of dolostone, with the underlaying Lower Triassic Jialingjiang Formation of anhydrite. The amplitude signals exhibit a strongly-disturbed texture around the

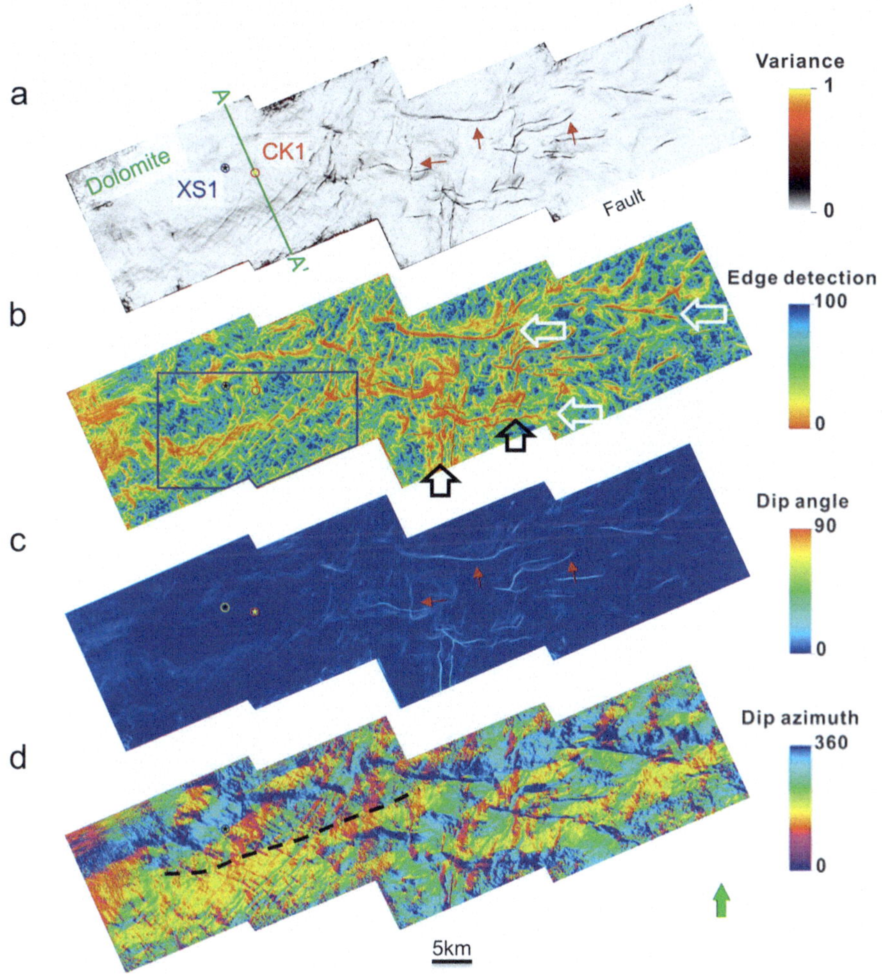

Fig. 6.3 **a** Strata slice (at layer of dolostone) from volumetric seismic attributes of variance, **b** edge detection, **c** dip, and **d** dip azimuth. A–A′: seismic profile; CK1, XS1: drilling wells; Solid blue box: the area of conjugate fault networks; Arrows: major fault damage zones; Dash line: fold axis

anhydrite layer, which is enhanced by the attribute of variance in shaded red zone (Fig. 6.5b). The anticline is comprised of several "Y" patterns indicative of reverse faults and their associated damage zones. However, the set of conjugate fault networks cannot be distinguished based on the profiles of seismic amplitude and attribute. Apparently, the vertical lengths of the conjugate fault networks are constrained by the thickness of the dolostone formation that is less than 300 m. The internal character of the damaged dolostone is discussed below via observations of thin sections.

Based on the above four seismic attributes (Figs. 6.3, 6.4), we recognize two sets of fault networks. One is the primary fault network that displays relatively large

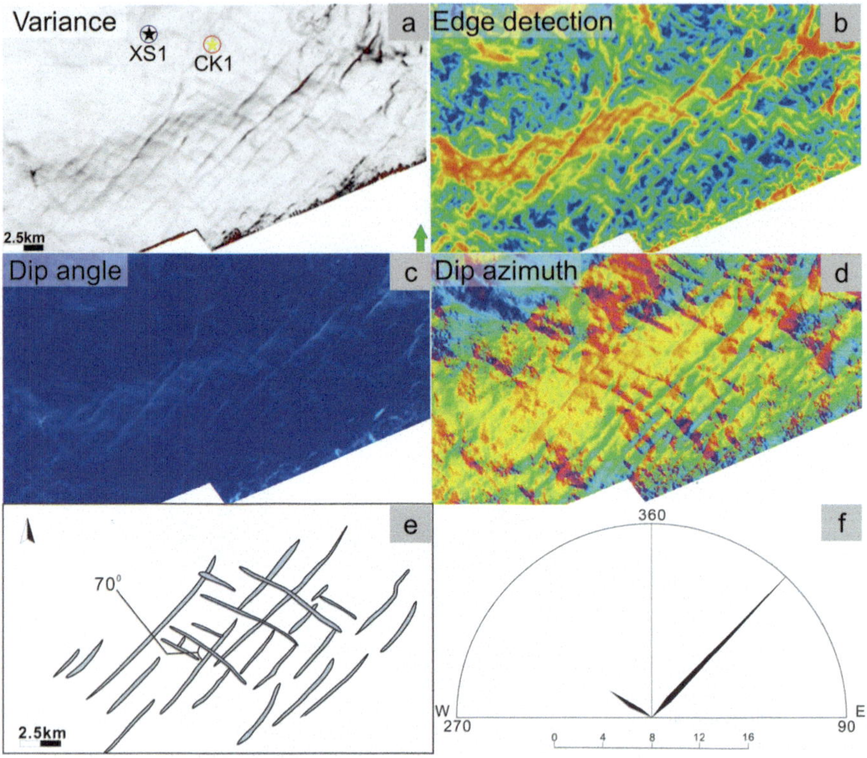

Fig. 6.4 a–d The conjugate fault networks depiction using volumetric seismic attributes (curvature, edge detection, dip/dip azimuth). Map location is shown in Fig. 6.3b. **e** Sketch of conjugate fault networks. **f** Rose diagram of the orientations of the conjugate fault networks

deformation intensity and rotation; the other set is the secondary conjugate fault network that is developed on the fold limb with a relatively smaller scale. The latter set of conjugate faults is associated with folding and are important for evaluation of reservoir conduits, which is the focus in this study.

6.4.3 Fractures Within Dolostone

The orthogonal geometry of the primary fault network and secondary fault network appear to be particularly developed in regions of relative compression characterized by appearance of anticlinal fold and twisting or rotation of the primary faults (e.g., Bergbauer & Pollard, 2004; Ghosh & Mitra, 2009; Stearns, 1967; Watkins et al., 2015). The formation of an anticline and reverse faults usually occurs with small-scale natural fractures in the brittle dolostone that have undergone compression. Based on the observations of the drilling core samples and thin sections, we found

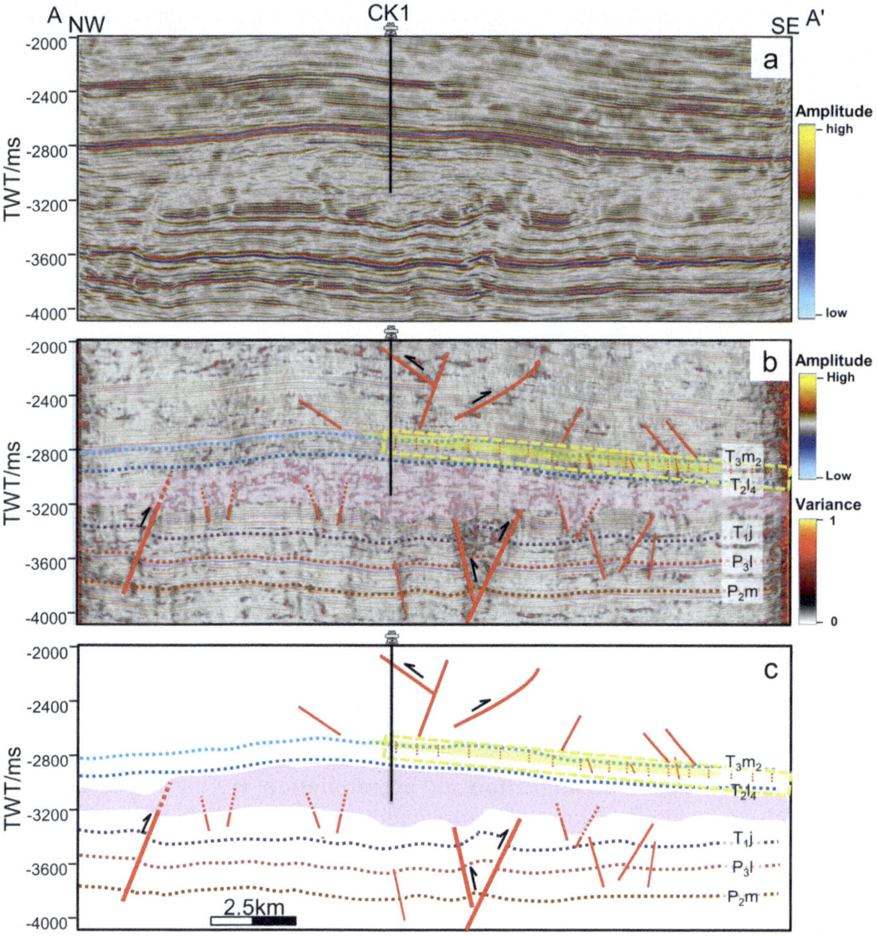

Fig. 6.5 Structural and geological interpretation of seismic profile A–A′, The **b, c** are the interpretations of the seismic profile (**a**). Dotted yellow rectangle depicts the zone of conjugate fault networks; thickness of the red line depicts the fault scale; shaded red zone depicts dolostone, limestone and anhydrite accumulation region

that two groups of the fractures: (1) large-fractures with apertures >0.1 mm, and (2) small-fractures with apertures <0.1 mm (Anders et al., 2014; Laubach, 2003). The large-fractures in cores and thin sections present average density of 1.4–2.2/mm (Fig. 6.6). These fractures are often filled with calcite with high dip-angles (50–90°). The small-fractures extend linearly with low dip-angles (0–30°, Fig. 6.6b–h). Some of them branch into several small sub-fractures. However, the wide-occurrence of low dip-angle fractures may be linked to overpressure of critically stressed fractures during natural resource extraction and storage (Fig. 6.6e–h; e.g., Ellsworth, 2013; Luo et al., 2020). The length of these two sets of fractures ranges from tens of microns to centimeters, forming a fracture network. The above analysis shows that the Triassic Leikoupo dolostone has a large number of fractures in addition to faults, consisted as the potential migration conduits in the subsurface.

6.5 Discussion

6.5.1 Implication of the Fault Networks in Hydrocarbon Accumulation Models

The previous results of reservoir characterization from the western Sichuan Basin show the Leikoupo Formation rock is composed of powder-crystalline dolostone and dolomicrite (He et al., 2019; Xiao et al., 2020). 32% of samples show moderate porosity of 2–5% while the permeability is low with a peak value 0.002–0.25 md (He et al., 2019). The observation of low permeability is not easily reconciled with the primary discoveries of exploration targets in this region (Yulin et al., 2018; Tian et al., 2018). If the fault networks and associated fractures represent optimum conduits for Leikoupo Formation reservoirs, we would expect that the conjugate fault networks control the hydrocarbon accumulation and is the key for understanding the hydrocarbon accumulations in strongly heterogeneous and low permeable rocks. Here we speculate on implication of fault networks for explaining the hydrocarbon accumulation models in Xinchang reservoir and their limitations.

(1) **Model 1: Self-sourced migration and accumulation.** If Xinchang reservoir is the self-sourced migration and accumulation model (Fig. 6.7a, b), its inherited structural high in the anticlinal is favorable for hydrocarbon accumulation, and the conjugate fault networks and associated fractures might be the controlling conduits facilitating gas charging into its current trap. In this view, faults and fractures near the anticline would be sealed by mudstone above and anhydrite below as a lithological-controlled network. It is capable of carrying gas over a large area of 200 km^2. As the principal source rocks are inferred within the Leikoupo Formation itself (Long et al., 2019; Keming, 2016), the proximal hydrocarbon supply is possible when the fault-fracture networks developed into their current configurations. A further evolving of the structure would create clustering of faults that allow gas to escape from the trap. This self-sourced model has the advantage of being consistent with our observations of the fault networks, and the fractures show the significant potential as pathways.

(2) **Model 2: Long-range upward migration and accumulation.** The long-range upward migration model (Fig. 6.7c) would connect the current dolomitic reservoir with a source that is likely to be Permian ~7500 m (based on Li et al. (2016) velocity model) below. For this to be so, the model requires large-scale faults for hydrocarbon migration across the impermeable, 500-m-thick anhydrite layer. This hypothesis would not apply as the Xinchang structure formed with only an anticline and fault networks that diminished into the underlaying

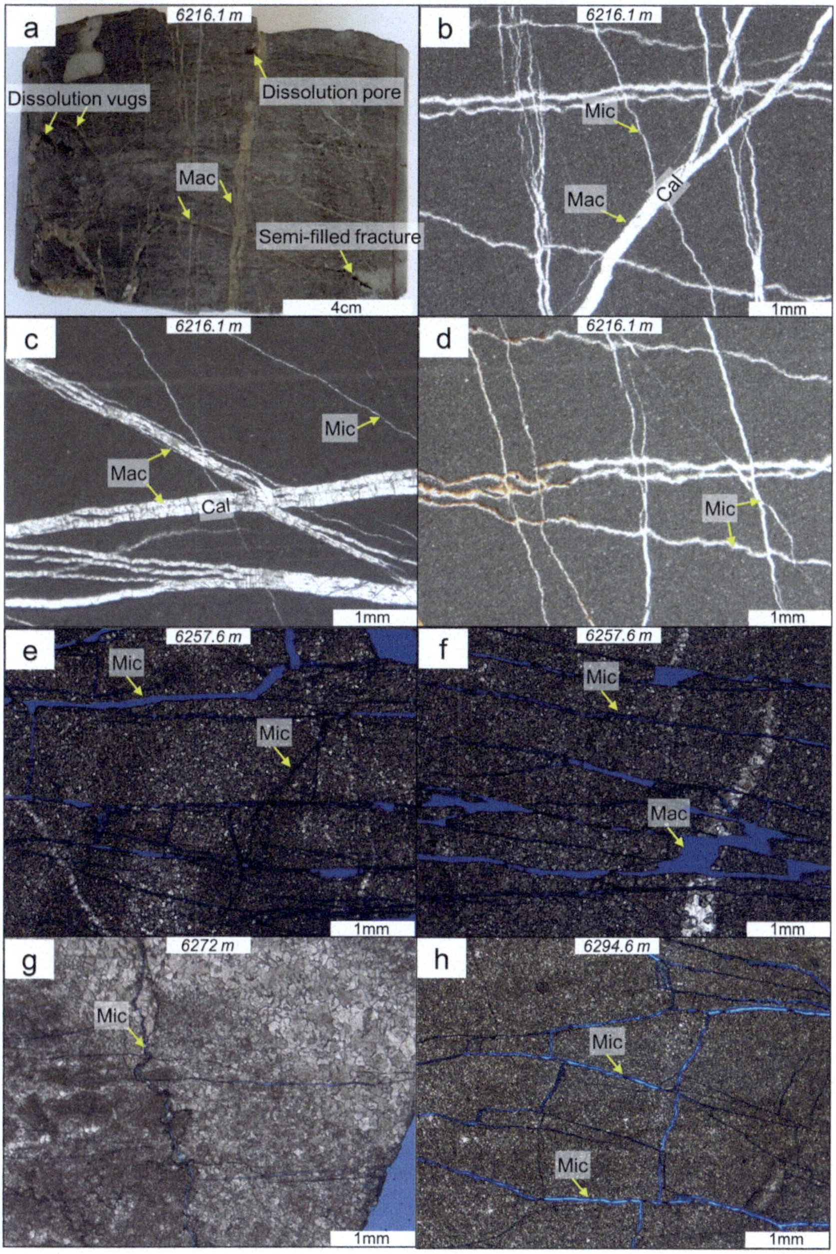

Fig. 6.6 Geological characteristics of calcite-filled/unfilled fractures system within the Middle Triassic Leikoupo Formation. **a** Core photo of macro-tectonic fractures partially filled by calcite from the CK1 well. **b–d** Thin sections of macro-tectonic fractures and micro-tectonic fractures filled with calcite from the CK1 well. **e–h** Thin sections of macro-tectonic fractures and micro-tectonic fractures from the LS1 well (13.7 km west of Xinchang block). Photomicrographs **b–h** were taken under plane-polarized light. *Abbreviations* Mac—Macro-fracture, Mic—Micro-fracture, Cal—Calcite

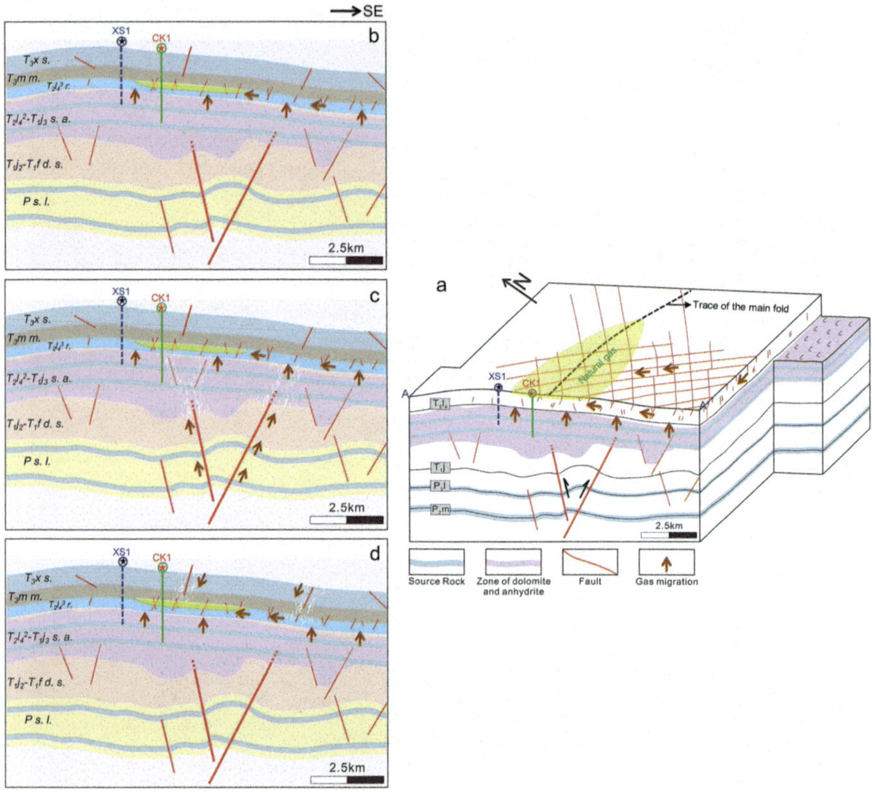

Fig. 6.7 **a** 3D hydrocarbon migration and accumulation models. The map displays: (1) structural interpretation map of the seismic profile A–A′; (2) motioning direction of the main faults; (3) location of main fold axis; (4) accumulation of natural gas; (5) drilling wells CK1 and XS1. **b–d** Three types of 2D hydrocarbon migration and accumulation models from the seismic profile **a** in Fig. 6.5. **b** Self-sourced migration and accumulation, **c** Long-range upward migration and accumulation, **d** Vertical downward migration and accumulation. *Abbreviations* P s. l.—Permian Formation source rock and limestone; T_1j_2–T_1f d. s.—Lower Triassic Jia-2 member to Lower Triassic Feixianguan Formation dolostone and shale; $T_2l_4{}^2$–T_1j_3 s. a.—Middle Triassic Lei-4-2 submember to Lower Triassic Jia-3 Member source rock and anhydrite; $T_2l_4{}^3$ r.—Middle Triassic Lei-4-3 submember reservoir; T_3m m.—Upper Triassic Ma'anatng Formation mudstone; T_3x s.—Upper Triassic Xujiahe Formation source rock

anhydrite. However, because of seismic resolution limitations, thin faults that partially propagate into anhydrite as secondary pathway would be possible.

(3) **Model 3: Vertical downward migration and accumulation.** Based on the previous studies on the origination of natural gas in the Leikoupo Formation and fault-scale identification, most of the source rock comes from Leikoupo Formation and overlying Ma'antang Formation in this model (Fig. 6.7d). If the Ma'antang Formation acted as source rock, the vertical downward migration model would play a role in the gas accumulation. He et al (2019) analyzed the

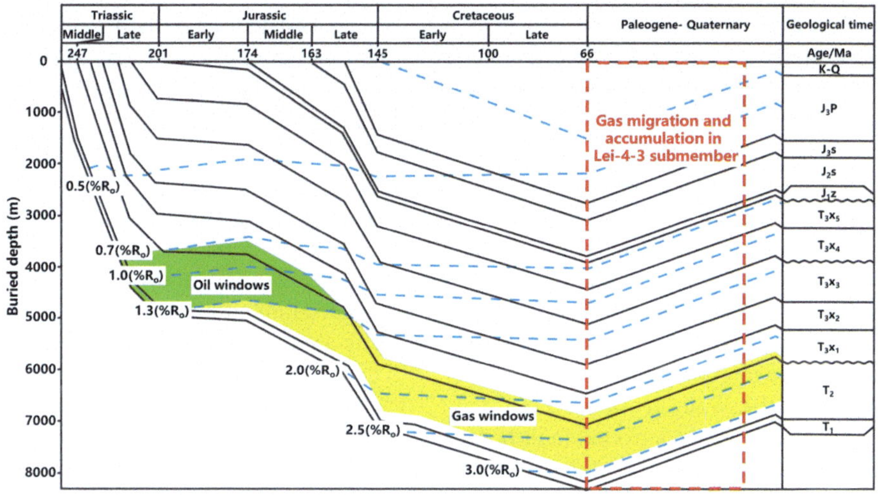

Fig. 6.8 A summary diagram showing the thermal evolution of organic matter in source rocks and hydrocarbon accumulation events at the CK1 well in the western Sichuan depression (modified after He et al., 2019). *Abbreviations* T_1—Lower Triassic; T_2—Middle Triassic; T_3—Upper Triassic; T_3x_1—Member 1 of the Upper Triassic Xujiahe Formation; T_3x_2—Member 2 of the Upper Triassic Xujiahe Formation; T_3x_3—Member 3 of the Upper Triassic Xujiahe Formation; T_3x_4—Member 4 of the Upper Triassic Xujiahe Formation; T_3x_5—Member 5 of the Upper Triassic Xujiahe Formation; J_1z—Lower Jurassic Ziliujing Formation; J_2s—Middle Jurassic Shaximiao Formation; J_3s—Upper Jurassic Suining Formation; J_3p—Upper Jurassic Penglaizhen Formation; K–Q—Cretaceous to Quaternary; R_o—vitrinite reflectance

burial history of the strata since Middle Triassic at the CK1 well and suggested that the period of natural gas generation was mainly from Lower Jurassic to Paleogene (~160–60 Ma, Fig. 6.8). The formation of fault networks was closely related with the uplift of the Longmen Shan range in the Late Miocene (Dai et al., 2013; Li et al., 2012; Wang et al., 2014), which was later than natural gas generation and would modify the Xinchang reservoir subsequently (Fig. 6.8). The tilted fold limb would enhance the downward migration of gas from Ma'antang source rock into the fracture system. Indeed, in this model the fault networks and associated fractures would behave as primary pathway for both resource sequences in above and below.

6.5.2 Exploration Plays in Xinchang

In this study, we propose that conjugate fault networks in the Xinchang fold limb would behave as Model 1 described above, although downward migration is a complementary possibility. In the model of self-sourced migration and accumulation, hydrocarbons accumulate in a stratigraphically-controlled reservoir in the Xinchang.

As the aforementioned migration model indicates, the distribution of faults and fractures would determine the favorable areas for investigation. The well of CK1 is a highly productive well near the anticlinal axis, with a cumulative gas production that exceeds 1.2×10^8 m^3 (Xu et al., 2011, 2015; Gangping, 2015; Hua-Jun et al., 2015). In contrast, another well, XS1, located 4 km to the east of CK1 in the same anticlinal high area, is proven as a dry well without production. The likely reason is that the well of CK1 is inside the conjugate fault networks, where the development of fault networks provide systematic pathways for natural gas migration and the accumulation of commercial gas. As to the well of XS1, the location that is far away from such fault networks would tend to prohibit any potential accumulation (Figs. 6.3, 6.7).

Previous studies on faults as migration conduits are mainly focused on seismic-scale single fault or regional fault systems (Billi et al., 2003; Karlsen & Skeie, 2006; Li et al., 2019; Wang et al., 2020; Xu et al., 2015; Zhang et al., 2019), while we focus on a lithologically-controlled fault network. Seismic-scale faults would cut through several layers in the vertical direction and communicate fluid between source rocks and reservoirs over a distance of several kilometers, e.g., San Andreas Parkfield (3 km, Bradbury et al., 2011; Rymer et al., 2006), EI Reno fault (1.6 km; Liao et al., 2019), Triunfo fault (0.2–0.6 km; Celestino et al., 2020) etc. These observations are probably limited in explaining lateral migration of hydrocarbon in a stratigraphic reservoir of ~100-m in thickness. In this interpretation, the small scale conjugate fault networks have an optimal geometry in the brittle dolostone and might be expected to control the lateral migration and accumulation of subsurface fluid horizontally. However, the slip directions of the fault planes are difficult to determine based on the seismic attribute maps, which require supplementary data and fault-plane solutions.

6.5.3 *Regional Stresses and Development of the Fault Networks*

Large plate-scale regions are generally subjected to a uniform stress field (Engelder & Geiser, 1980; Heidbach et al., 2010; Li et al., 2016; Zoback, 1992). As the eastern boundary of the Tibetan Plateau, the Longmen Shan is a steep transition zone between the Tibetan Plateau and the Sichuan Basin. The Longmen Shan has experienced several stages of rapid uplift with the latest and most important stage starting in the Late Miocene (Dai et al., 2013; Li et al., 2012; Wang et al., 2014). It is at this stage that Xinchang region is compressed and the Longquan Shan is uplifted as a response to the rapid uplift of the Longmen Shan about 10–15 Ma ago (Dai et al., 2013; Li et al., 2012; Richardson et al., 2008). Additionally, the direction of the maximum principle compressional stress of Longmen Shan and adjacent western Sichuan Basin was nearly on E-W during the Miocene-Pliocene, which is proximately parallel to the acute bisector of conjugate fault network in Figs. 6.3, 6.4 (Burchfiel et al., 1995; Li et al., 2014). Faults tend to form within the brittle layer of dolostone while the weak/ductile layer of anhydrite can deform by distributed shearing and inflation that accommodate the overall strain (Davies et al., 2012; Ferrill et al., 2017a, b; Vasquez et al., 2018). The fault networks in the dolostone layer are more developed than underlaying anhydrite (Figs. 6.3, 6.4), which we speculate is based on mechanical contrast that controls the lithologically-controlled conjugate faults. However, the primary fault network is not constrained by the lithologies.

6.6 Conclusions

Based on high-resolution 3D seismic data, we have interpreted fault networks in the Triassic Leikoupo Formation, composed of carbonates and evaporates, with implications for understanding gas migration and accumulation in carbonate reservoirs. We found that seismic attributes could be applied to identify the fault networks in the subsurface. The results show an orthogonal network of primary faults in the Xinchang region and a secondary set of conjugate fault networks in the fold limb. The observation of the conjugate fault networks integrated with fractures from core samples illustrates a lithologically-controlled conduit system for natural gas in the Leikoupo Formation. The characteristics of the conjugate fault networks favor the model of self-sourced migration and accumulation in the Xinchang reservoir, where the dolostone of the Leikoupo Formation, that contains such fault networks, as a potential exploration target.

References

Al-Dossary, S., & Marfurt, K. J. (2006). 3D volumetric multispectral estimates of reflector curvature and rotation. *Geophysics, 71*(5), P41–P51. https://doi.org/10.1190/1.2242449

Ameen, M. S. (2003). Fracture and in-situ stress characterization of hydrocarbon reservoirs: Definitions and introduction. *Geological Society, London, Special Publications, 209*(1), 1–6. https://doi.org/10.1144/GSL.SP.2003.209.01.01

Anders, M. H., Laubach, S. E., & Scholz, C. H. (2014). Microfractures: A review. *Journal of Structural Geology, 69*, 377–394. https://doi.org/10.1016/j.jsg.2014.05.011

Bahorich, M., & Farmer, S. (1995). 3-D seismic discontinuity for faults and stratigraphic features: The coherence cube. *The Leading Edge, 14*(10), 1053–1058. https://doi.org/10.1190/1.1437077

Barchi, M. R. (2002). Lithological and structural controls on the seismogenesis of the Umbria region: observations from seismic reflection profiles. *Bollettino Della Società Geologica Italiana, 121*(1), 855–864.

Bergbauer, S., & Pollard, D. D. (2004). A new conceptual fold-fracture model including prefolding joints, based on the Emigrant Gap anticline, Wyoming. *Geological Society of America Bulletin, 116*(3–4), 294–307. https://doi.org/10.1130/B25225.1

Bhattacharya, S., & Verma, S. (2019). Application of volumetric seismic attributes for complex fault network characterization on the North Slope, Alaska. *Journal of Natural Gas Science and Engineering, 65*, 56–67. https://doi.org/10.1016/j.jngse.2019.02.002

Billi, A., Salvini, F., & Storti, F. (2003). The damage zone-fault core transition in carbonate rocks: Implications for fault growth, structure and permeability. *Journal of Structural Geology, 25*(11), 1779–1794. https://doi.org/10.1016/S0191-8141(03)00037-3

Bradbury, K. K., Evans, J. P., Chester, J. S., Chester, F. M., & Kirschner, D. L. (2011). Lithology and internal structure of the San Andreas fault at depth based on characterization of Phase 3 whole-rock core in the San Andreas Fault Observatory at Depth (SAFOD) borehole. *Earth and Planetary Science Letters, 310*(1–2), 131–144. https://doi.org/10.1016/j.epsl.2011.07.020

Burberry, C. M., Cannon, D. L., Cosgrove, J. W., & Engelder, T. (2020). Fracture patterns associated with the evolution of the Teton anticline, Sawtooth Range, Montana, USA. *Geological Society, 487*(1), 229–261. https://doi.org/10.1144/SP487.12

Burchfiel, B. C., Zhiliang, C., Yupinc, L., & Royden, L. H. (1995). Tectonics of the Longmen Shan and adjacent regions, central China. *International Geology Review, 37*(8), 661–735. https://doi.org/10.1080/00206819509465424

Caine, J. S., Evans, J. P., & Forster, C. B. (1996). Fault zone architecture and permeability structure. *Geology, 24*(11), 1025–1028. https://doi.org/10.1130/0091-7613

Carter, N., & Lines, L. (2001). Fault imaging using edge detection and coherency measures on Hibernia 3-D seismic data. *The Leading Edge, 20*(1), 64–69. https://doi.org/10.1190/1.1438880

Celestino, M. A. L., de Miranda, T. S., Mariano, G., de Lima Alencar, M., de Carvalho, B. R. B. M., da Cruz Falcão, T., et al. (2020). Fault damage zones width: Implications for the tectonic evolution of the northern border of the Araripe Basin, Brazil, NE Brazil. *Journal of Structural Geology, 138*, 104116. https://doi.org/10.1016/j.jsg.2020.104116

Chester, F. M., & Chester, J. S. (1998). Ultracataclasite structure and friction processes of the Punchbowl fault, San Andreas system. *California. Tectonophysics, 295*(1–2), 199–221. https://doi.org/10.1016/S0040-1951(98)00121-8

Chester, F. M., Evans, J. P., & Biegel, R. L. (1993). Internal structure and weakening mechanisms of the San Andreas fault. *Journal of Geophysical Research: Solid Earth, 98*(B1), 771–786. https://doi.org/10.1029/92JB01866

Chilingarian, G. V., Torabzadeh, J., Rieke, H. H., Metghalchi, M., & Mazzullo, S. J. (1992). Interrelationships among surface area, permeability, porosity, pore size, and residual water saturation. In *Developments in petroleum science* (Vol. 30, pp. 379–397). Elsevier. https://doi.org/10.1016/S0376-7361(09)70130-X

Chopra, S., & Marfurt, K. J. (2007). Seismic attributes for prospect identification and reservoir characterization. In *Society of exploration geophysicists and European association of geoscientists and engineers.*

Crider, J. G., & Peacock, D. C. (2004). Initiation of brittle faults in the upper crust: A review of field observations. *Journal of Structural Geology, 26*(4), 691–707. https://doi.org/10.1016/j.jsg.2003.07.007

Dai, J., Wang, C., Hourigan, J., & Santosh, M. (2013). Insights into the early Tibetan Plateau from (U–Th)/He thermochronology. *Journal of the Geological Society, 170*(6), 917–927. https://doi.org/10.1144/jgs2012-076

Davies, R. K., Knipe, R. J., & Welch, M. J. (2012). The role of vertical mechanical heterogeneity in predicting fault zone architecture. In *Proceedings of the 3rd EAGE international conference on fault and top seals* (pp. cp-311). European Association of Geoscientists and Engineers.https://doi.org/10.3997/2214-4609.20143050

De Paola, N., Collettini, C., Faulkner, D. R., & Trippetta, F. (2008). Fault zone architecture and deformation processes within evaporitic rocks in the upper crust. *Tectonics, 27*(4), 2230. https://doi.org/10.1029/2007TC002230

Dominguez, G. C. (1992). *Carbonate reservoir characterization: A geologic-engineering analysis.* Elsevier.

Ellsworth, W. L. (2013). Injection-induced earthquakes. *Science, 341*(6142), 1225942. https://doi.org/10.1126/science.1225942

Eng, C., & Tsuji, T. (2019). Influence of faults and slumping on hydrocarbon migration inferred from 3D seismic attributes: Sanriku-Oki forearc basin, northeast Japan. *Marine and Petroleum Geology, 99*, 175–189. https://doi.org/10.1016/j.marpetgeo.2018.10.013

Engelder, T., & Geiser, P. (1980). On the use of regional joint sets as trajectories of paleostress fields during the development of the Appalachian Plateau, New York. *Journal of Geophysical Research: Solid Earth, 85*(B11), 6319–6341. https://doi.org/10.1029/JB085iB11p06319

Ferrill, D. A., Evans, M. A., McGinnis, R. N., Morris, A. P., Smart, K. J., Wigginton, S. S., et al. (2017a). Fault zone processes in mechanically layered mudrock and chalk. *Journal of Structural Geology, 97*, 118–143. https://doi.org/10.1016/j.jsg.2017.02.013

Ferrill, D. A., Morris, A. P., McGinnis, R. N., Smart, K. J., Wigginton, S. S., & Hill, N. J. (2017b). Mechanical stratigraphy and normal faulting. *Journal of Structural Geology, 94*, 275–302. https://doi.org/10.1016/j.jsg.2016.11.010

Flodin, E. A., & Aydin, A. (2004). Evolution of a strike-slip fault network, Valley of Fire State Park, southern Nevada. *Geological Society of America Bulletin, 116*(1–2), 42–59. https://doi.org/10.1130/B25282.1

Gangping, X. (2015). Source of gas reservoirs in the fourth member of the Middle Triassic Leikoupo Formation in Western Sichuan Depression. *Petroleum Geology and Experiment, 37*(4), 418–422.

Gao, D. (2013). Integrating 3D seismic curvature and curvature gradient attributes for fracture characterization: Methodologies and interpretational implications. *Geophysics, 78*(2), O21–O31. https://doi.org/10.1190/geo2012-0190.1

Ghosh, K., & Mitra, S. (2009). Structural controls of fracture orientations, intensity, and connectivity, Teton anticline, Sawtooth Range, Montana. *AAPG Bulletin, 93*(8), 995–1014. https://doi.org/10.1306/04020908115

Guo, J. L., Gao, S., Wu, Y. B., Li, M., Chen, K., Hu, Z. C., et al. (2014). 3.45 Ga granitic gneisses from the Yangtze Craton, South China: Implications for Early Archean crustal growth. *Precambrian Research, 242*, 82–95. https://doi.org/10.1016/j.precamres.2013.12.018

Harmon, D. (2019). *Structural mini-atlas for physical geology lecture [EB/OL].* http://maps.unomaha.edu/Maher/geo117/part3/structures/structureatlas.html

He, D., Ma, Y., Li, Y., & Wu, S. (2019). New directions in an established gas play: Promising dolomite reservoirs in the Middle Triassic Leikoupo Formation of the Sichuan Basin, China. *AAPG Bulletin, 103*(1), 1–29. https://doi.org/10.1306/05111816502

Heidbach, O., Tingay, M., Barth, A., Reinecker, J., Kurfeß, D., & Müller, B. (2010). Global crustal stress pattern based on the World Stress Map database release 2008. *Tectonophysics, 482*(1–4), 3–15. https://doi.org/10.1016/j.tecto.2009.07.023

Hua-Jun, W., Zhi-Ru, Y., & Bing, H. (2015). Superposed evolution of Sichuan Basin and its petroleum accumulation. *Earth Science Frontiers, 22*(3), 161.

Iacopini, D., & Butler, R. W. (2011). Imaging deformation in submarine thrust belts using seismic attributes. *Earth and Planetary Science Letters, 302*(3–4), 414–422. https://doi.org/10.1016/j.epsl.2010.12.041

Karlsen, D. A., & Skeie, J. E. (2006). Petroleum migration, faults and overpressure, Part I: Calibrating basin modelling using petroleum in traps—a review. *Journal of Petroleum Geology, 29*(3), 227–256. https://doi.org/10.1111/j.1747-5457.2006.00227.x

Keming, Y. (2016). Hydrocarbon potential of source rocks in the Middle Triassic Leikoupo Formation in the Western Sichuan depression. *Petroleum Geology and Experiment, 38*(3), 366–374.

Kim, Y. S., Peacock, D. C. P., & Sanderson, D. J. (2003). Mesoscale strike-slip faults and damage zones at Marsalforn, Gozo Island, Malta. *Journal of Structural Geology, 25*(5), 793–812. https://doi.org/10.1016/S0191-8141(02)00200-6

Laubach, S. E. (2003). Practical approaches to identifying sealed and open fractures. *AAPG Bulletin, 87*(4), 561–579. https://doi.org/10.1306/11060201106

Li, M., Wu, G., Xia, B., Huang, T., Ni, B., Pang, S., & Long, X. (2019). Controls on hydrocarbon accumulation in clastic reservoirs of the Tarim Craton, NW China. *Marine and Petroleum Geology, 104*, 423–437. https://doi.org/10.1016/j.marpetgeo.2019.04.008

Li, P., Hao, F., Guo, X., Zou, H., Yu, X., & Wang, G. (2015). Processes involved in the origin and accumulation of hydrocarbon gases in the Yuanba gas field, Sichuan Basin, southwest China. *Marine and Petroleum Geology, 59*, 150–165. https://doi.org/10.1016/j.marpetgeo.2014.08.003

Li, Z., Jia, D., Chen, W., Yin, H., Shen, L., Sun, C., et al. (2014). Late Cenozoic east–west crustal shortening in southern Longmen Shan, eastern Tibet: Implications for regional stress field changes. *Tectonophysics, 623*, 169–186. https://doi.org/10.1016/j.tecto.2014.03.033

Li, Z. W., Liu, S., Chen, H., Deng, B., Hou, M., Wu, W., & Cao, J. (2012). Spatial variation in Meso-Cenozoic exhumation history of the Longmen Shan thrust belt (eastern Tibetan Plateau) and the adjacent western Sichuan basin: Constraints from fission track thermochronology. *Journal of Asian Earth Sciences, 47*, 185–203. https://doi.org/10.1016/j.jseaes.2011.10.016

Li, Z., Liu-Zeng, J., Jia, D., Sun, C., Wang, W., Yuan, Z., & Liu, B. (2016). Quaternary activity of the range front thrust system in the Longmen Shan piedmont, China, revealed by seismic imaging and growth strata. *Tectonics, 35*(12), 2807–2827. https://doi.org/10.1002/2015TC004093

Liao, Z., Liu, H., Carpenter, B. M., Marfurt, K. J., & Reches, Z. E. (2019). Analysis of fault damage zones using three-dimensional seismic coherence in the Anadarko Basin, Oklahoma. *AAPG Bulletin, 103*(8), 1771–1785. https://doi.org/10.1306/1219181413417207

Liao, Z., Liu, H., Jiang, Z., Marfurt, K. J., & Reches, Z. E. (2017). Fault damage zone at subsurface: A case study using 3D seismic attributes and a clay model analog for the Anadarko Basin, Oklahoma. *Interpretation, 5*(2), T143–T150. https://doi.org/10.1190/INT-2016-0033.1

Liu, S. H., Jin, X. B., Hu, Y. Z., & Yang, Z. B. (2010). Exploration prospect of lower Triassic in western Sichuan basin: An application of Chuanke 1 well. *Natural Gas Exploration and Development, 4*, 21–24.

Long, K., Chen, Z. Y., Song, X. B., & Yuan, H. (2019). Analysis on natural gas source of Middle Triassic Leikoupo Formation in western Sichuan Basin, China. *Natural Gas Industry, S1*, 48–53.

Lu, Y., Lesher, C. M., & Deng, J. (2019). Geochemistry and genesis of magmatic Ni-Cu-(PGE) and PGE-(Cu)-(Ni) deposits in China. *Ore Geology Reviews, 107*, 863–887. https://doi.org/10.1016/j.oregeorev.2019.03.024

Lunn, R. J., Willson, J. P., Shipton, Z. K., & Moir, H. (2008). Simulating brittle fault growth from linkage of preexisting structures. *Journal of Geophysical Research: Solid Earth, 113*, 5388. https://doi.org/10.1029/2007JB005388

Luo, Y., Higgs, W. G., & Kowalik, W. S. (1996). Edge detection and stratigraphic analysis using 3D seismic data. In *SEG technical program expanded abstracts 1996* (pp. 324–327). Society of Exploration Geophysicists. https://doi.org/10.1190/1.1826632

Luo, Y., Wang, Y., Liu, H., Wang, G., & Zhao, Y. (2020). Overpressure controlling factors for tectonic fractures in near-source tight reservoirs in the southwest Ordos Basin, China. *Journal of Petroleum Science and Engineering, 188*, 106818. https://doi.org/10.1016/j.petrol.2019.106818

Marfurt, K. J., Kirlin, R. L., Farmer, S. L., & Bahorich, M. S. (1998). 3-D seismic attributes using a semblance-based coherency algorithm. *Geophysics, 63*(4), 1150–1165. https://doi.org/10.1190/1.1444415

Marfurt, K. J., Sudhaker, V., Gersztenkorn, A., Crawford, K. D., & Nissen, S. E. (1999). Coherency calculations in the presence of structural dip. *Geophysics, 64*(1), 104–111. https://doi.org/10.1190/1.1444508

Miller, S. A., Collettini, C., Chiaraluce, L., Cocco, M., Barchi, M., & Kaus, B. J. (2004). Aftershocks driven by a high-pressure CO_2 source at depth. *Nature, 427*(6976), 724–727. https://doi.org/10.1038/nature02251

Nixon, C. W., Sanderson, D. J., Dee, S. J., Bull, J. M., Humphreys, R. J., & Swanson, M. H. (2014). Fault interactions and reactivation within a normal-fault network at Milne Point, Alaska. *AAPG Bulletin, 98*(10), 2081–2107. https://doi.org/10.1306/04301413177

Prentice, C. S., Scotchmoor, J. G., Moores, E. M., & Kiland, J. P. (2006). *1906 San Francisco earthquake centennial field guides: Field trips associated with the 100th Anniversary Conference, 18–23 April 2006*. Geological Society of America. https://doi.org/10.1130/0-8137-0007-8

Putz-Perrier, M. W., & Sanderson, D. J. (2010). Distribution of faults and extensional strain in fractured carbonates of the North Malta Graben. *AAPG Bulletin, 94*(4), 435–456. https://doi.org/10.1306/08260909063

Richardson, N. J., Densmore, A. L., Seward, D., Fowler, A., Wipf, M., Ellis, M. A., ... & Zhang, Y. (2008). Extraordinary denudation in the Sichuan Basin: Insights from low-temperature thermochronology adjacent to the eastern margin of the Tibetan Plateau. *Journal of Geophysical Research: Solid Earth, 113*(B4). https://doi.org/10.1029/2006JB004739

Roberts, A. (2001). Curvature attributes and their application to 3 D interpreted horizons. *First Break, 19*(2), 85–100.

Rymer, M. J., Hickman, S. H., & Stoffer, P. W. (2006). *A field guide to the central, creeping section of the San Andreas fault and the San Andreas Fault Observatory at Depth.* https://doi.org/10.1130/2006.1906SF(16)

Sarkar, S., Marfurt, K. J., & Slatt, R. M. (2010). Generation of sea-level curves from depositional pattern as seen through seismic attributes-seismic geomorphology analysis of an MTC-rich shallow sediment column, northern Gulf of Mexico. *The Leading Edge, 29*(9), 1084–1091. https://doi.org/10.1190/1.3485769

Segall, P., & Pollard, D. D. (1983). Nucleation and growth of strike slip faults in granite. *Journal of Geophysical Research: Solid Earth, 88*(B1), 555–568. https://doi.org/10.1029/JB088iB01p00555

Stearns, D. W. (1967). Certain aspects of fracture in naturally deformed rock. In *National sciences foundation advanced science seminar in rock mechanics* (pp. 97–118).

Sun, T., Luo, X., Qing, H., Kou, X., Sheng, Z., Xu, G., & Zuo, Y. (2020). Characteristics and natural gas origin of middle–late triassic marine source rocks of the Western Sichuan depression, SW China. *Acta Geologica Sinica-English Edition, 94*(2), 376–398. https://doi.org/10.1111/1755-6724.14341

Tian, H., Tang, S., Zhang, J. Y., Xin, Y., Wang, X., & Li, W. (2018). Characteristics and formation conditions of carbonate reservoir in Leikoupo Formation of western Sichuan Basin. *Natural Gas Geoscience, 29*(11), 1585–1594.

Vasquez, L., Nalpas, T., Ballard, J. F., De Veslud, C. L. C., Simon, B., Dauteuil, O., & Du Bernard, X. (2018). 3D geometries of normal faults in a brittle-ductile sedimentary cover: Analogue modelling. *Journal of Structural Geology, 112*, 29–38. https://doi.org/10.1016/j.jsg.2018.04.009

Wang, C., Dai, J., Zhao, X., Li, Y., Graham, S. A., He, D., et al. (2014). Outward-growth of the Tibetan Plateau during the Cenozoic: A review. *Tectonophysics, 621*, 1–43. https://doi.org/10.1016/j.tecto.2014.01.036

Wang, F., Chen, D., Wang, Q., Shi, X., Xie, G., Wang, Z., et al. (2020). Evolution characteristics of transtensional faults and their impacts on hydrocarbon migration and accumulation: A case study from the Huimin Depression, Bohai Bay Basin, eastern China. *Marine and Petroleum Geology, 120*, 104507. https://doi.org/10.1016/j.marpetgeo.2020.104507

Watkins, H., Butler, R. W., Bond, C. E., & Healy, D. (2015). Influence of structural position on fracture networks in the Torridon Group, Achnashellach fold and thrust belt, NW Scotland. *Journal of Structural Geology, 74*, 64–80. https://doi.org/10.1016/j.jsg.2015.03.001

Weimer, P., & Davis, T. L. (Eds.) (1996). *Applications of 3-D seismic data to exploration and production* (Vol. 5). Published jointly by the American Association of Petroleum Geologists and the Society of Exploration Geophysics.

Wilson, T. H., Smith, V., & Brown, A. (2015). Developing a model discrete fracture network, drilling, and enhanced oil recovery strategy in an unconventional naturally fractured reservoir using integrated field, image log, and three-dimensional seismic data. *AAPG Bulletin, 99*(4), 735–762. https://doi.org/10.1306/10031414015

Wu, X. (2017). Directional structure-tensor-based coherence to detect seismic faults and channels. *Geophysics, 82*(2), 13. https://doi.org/10.1190/geo2016-0473.1

Xiao, K., Li, H., Duan, Y., Zhang, Y., & Liu, L. (2020). Reservoir characteristics and main controlling factors of the Leikoupo gas pools in the western Sichuan Basin. *Natural Gas Industry B, 7*(1), 13–23. https://doi.org/10.1016/j.ngib.2020.01.002

Xu, G. M., Feng, X., Zhu, L., & Wang, Q. X. (2011). Exploration suggestions on Leikoupo to Maantang Formations in Xinchang structural belt, west Sichuan Basin. *Natural Gas Technology and Economy, 5*(5), 16–20.

Xu, S., Hao, F., Xu, C., Wang, Y., Zou, H., & Gong, C. (2015). Differential compaction faults and their implications for fluid expulsion in the northern Bozhong Subbasin, Bohai Bay Basin, China. *Marine and Petroleum Geology, 63*, 1–16. https://doi.org/10.1016/j.marpetgeo.2015.02.013

Yong, L., Allen, P. A., Densmore, A. L., & Qiang, X. (2003). Evolution of the Longmen Shan foreland basin (western Sichuan, China) during the Late Triassic Indosinian orogeny. *Basin Research, 15*(1), 117–138. https://doi.org/10.1046/j.1365-2117.2003.00197.x

Yong, L., Yunfu, Z., & Haisheng, Y. (1995). *Sedimentary and tectonic evolution of the Longmenshan foreland basin, western Sichuan.* Press of Chengdu University of Science and Technology.

Yulin, C., Yan, Z., Yongming, D., & Qiongxian, W. (2018). Pore structure characteristics and reservoir classification of dolomite reservoirs in fourth member of Leikoupo formation, Longmen mountain front, western Sichuan Basin. *Petroleum Geology and Experimental, 40*(5), 621–631.

Zhang, J., Cao, J., Wang, Y., Hu, G., Zhou, N., & Shi, T. (2019). Origin of giant vein-type bitumen deposits in the northwestern Junggar Basin, NW China: Implications for fault-controlled hydrocarbon accumulation. *Journal of Asian Earth Sciences, 179*, 287–299. https://doi.org/10.1016/j.jseaes.2019.05.009

Zoback, M. L. (1992). First-and second-order patterns of stress in the lithosphere: The World Stress Map Project. *Journal of Geophysical Research: Solid Earth, 97*(B8), 11703–11728. https://doi.org/10.1029/92JB00132

Chapter 7
Caprock Integrity of Carbonate and Evaporite

Abstract In geologic time scales, the slow, subcritical fracture propagation can contribute considerably to the fault and fracture system development, especially under reactive fluid environments during petroleum migration and accumulation. Using the double torsion technique, we measure the subcritical fracture growth properties of carbonate and anhydrite caprock core samples in Sichuan Basin, China, under different environments of ambient air, de-ionized water, and formation water. Experimental results show that the aqueous fluids significantly promote subcritical fracture growth: (1) Compared to ambient air, the presence of aqueous fluids dramatically reduces both the fracture toughness (K_{IC}) and subcritical fracture growth index (SCI) of the caprocks; (2) The K_{IC} reduction is more prominent for the anhydrite (50%) caprock than the carbonate (30%) caprock, which indicates that mineral solubility is responsible for the fluid-weakening effect; (3) A systematic change of SCI with varying fluid conditions is observed: SCI is the largest in air, smallest in formation water, and intermediate in de-ionized water. These observations provide insights into the mechanisms controlling fracture growth in carbonate and evaporite caprocks, which could extend to brittle failure and fluid-rock interactions occurring during the seismic cycle in the upper crust.

7.1 Introduction

Caprock is a crucial component in subsurface storage reservoirs. For effective trapping of the oil and gas in petroleum systems, a qualified caprock should meet a suite of requirements in its petro-physical and rock mechanical properties. For intact caprocks, low permeability is essential for permeability barrier establishment to trap the fluid from migrating upward through capillary sealing (e.g., Downey, 1984; Watts, 1987). However, rocks frequently fail by fracturing and faulting under subsurface stress and temperature conditions which significantly threatens caprock sealing performance. For example, unsealed faults are frequently identified in petroleum exploration practices as examples of seal failure (Allan, 1989; Fisher et al., 2001; Finkberner et al., 2001; Hao et al., 2015). For sealed faults, risk assessment for seal

Z. Liao, *Applied Seismic Geostructures*, https://doi.org/10.1007/978-981-97-7079-3_7

failure is also necessary because of the potential for fault reactivation (Fulljames et al., 1997; Smith, 1966; Yielding et al., 1997) within the critically stressed crust (Zoback, 2010). Fractures within tight formations can provide fracture porosity for storage and pathway for fluid migration, which significantly enhances the storage and flow properties of the host rocks (Cipolla et al., 2010; Olson et al., 2009). On the contrary, fracture growth in caprocks could be a deteriorating process that risks the caprock integrity (Cosgrove, 2001; Hao et al., 2015; Ingram & Urai, 1999; Tingay et al., 2009). As a result, the overall caprock sealing efficiency is related to both how impermeable the caprocks are and how strong the caprocks can withstand failure by faulting and fracturing under in-situ stress, temperature, and fluid conditions.

Rock failure characteristics are commonly investigated experimentally through rock mechanical and fracture mechanical testing under relevant stress conditions (e.g., Lee et al., 2015; Zhang et al., 2017), different fluid chemical environments (e.g., Nara et al., 2012; Erguler & Ulusay, 2009; Chen et al., 2019), and variable temperatures (e.g., Atkinson & Meredith, 1987; Handin et al., 1967; Meredith & Atkinson, 1985). Rock mechanics testing can determine rock strength, failure modes, failure envelope, and failure microstructures, along with other rock physical properties (Chen et al., 2017; Sone & Zoback, 2013). Fracture mechanical testing, on the other hand, explores the fracture growth properties within the rocks. For instance, fracture toughness, or K_C with the unit of MPa $\sqrt{m}$, is a measure of the ability for materials to resist fracture (Anderson, 2005). When the stress state near a fracture tip, named stress intensity factor K, which is a function of applied stress and fracture geometry, approaches the fracture toughness of the material, critical fracture propagation is expected, e.g., in dynamic rupture propagation (Reches & Lockner, 1994) and hydraulic fracturing (Cosgrove, 2001; Tingay et al., 2009) processes. Fractures can also propagate subcritically when $K < K_C$, by chemically assisted processes such as stress corrosion (Atkinson, 1982, 1984) and mineral dissolution (Major et al., 2018). Subcritical fracture growth is related to the long-term, environmentally assisted processes such as joint system development, static fatigue, and creep deformation (Olson, 1993).

The recent interest in unconventional petroleum resource development, subsurface waste storage, and hydrothermal systems resulted in increased interest in geomechanical properties in tight formations (e.g., Callahan et al., 2019; Chen et al., 2019; Major et al., 2018; Meléndez-Martínez & Schmitt, 2016). Carbonates generally deform brittly under upper crustal conditions, but they can also have ductile failure under elevated temperature and confine pressure (Handin et al., 1963) or assisted by pore fluids (Renner & Rummel, 1996) with considerable porosity, small grain size, and the presence of clay minerals (Chen et al., 2019). Despite with similar evaporative origin, halite, gypsum, and anhydrite are often thought as ideal caprocks due to their dominant ductile flow characteristic even under shallow depth stress and temperature conditions (Chester, 1988; Fu et al., 2016; Handin & Hager, 1958; Shimamoto, 1986) and therefore their low permeability and porosity. Fracture toughness and brittleness measurements are routinely conducted to guide hydraulic fracturing operations (e.g., Jin et al., 2014a, 2014b). As a result, a large amount of literature reported fracture toughness values in tight formations such as shales (Chen et al., 2017, 2019; Swanson,

1984) and tight carbonates (Brantut et al., 2013; Piane et al., 2017). However, there is a scarce of studies on slow subcritical fracture growth properties in tight formations, especially for rock salts under reactive fluids.

Previous studies show that fluid pH and cation concentration have impact on subcritical fracture growth (Billi et al., 2003; Rostom et al., 2013). However, no previous study focused on the fracture growth properties in rock salts such as halite, gypsum, and anhydrite primarily due to the abovementioned strong flow properties of these evaporites, which made fractures as a common brittle deformation feature unlikely to occur or quickly healed/sealed during ductile flow. Despite their strong potential to withstand large deformation, rock salts are vulnerable under fluids due to their much more significant solubility than carbonates and silicates. Since mineral dissolution is thought to be an effective mechanism of subcritical fracture growth in calcites/carbonates (Bergsaker et al., 2016; Chen et al., 2019) and silicates (Atkinson & Meredith, 1981), it is reasonable to assume that gypsum/anhydrite dissolution can also be a significant contributing factor for subcritical fracture growth. In addition, the massive swelling potential of anhydrite-gypsum transformation could make initial cracks deteriorating factor for fracture system development (Bildstein et al., 2001; Fu et al., 2016; Machel, 2001). For petroleum systems, subcritical fracture growth may control brittle deformation processes at tectonic loading rates, at slow pore pressure buildup during burial and maturation of organic-rich sediments, and at the interfaces of changing fluid composition during migration of hydrocarbons (Chen et al., 2017). Therefore, assessment of the interaction between fluid and fractures in hydrocarbon systems is necessary.

In this study, we focus on the fracture characteristics of the carbonate and anhydrite caprocks by fracturing, specifically in a deep gas reservoir in Sichuan Basin, China. We conduct double torsion fracture mechanical testing on core caprock samples of carbonates and anhydrite under various fluid chemical environments of air, deionized water, and formation water to get fracture growth parameters of fracture toughness, subcritical fracture growth index (SCI) and stress intensity versus fracture velocity curves. We also compare the microstructures between natural fractures and those formed during the fracture mechanics testing. The controlling mechanisms for fracture development in marine evaporites under subsurface conditions are discussed, and the general implication for caprock failure in deep gas reservoirs is addressed.

7.2 Geologic Setting

The Sichuan Basin is the largest nature gas-rich basin in Southwest China (Li et al., 2015). Northeast Sichuan Basin is located at the northern margin of the Yangtze plate, surrounded by the Micangshan uplift in the north, Dabashan thrust belt in the northeast, and East Sichuan fold belt in the east (Fig. 7.1a). The major petroleum systems are distributed in three sub-divisions: the Tongnanba anticline in Dabashan forebulge, the Yuanba area in the slope of the Micangshan uplift, and the Xuanhan-Daxian area

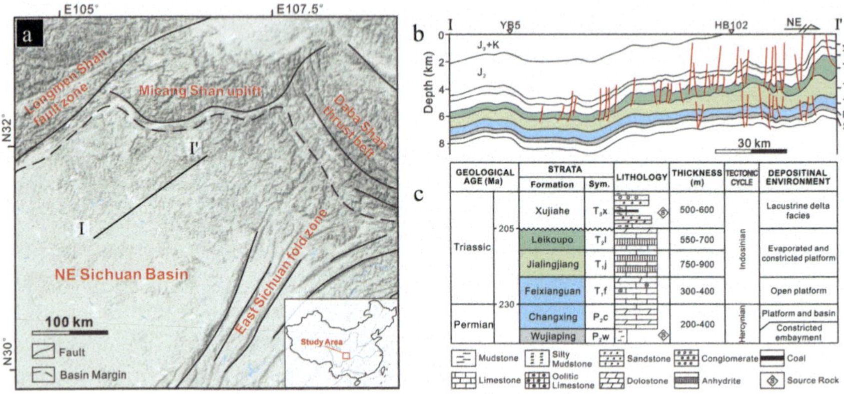

Fig. 7.1 Geologic setting of the study area. (**a**) Major structures are bounding the NE Sichuan Basin. Inset marks the location of the Sichuan Basin (green shade) and the mapped area in A (square). (**b**) Stratigraphic units crossing the blue line in (**a**) constructed from well-logs and seismic data. Major faults marked in red. (**c**) Typical stratigraphic columns for the present petroleum system. Adapted from Li et al., (2015)

in the edge of East Sichuan fold belt (Fig. 7.1a). Two large deposition stages associated with the petroleum systems are identified in the Northeast Sichuan Basin: marine carbonate deposition from Permian to Middle Triassic periods, terrestrial clastic rock deposition from Late Triassic to Middle Paleogene (Fig. 7.1b, Fig. 7.1c).

This study focuses on the petroleum systems in marine stratum from Permian to Middle Triassic periods. The source rocks are coaly mudstone and dark marlite in the Permian Wujiaping Formation (P_2w). The main reservoir rocks are the striped reef-bank dolostone along the platform margin in the Upper Permian Changxing Formation (P_2c) and the shallow shoal oolitic dolostone and limestone in the Lower Triassic Feixianguan Formation (T_1f). The carbonates (P_2c, T_1f) underwent intense dolomitization, dissolution, and fractures, which made them good reservoir rocks. However, the tight and impermeable carbonates without dolomitization, dissolution, or fracture networks made them direct caprocks overlaying the carbonate reservoir rocks. The dense limestone usually formed at platform slopes and outer shelf overlaying porous shoal facies reservoir rocks (Jin, 2012; Jin et al., 2014a, 2014b). From Late Feixianguan Age, the open platforms gradually transitioned to restricted platforms, and eventually to evaporative platforms as sea level regressed. Anhydrite layers started to deposit, interbedded with mud-grade carbonate rocks in the Upper Feixianguan Formation. Thick anhydrite, gypsum, and halite rocks compose most of the Jialingjiang Formation (T_2j). Anhydrite interbeds with micritic dolostone in Leikoupo Formation (T_2l). These evaporites gradually thickened from the edge toward the center of the basin (NE to SW) and covered nearly the whole Northeast Sichuan Basin with an average stacking thickness of over 350 m (Wei et al., 2011; Zhang et al., 2012). Seismic and logging evidence confirmed that major faults died out in these evaporite-bearing formations (Fig. 7.1b), suggesting that the evaporites are ideal regional caprocks for the underlying petroleum resources (Ma et al., 2008).

Fig. 7.2 Recovered core samples (**a-d**) from the caprocks of deep gas reservoirs in NE Sichuan Basin and their thin-section images (**e–h**). Thin-section planes were parallel to bedding and were imaged under cross-polarized light (**e**, **f**) and plane-polarized light (**g**, **h**). See the text for descriptions

Fractures play an essential role in the current petroleum system. On the one hand, fracture network within the carbonate rocks improves the reservoir storage capacity (Guo, 2018; Ma et al., 2007). On the other hand, field exploration data suggested that whether fractures can develop within the dense impermeable carbonate direct caprocks overlaying the reservoir rocks is critical for caprock sealing efficiency. For instance, fractures developed well in the tight, fold-modified Middle-Upper Feixianguang Formation, making it a fractured reservoir. While in the same Feixianguang Formation, tight carbonate rocks in Yuanba area served as locally direct caprock due to the lack of fractures. The indirect caprocks of anhydrite and gypsum did not show clear evidence of seal failure from seismic data (Fig. 7.2b), but there is a need to evaluate the effect of fluids on compromising the integrity of these evaporites as the likelihood of direct caprock failure in the current petroleum system is high.

7.3 Materials and Methods

7.3.1 Materials

The materials used are core samples from four wells within three formations, with the depth range of 4142–6705 m (13,589–21,998 ft) in NE Sichuan Basin (Table 7.1). Powder X-ray diffraction (XRD) analysis determines bulk mineralogy of the four samples (Table 7.1). The fragmented core samples were first crushed and pulverized. The sieved portion was then ground in a micronizing mill for a few minutes with ethanol. The slurry was then oven-dried at 70 °C (well below the gypsum-anhydrite transition temperature of 200 °C at normal atmospheric condition as in Deer et al. (1992)), and the dry powder was then mounted on the glass sample holder and inserted into the Rigaku Ultima IV diffractometer for XRD analysis using Cu Kα radiation, 2–70 °C 2θ scan range, and 0.02° step size. We used JADE software for peak identification. T_1j and T_2l are anhydrite-bearing regional caprocks. T_1j from well HB102 is composed of predominantly anhydrite, while T_2l from well YB204 is dolomitic anhydrite. T_1f is a carbonate-rich direct caprock. T_1f from Well YB5 is limestone with 94% calcite, while T_1f from well MB2 is muddy as it has less calcite but contains small amount of dolomite, clay, and pyrite.

The recovered core samples and the thin-section images display the general appearance and texture of the caprocks. T_1j from well HB102 has fine planer lamination (Fig. 7.2a), with tortuously striped anhydrite crystals as large as hundreds of μm tightly aligned with neighboring grains (Fig. 7.2e). T_2l from well YB204 shows wavy and thicker laminations, with the dark dolomitic layers interbedded with the lighter anhydrite layers (Fig. 7.2b). The thin-section image reveals the interconnected anhydrite crystals and the muddy textured dolostone (Fig. 7.2f). T_1f from well YB5 is homogeneous with fine lamination (Fig. 7.2c), and a tight microcrystal calcite matrix as displayed in the thin-section image (Fig. 7.2g). T_1f from well MB2 has more impurities as indicated by the wavy and lighter bands in the core sample (Fig. 7.2d) and the dark spots in the thin section image (Fig. 7.2h). Except for a few cracks along bedding contrasts, all three formations hardly display natural fractures in the core samples, and all formations reveal tight matrix with hardly any pore spaces in the thin sections.

The basic physical properties of these rocks are compiled in Table 7.2. Porosity was calculated based on mineralogy data from XRD analysis. Dynamic Young's modulus and Poisson's ratio were calculated using measured sample density and ultrasonic P-wave and S-wave velocities at zero confine pressure under elastic isotropic assumption. Wave speeds were transmitted using core plugs with the axial direction perpendicular to bedding. Static Young's modulus and Poisson's ratio are derived from unconfined compressive tests with compression normal to bedding. The static Poisson's ratio values were used for later stress intensity factor calculations.

Table 7.1 Mineralogy of the caprock core samples determined from XRD analysis

Well	Formation	Depth (m)	Mineral type (%)[a]					
			Anhydrite	Gypsum	Calcite	Dolomite	Quartz	Others
HB102	T_1j	4705–4712	98	–	–	–	2	–
YB204	T_2l	4891–4893	62	4		30	4	–
YB5	T_1f	6703–6705	–	–	94	–	6	–
MB2	T_1f	4142–4145	–	–	70	11	6	13[b]

[a] Mineral composition is in weight percent
[b] 12% chlorite, and 1% pyrite

Table 7.2 Basic Physical properties of the caprocks tested

Well	Density (g/cm^3)		Wave speed (m/s)		E (GPa)		Poisson's ratio		Φ (%)
	Measured	Calculated	P	S	Dynamic	Static	Dynamic	Static	
HB102	2.878	2.942	5052	2791	57.2	33.4	0.28	0.51	0.72
YB204	2.893	2.859	4937	2767	57.6	15.9	0.27	0.75	1.15
YB5	2.709	2.694	5544	2733	54.1	44.5	0.34	0.22	0.11
MB2	2.787	2.745	–	–	–	34.3	–	0.20	0.33

7.3.2 Experimental Methods

A double torsion (DT) test is suitable to measure fracture behavior, especially for investigating subcritical fracture growth under reactive fluids (Atkinson, 1984; Chen et al., 2019). The loading configuration consists of a symmetric four-point support at the corners of a rectangular plate with loading applied from the two points at the top edge of the plate (Fig. 7.3a). The stress intensity factor K_I and fracture velocity are calculated as (Williams & Evans, 1973):

$$K_I = PW_m\left[\frac{3(1+v)}{Wd^3 d_n \psi}\right]^{1/2} \tag{7.1}$$

$$V = -\varphi \frac{P_0 a_0}{P^2}\frac{dP}{dt} \tag{7.2}$$

In Eq. (7.1), P is the applied load, W_m is bending moment arm, W is specimen width, d is specimen thickness, d_n is the reduced specimen thickness in the groove (Fig. 7.3a), v is Poisson's ratio, and ψ is the thickness correction factor expressed as (Fuller, 1979): $\psi = 1-1.26(d/W) + 2.4(d/W)\exp(-\pi W/2d)$. In Eq. (7.2), ϕ is a geometric correction factor taking into account the curved crack front, P_0, and a_0 are the load and crack length at the beginning of the relaxation period, respectively. We chose $\phi = 0.2$, as suggested by Williams and Evans (1973) and a constant $a_0 =$ 13 mm (Chen et al., 2017). With both K_I and V, the K-V curves can be constructed and the power law fitting of the K-V curves yields the SCI as (Atkinson, 1984; Brantut et al., 2013): $V = AK_I^n$, where A is an experimental constant, n is the SCI measuring the susceptibility of subcritical fracture growth for materials in the specific environments.

Core samples were sliced parallel to bedding and smoothed to rectangular wafers of 75 mm by 25 mm by 2 mm. Rectangular fracture guiding grooves of 0.4 mm in depth and 1 mm in width were cut on one side of each specimen (Fig. 7.3b). The specimens were tested in three conditions: ambient air (at room humidity of ~50%), DI water, and formation water at room temperature of ~20 °C. The specimen wafers were then soaked in relative fluids of DI water and formation water for 24 h before the DT tests to ensure complete fluid saturation. Laboratory formation

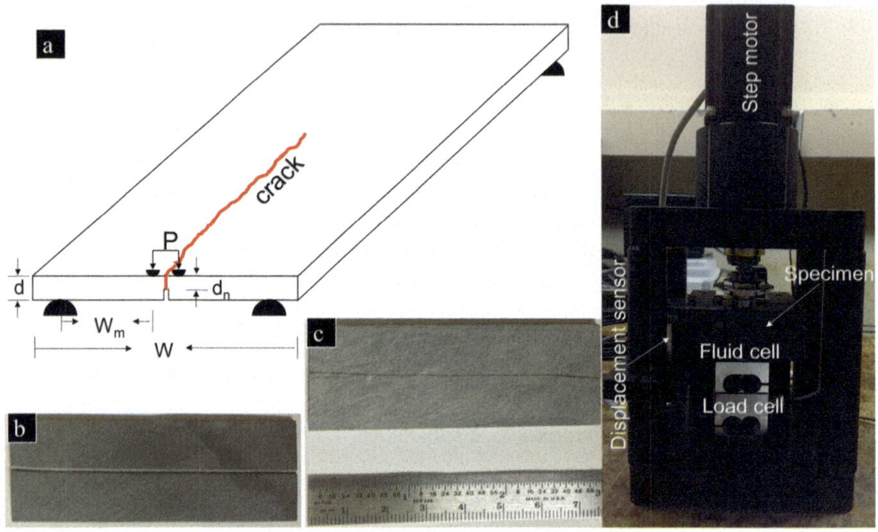

Fig. 7.3 The double torsion experimental method. (**a**) Loading configuration of the Double Torsion (DT) setup with useful specimen geometric parameters. (**b**) An intact MB2 specimen before DT testing with the fracture guiding groove facing upward. (**c**) A fractured YB5 specimen showing a fracture path along the central axis. (**d**) Experimental apparatus with main components

water composition was based on ion chromatography results of formation water at relevant layers (Table 7.3). Intact wafers were first pre-cracked under a specific environment under a slow loading rate of 0.05 mm/s to introduce the starting fracture (Chen et al., 2017; Holder et al., 2001). The pre-cracked specimens were loaded at 0.5 mm/s to load level of approximately 80–90% K_{IC}. The load point was then hold still to allow load relaxation for ~10 min. In a typical load relaxation test, ~4 load-relaxation cycles can be applied before the fracture propagates to the end of the specimen. The maximum load was used to calculate K_{IC} (termed K_{IC}^*) using Eq. (7.1). Due to sample inhomogeneity, such as existing fractures, layering contrast, and uneven crystal domains, some fractures can propagate deviating from the central axis, which was categorized as failed tests and was not used for further analysis. The DT test apparatus (Fig. 7.3d) consists of a step motor (Automation direct STP-MTR-23079) for displacement control, a S-beam load cell (Omega LCCD-50) with maximum range of ~50 lbs (23 kg) for load measurement, a non-contact linear displacement sensor (Omega LD701-5/10) for displacement measurement, and a National Instrument USB-6001 with Labview software for data acquisition under sampling rate of 5 Hz.

Original core samples were made into thin-section slides for microstructural observations. Thin-section slides containing natural fractures were further used for scanning electron microscopy (SEM) analysis. Fractures formed during DT tests were also preserved for SEM imaging of both fracture surfaces and fracture traces.

Table 7.3 Formation water composition in mol/L

Well	NaCl	KCl	$CaCl_2$	$MgCl_2$	Na_2SO_4	Salinity (‰)
HB102	0.76	–	0.56	0.13	–	104.99
YB204	1.278	0.196	0.423	0.364	–	141.21
YB5	0.508	–	0.258	0.029	0.014	57.29
MB2	0.783	0.135	–	–	0.115	67.27

7.4 Results

7.4.1 Fracture Mechanical Behavior

An example of the K-V curves for the anhydrite caprock T_1j from well HB102 is displayed in Fig. 7.4. Some general features of the K-V curves for all the caprock samples can be drawn. First, the K-V curves generally follow the empirical power-law relation under all conditions tested, as represented by straight lines in the log–log space (Fig. 7.4). Second, the test condition strongly influences the K-V curves. From air towards formation water to water, the K-V curves shift toward lower stress intensity factor and higher fracture velocity (the upper-left corner in the K-V space). This shift marks an enhanced subcritical fracture growth: fractures propagate under lower K as the K-V curves shift leftward, and fractures move faster as the K-V curves shift upward. From the K-V curves, we derive the maximum stress intensity factor K_{IC}^* as an approximation to K_{IC} (Callahan et al., 2019), the specific K values at $V = 10^{-5}$ m/s (termed K_a) to measure the scatter of K under each test condition and the shift of K under different test conditions (Chen et al., 2017, 2019; Nara et al., 2012), and the SCI (slope of the K-V curve) to measure the rate-dependency of the stress intensity factor.

We now integrate the measured fracture mechanical properties of all samples. The synthesized K-V curves (Fig. 7.5a–d) were simplified based on the raw data (Fig. 7.4). Each curve was drawn based on its mean SCI (E–H) and mean K_a data (I-L), and then extended to the velocity range of 10^{-7}–10^{-4} m/s. In ambient air, the anhydrite caprock from well HB102 has the largest fracture toughness, about 1.47 MPa $m^{1/2}$; while the other three caprocks are weaker, with K_{IC}^* around 1.2 MPa $m^{1/2}$. SCI spreads between 17 and 75. The systematic environmental effect on the shift of the K-V curves, SCI, K_{IC}^*, and K_a is maintained for all the samples. Compared to the air environment, DI water has the strongest effect on the shift of the K-V curve and the reduction of K_{IC}^* and K_a, while such weakening effect is smaller in formation water. The only exception is for the dolomitic anhydrite caprock from well YB204 (Fig. 7.5b, j), where both formation water and water have similar weakening effect on the shift of K-V curve, K_{IC}^*, and K_a. The introduction of fluids (water and formation water) induces stronger shift of the K-V curves and stronger reduction of K_{IC}^* and K_a for the two anhydrite-rich caprocks (HB102 and YB204) than the carbonate caprocks (YB5 and MB2). For example, K_{IC}^* and K_a reduced over 50% for HB102 and YB204

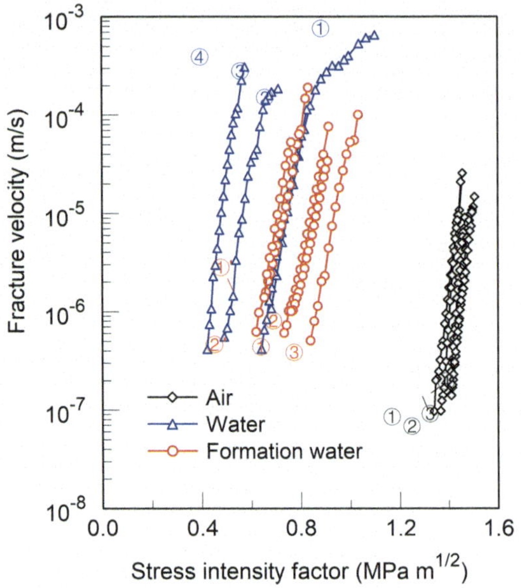

Fig. 7.4 K-V curves for the anhydrite caprock (HB102) tested under conditions of ambient air (*black*), De-ionized water (*blue*), and formation water (*red*). The load-relaxation sequences are also labeled for each fluid environment

(Fig. 7.5i, j) under the two aqueous fluids of water of formation water, while such reduction is ~30% for the carbonate caprocks (Fig. 7.5k, l). A systematic change of the SCI is also observed (except YB204): SCI is the largest in air (~60), smallest in formation water (20–30), and intermediate in water (30–40).

7.4.2 *Fracture Microstructure*

Fracture microstructures demonstrate both similarities and differences between natural fractures and fractures formed by DT tests. Only a few natural fractures present in these caprock samples (Fig. 7.2a–d). We did not succeed in finding any natural fractures in the thin-section slides from anhydrite caprock from well HB102. The example in Fig. 7.6a is a mud band impurity within the anhydrite matrix. YB204 holds natural fractures along the evaporite-carbonate boundary. These fractures extend into the carbonate matrix but disappear in the evaporite matrix (Fig. 7.6b). A thin natural fracture is captured in the carbonate caprock YB5, with the fracture space filled with calcite. The natural fracture example in MB2 has a bifurcation structure. This fracture was first filled with quartz (white crystals on the sidewall of the fracture), and then this fracture re-opened for additional calcite filling (black zone in the middle of the fracture) as confirmed by SEM energy dispersive spectrum analysis. Some of the natural fractures in MB2 also contains bitumen (not shown here).

Despite that natural fractures rarely exist in anhydrite (Fig. 7.6a, b), the induced fractures propagate within the anhydrite matrix, and can both cut across individual anhydrite grains and along the boundaries of two crystal domains (Fig. 7.6e, f).

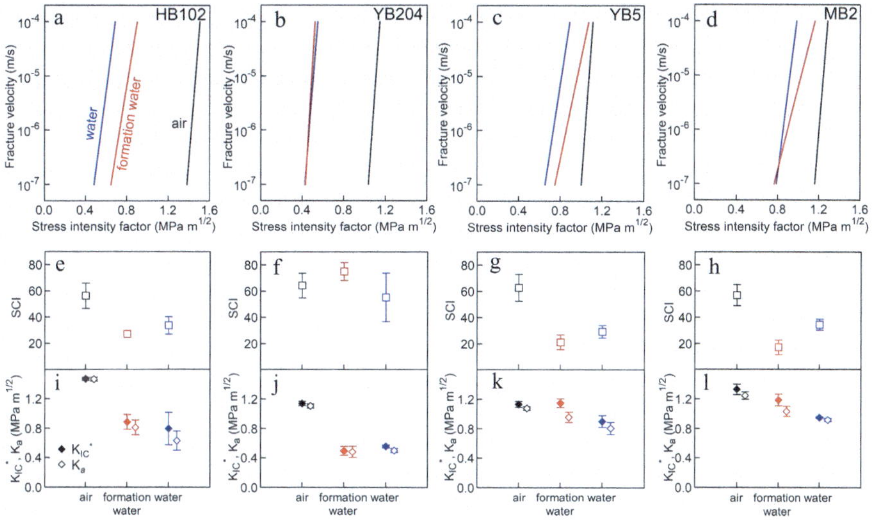

Fig. 7.5 Synthesize the measured fracture mechanical parameters. (**a-d**) Constructed K-V curves for the four wells under three test conditions of air (*black*), water (*blue*), and formation water (*red*). The color codes of black (air), blue (water), and red (formation water) apply to all figures here. (**e–h**) Mean SCI with 1σ error bars. (**i–l**) Mean fracture toughness (solid diamonds) and K_a (hollow diamonds) with 1σ error bars. K = Stress intensity factor, V = Fracture velocity, SCI = Subcritical fracture growth index

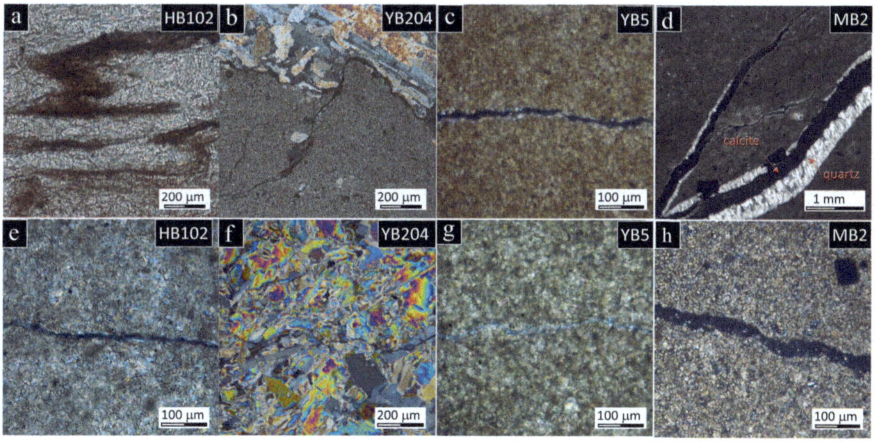

Fig. 7.6 Thin-section images of both natural fractures (**a-d**) and induced fractures during Double Torsion tests (**e–h**) under cross-polarized light. The quartz and calcite filling in the fracture in (**d**) were identified with energy dispersive spectrum analysis. See the text for details

Because of the micritic texture, intergranular fracture propagation is dominant in the carbonate caprocks as displayed in Fig. 7.6g, h. The fracture paths can be either straight (Fig. 7.6e, h) or twisting (Fig. 7.6f, g). All these natural and induced fractures are predominantly opening mode, as can be identified from the matching shapes on fracture walls.

SEM images display fracture microstructures at finer scales. The fracture surfaces from anhydrite caprock from well HB102 expose abundant elongated anhydrite crystal grains (Fig. 7.7a, b). These grains have smooth surfaces, with only a few grains show breakage surfaces (red arrows in Fig. 7.7a, b). The variation of test environments from air to water has a negligible effect on fracture surfaces for anhydrite caprock from well HB102. The anhydrite grains in the dolomitic anhydrite caprock from well YB204 are larger than those in HB102, with the fine-grain dolomite scatter between anhydrite domains (Fig. 7.7c, d). Sharp anhydrite grain edges are well preserved in the fracture surface formed in air (Fig. 7.7c), while the grain edges are less sharp in the fracture surface formed in formation water (Fig. 7.7d). Fracture surfaces in the two carbonate caprocks (YB5 and MB2) are similar in their fine-grain texture (Fig. 7.7e, h). Furthermore, fluid chemistry has similar effects in altering fracture surface morphology of these carbonate caprocks: the spotty surfaces formed in air (Fig. 7.7e, h) evolved to flatter surfaces formed in water (Fig. 7.7f, i), and to the smoothest fracture surfaces formed in formation water (Fig. 7.7g, j).

Fracture traces between the anhydrite and carbonate caprocks show contrasting features. On one hand, fractures show step-like traces in the anhydrite caprock

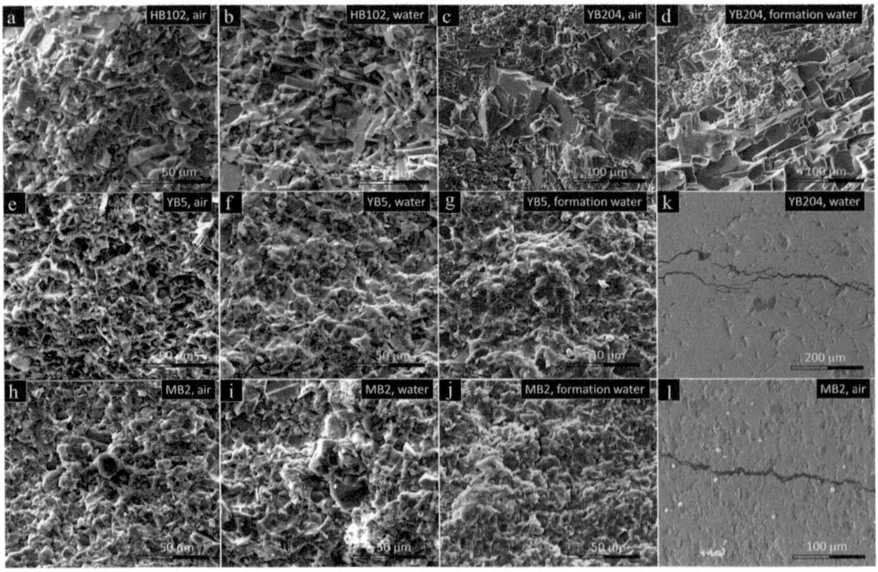

Fig. 7.7 SEM images of fracture surfaces (**a–j**) and fracture traces (**k–l**) formed during Double Torsion tests. See the text for details

(Fig. 7.7k), suggesting cleavage along preferred crystal planes as the main fracture growth process. The anhydrite fractures also have a wide damage zone around the fracture tip as seen by multiple bifurcated fracture traces (Fig. 7.7k). On the other hand, because of the fine-grain texture, the calcite grains have little influence on fracture traces along the carbonate caprock, and the damage zone is concentrated around the main fracture wall, as can be seen from the loose grains close to the fracture wall (Fig. 7.7l).

7.5 Discussion and Implications

7.5.1 Comparison to Other Studies

For the carbonate caprocks (YB5, MB2), our measured fracture toughness data in air are ~1.2 MPa $m^{1/2}$ (Fig. 7.5k, Fig. 7.5l), which is comparable to reported values for micritic limestone as summarized in Atkinson (1984) and in agreement with that for carbonate-rich Marcellus Shale in recent study by Chen et al. (2019). The reduction in fracture toughness, when exposed to water and formation water (Fig. 7.5k, l), is also in agreement with the well-documented water-weakening effect on rock strength and toughness (Millero & Roy, 1997; Mucci, 1983). We also observe a substantial reduction in SCI when exposed to water and DI water. SCI values in air are ~ 60, comparable to those from the carbonate-rich Marcellus Shale (Chen et al., 2019), and the lower SCI values of 20–40 for wet carbonates are also similar to those for carbonates exposed to aqueous fluids as documented in Atkinson (1984). For the anhydrite caprocks (HB102, YB204), their fluid-weakening effect on K_{IC}^* and SCI is similar to that for carbonate caprocks. However, anhydrite caprocks have stronger water-weakening effects on K_{IC}^*, and water and formation water have similar effects (Fig. 7.5a, b). Among a few reported K_{IC} values for anhydrite/gypsum, our measured K_{IC}^* of 1.14 MPa $m^{1/2}$ for YB204 and 1.47 MPa $m^{1/2}$ for HB102 in air are larger than K_{IC} of ~1.07 MPa $m^{1/2}$ for gypsum rock (60–85% gypsum) measured using Cracked Chevron Notched Brazilian Disc (CCNBD) specimens (Meng et al., 2015), which is consistent with the fact that anhydrite is stronger than gypsum (Bell, 1981, 1994). Moreover, the high K_{IC}^* values of ~1.5 MPa $m^{1/2}$ for the purer anhydrite in HB102 in air is more than 4 times larger than theoretically calculated values of intergranular and grain boundary fractures for anhydrite crystal (Tromans & Meech, 2002), which is similar to observed stronger toughness in carbonate rocks than in single crystal calcite (Chen et al., 2019). Except for the YB204, consistent K_{IC}^* and SCI dependency on fluid conditions are observed: air > formation water > water for K_{IC}^* (Fig. 7.5i–l), air > water > formation water for SCI (Fig. 7.5e–h). These observations provide insights into the mechanisms controlling fracture growth as will be discussed in the following section.

7.5.2 *Fracture Growth Mechanisms*

We now explore the possible mechanisms associated with our experimental observations. At ambient pressure/temperature conditions, calcite solubility increases with increasing water salinity (Millero & Roy, 1997; Mucci, 1983), while the anhydrite solubility increases first with water salinity then decreases at higher salt concentration (Klimchouk, 1996). The consistent SCI reduction from air to water toward formation water (Fig. 7.5e, g, h) can be linked to the mineral dissolution rates. Both anhydrite and calcite dissolve faster in saline water than in DI water. Existing data suggest that the faster the mineral dissolution rate, the lower the SCI (see Atkinson (1984) for carbonates and Chen et al. (2019) for Marcellus Shale in HCl solution).

The consistent K_{IC}^* reduction from air to aqueous conditions for all the caprock samples is most likely resulted from the soaking process before the DT tests. Fracture toughness is a material property. Like in the case of glass, the variation in fluid composition influences SCI through surficial processes such as stress-corrosion and dissolution reactions at the fracture surfaces, but K_{IC}^* marking the initiation of critical crack growth in the material is unaffected (Wiederhorn & Johnson, 1973). Indeed, the reduction in K_{IC}^* is linked to the physical weakening of the material in strength and elastic modulus, as in the case of water weakening in clay-bearing rocks (Chen et al., 2017, 2019; Nara et al., 2012). The large surface area of the thin specimen slices (Fig. 7.3) enables more effective fluid-rock interaction than bulky samples. Anhydrite solubility is ~2.5 g/L in water at ambient condition, which is more than four orders of magnitude larger than calcite solubility of ~1.5 mg/L (Klimchouk, 1996). This explains why anhydrite caprocks have a stronger K_{IC}^* reduction than carbonate caprocks (Fig. 7.5a–d, i–l). The stronger reduction in K_{IC}^* for anhydrite caprocks than carbonates is also in line with water-enhanced creep during salt deformation process (Urai et al., 1986). In addition, at ambient conditions, the irreversible anhydrite dissolution reaction can quickly transform anhydrite into gypsum with substantial volumetric expansion, which might also contribute to their larger K_{IC}^* reduction in solutions. The strong K_{IC}^* weakening in water than in formation water could not be explained by the dissolution rates since both anhydrite and calcite dissolves faster in saline water than in DI water. We suspect that the DI water with lower density can diffuse into the rock matrix more easily than the denser formation water, which causes stronger weakening effect.

It is not surprising that for both anhydrite and carbonate caprocks their K_{IC}^* values are much higher than their single-crystal counterparts because cracking multiple grain domains with different crystal orientations is harder than cracking a single crystal along the weak cleavage planes. The less preserved cleavage features along fracture surfaces of HB102 than YB204 (Fig. 7.7a–d) is likely due to faster dissolution reaction rates for smaller grains. Ambient humidity may also impact fresh anhydrite fracture surfaces, dissolving cleavage signatures (Fig. 7.7a). The lower K_{IC}^* for YB204 in air than HB102 indicates that multi-crystalline materials with smaller grain sizes are fracture mechanically stronger than those with bigger grains, most likely due to the less contribution of cleavage to fracture processes than rocks with larger grains. Grain size also influences fracture trace shapes. Fractures in multi-domain crystalline rocks have zig-zag or step-like shapes and wider damage zone (Fig. 7.7k) due to the combined effect of stress concentration (grooving and applied stress) and preferred cleavage planes, whereas fine-grain limestone is similar to a homogeneous material for fracture growth where individual grain has little influence on fracture trace (Fig. 7.7l).

7.5.3 Implications for Reservoir Sealing Performance

Subsurface reservoirs need to meet certain seal performance in applications such as petroleum systems (here), carbon sequestration, and waste disposal. In general, anhydrite caprock is better than carbonate caprock, in its larger dry toughness (Fig. 7.5), lack of natural fractures (Fig. 7.7), and the strong ductility under subsurface stresses (Hangx et al., 2010). Although experimental opening-mode fractures have been generated for anhydrite under atmospheric pressure, the brittle-ductile transition under higher confining stresses and higher temperatures with increasing burial depth gradually suppresses brittle fracture growth as verified by the lack of natural fractures in anhydrite. Under higher confining stresses, brittle failure in anhydrite was taken over by plastic flow in the forms of grain boundary sliding, dislocation creeps, and solution transfer creep (Urai et al., 1986). One technical limitation of the current study is that the DT tests did not consider could only be performed under atmospheric stress instead of the in-situ subsurface stress conditions such as confining stress and pore pressure (see method). Unlike rock mechanics tests, most fracture mechanical tests of rocks were performed without confining stresses (e.g., Kuruppu et al., 2014; ISRM, 1988). The reason is that the effective stress law controls fracture growth: Although fractures should grow under positive driving stress, e.g., pore pressure P_p larger than minimum stress S_{hmin} for opening mode fractures. Even though our tests were performed under atmospheric pressure, our results can be applied to subsurface conditions where positive driving stresses exist, such as, iin scenarios such aswith significant overpressure (large P_p with hydrocarbon accumulation) and or tectonic uplift (small S_{hmin} active tectonics of the Himalaya orogeny), which could be satisfied at Sichuan basin resulting from active tectonics of the Himalaya orogeny (Fig. 7.1a). Indeed, the positive driving stress condition ($P_p > S_{hmin}$) does occur widely in nature as confirmed by the wide occurrence of opening-mode natural fractures in carbonate caprocks here (Fig. 7.6) and in other studies (Fall et al., 2015; Gale et al., 2014).

Temperature is another important parameter affecting rock failure processes. Though our current tests were not designed to investigate the temperature effect on fracture growth (all tests were at room temperature), studies have shown that higher temperature can have greater impact on subcritical fracture growth than fluid salinity and pH level variations (Atkinson, 1984; Chen et al., 2019). In general, fractures grow faster and the rocks are weaker at higher temperatures due to enhanced physical and chemical interaction and reactions. However, caution should be made when applying the present results to higher temperature subsurface conditions with extreme cases such as the complex calcite dissolution kinetics (e.g., Naviaux et al., 2019).

Fluid chemistry is another main factor for fluid assisted fracture growth (Atkinson, 1984; Chen et al., 2019), though not targeted in this study. Many fluid-rock interactions and reactions are functions of fluid chemistry such as the salinity-modulated clay hydration reactions (Chen et al., 2019) and calcite dissolution reactions (Rostom et al., 2013), pH-adjusted quartz and calcite dissolution reactions (Atinson & Meredith, 1981; Chen et al., 2019; Wiederhorn & Johnson, 1973), to name a few. Oftentimes, the fluid chemistry and temperature act simultaneously in natural cases, making this problem even more complicated.

Our laboratory scale study focused on single fracture behavior under controlled fluid and temperature conditions. From the lab-scale single-crack case to the reservoir-scale cases with multiple interacting fractures, one need to perform upscaling geomechanical modeling that considers interactions among fracture mechanical properties (fracture network evolution, stress condition variation), hydrologic properties (transport properties evolutions), and fluid chemo-mechanical reactions (dissolution–precipitation, hydration, etc.) over relevant time scales. Nonetheless, experimental results demonstrate that aqueous fluids (water and formation water), rock composition (carbonate, anhydrite/gypsum), and rock texture (grain size) strongly influence fracture propagation for both carbonate and anhydrite caprocks, and such effect is stronger for anhydrite caprocks. If aqueous fluids exist at the reservoir-caprock interface, the drying effect during the charging process in a gas reservoir like the present case drives out formation water away from the interface, which can potentially strengthen the caprocks at the reservoir-caprock interface. Another conclusion can be drawn from the different effects between water and formation water: caprocks have larger K_{IC}^* but smaller SCI in formation water (Fig. 7.5). This infers fracture growth in formation water is more difficult at high stress intensity factor (large K_{IC}^*) but easier under lower stress intensity factor (smaller SCI) than in water or low salinity solutions. Over short time scales such as during hydraulic fracturing operations (high stress), fracture growth in water is more natural. Nevertheless, over geologic time scales, faster subcritical fracture growth is expected in formation water environments.

7.6 Summary and Conclusion

We conducted double torsion experiments on carbonate and evaporite caprocks in Sichuan Basin, China, under environments of air, DI-water, and formation water to characterize the mechanical properties of fracture growth. The measured fracture toughness and SCI agree with test data from previous research. In ambient air, the anhydrite caprock (well HB102) has the largest fracture toughness of 1.47 MPa $m^{1/2}$, while the other three caprocks are weaker with toughness of ~1.2 MPa $m^{1/2}$. Both fracture toughness and SCI of the caprocks are systematically reduced from ambient air condition to water environments: K_{IC}^* reduction is most active in the water, while

SCI reduction is the strongest in formation water. The mechanical weakening process and microstructural observations suggest strong fluid-rock interactions (primarily dissolution reaction) during fracture growth. The larger dry toughness and lack of natural fractures suggest that anhydrite caprocks have better sealing performance in addition to its strong ductility under subsurface confine stress. Our results also suggest a higher increased risk of failure fracture propagation in anhydrite caprocks than carbonates with the presence of excess fluids. This approach and data extend the mechanical evaluation of caprock integrity and improve the understanding of subcritical fracture growth, which could also be helpful for understanding brittle failure in the upper crust over geologic time scales.

References

Allan, U. S. (1989). Model for hydrocarbon migration and entrapment within faulted structures. *AAPG Bulletin, 73*(7), 803–811. https://doi.org/10.1306/44B4A271-170A-11D7-8645000102C1865D

Anderson, T. L. (2005). *Fracture mechanics: Fundamentals and applications*. Taylor & Francis.

Atkinson, B. K. (1982). Subcritical crack propagation in rocks: Theory, experimental results and applications. *Journal of Structural Geology, 4*(1), 41–56. https://doi.org/10.1016/0191-8141(82)90005-0

Atkinson, B. K. (1984). Subcritical crack growth in geological materials. *Journal of Geophysical Research: Solid Earth, 89*(B6), 4077–4114. https://doi.org/10.1029/JB089iB06p04077

Atkinson, B. K., & Meredith, P. G. (1981). Stress corrosion cracking of quartz: A note on the influence of chemical environment. *Tectonophysics, 77*(1–2), T1–T11. https://doi.org/10.1016/0040-1951(81)90157-8

Bell, F. (1981). Geotechnical properties of some evaporitic rocks. *Bulletin of Engineering Geology & the Environment, 24*(1).

Bell, F. G. (1994). A survey of the engineering properties of some anhydrite and gypsum from the north and midlands of England. *Engineering Geology, 38*(1–2), 1–23. https://doi.org/10.1016/0013-7952(94)90021-3

Bergsaker, A. S., Røyne, A., Ougier-Simonin, A., Aubry, J., & Renard, F. (2016). The effect of fluid composition, salinity, and acidity on subcritical crack growth in calcite crystals. *Journal of Geophysical Research: Solid Earth, 121*(3), 1631–1651. https://doi.org/10.1002/2015JB012723

Bildstein, O., Worden, R. H., & Brosse, E. (2001). Assessment of anhydrite dissolution as the rate-limiting step during thermochemical sulfate reduction. *Chemical Geology, 176*(1–4), 173–189. https://doi.org/10.1016/S0009-2541(00)00398-3

Billi, A., Salvini, F., & Storti, F. (2003). The damage zone-fault core transition in carbonate rocks: Implications for fault growth, structure and permeability. *Journal of Structural Geology, 25*(11), 1779–1794. https://doi.org/10.1016/S0191-8141(03)00037-3

Brantut, N., Heap, M. J., Meredith, P. G., & Baud, P. (2013). Time-dependent cracking and brittle creep in crustal rocks: A review. *Journal of Structural Geology, 52*, 17–43. https://doi.org/10.1016/j.jsg.2013.03.007

Callahan, O. A., Eichhubl, P., Olson, J. E., & Davatzes, N. C. (2019). Fracture mechanical properties of damaged and hydrothermally altered rocks, Dixie Valley-Stillwater fault zone, Nevada, USA. *Journal of Geophysical Research: Solid Earth, 124*(4), 4069–4090. https://doi.org/10.1029/2018JB016708

Chen, X., Eichhubl, P., & Olson, J. E. (2017). Effect of water on critical and subcritical fracture properties of Woodford shale. *Journal of Geophysical Research: Solid Earth, 122*(4), 2736–2750. https://doi.org/10.1002/2016JB013708

Chen, X., Eichhubl, P., Olson, J. E., & Dewers, T. A. (2019). Effect of water on fracture mechanical properties of shales. *Journal of Geophysical Research: Solid Earth, 124*(3), 2428–2444. https://doi.org/10.1029/2018JB016479

Chester, F. M. (1988). The brittle-ductile transition in a deformation-mechanism map for halite. *Tectonophysics, 154*(1–2), 125–136. https://doi.org/10.1016/0040-1951(88)90230-2

Cipolla, C. L., Warpinski, N. R., Mayerhofer, M., Lolon, E. P., & Vincent, M. (2010). The relationship between fracture complexity, reservoir properties, and fracture-treatment design. *SPE Production & Operations, 25*(4), 438–452. SPE-115769-PA. https://doi.org/10.2118/115769-PA

Cosgrove, J. W. (2001). Hydraulic fracturing during the formation and deformation of a basin: A factor in the dewatering of low-permeability sediments. *AAPG Bulletin, 85*(4), 737–748. https://doi.org/10.1306/8626C997-173B-11D7-8645000102C1865D

Deer, W. A., Howie, R. A., & Zussmanm J. (1992). An introduction to the rock-forming minerals. Longman Scientific & Technical.

Downey, M. W. (1984). Evaluating seals for hydrocarbon accumulations. *AAPG Bulletin, 68*(11), 1752–1763. https://doi.org/10.1306/AD461994-16F7-11D7-8645000102C1865D

Erguler, Z. A., & Ulusay, R. (2009). Water-induced variations in mechanical properties of clay-bearing rocks. *International Journal of Rock Mechanics and Mining Sciences, 46*(2), 355–370. https://doi.org/10.1016/j.ijrmms.2008.07.002

Fall, A., Eichhubl, P., Bodnar, R. J., Laubach, S. E., & Davis, J. S. (2015). Natural hydraulic fracturing of tight-gas sandstone reservoirs, Piceance Basin, Colorado. *Bulletin, 127*(1–2), 61–75. https://doi.org/10.1130/B31021.1

Finkbeiner, T., Zoback, M., Flemings, P., & Stump, B. (2001). Stress, pore pressure, and dynamically constrained hydrocarbon columns in the South Eugene Island 330 field, northern Gulf of Mexico. *AAPG Bulletin, 85*(6), 1007–1031. https://doi.org/10.1306/8626CA55-173B-11D7-8645000102C1865D

Fisher, M. A., Normark, W. R., Bohannon, R. G., Sliter, R. W., & Calvert, A. J. (2001, December). Geology of the continental margin beneath Santa Monica Bay, southern California, from Small-Airgun, seismic-reflection data. In *AGU fall meeting abstracts* (Vol. 2001, pp. S11A-0544).

Fu, Y., van Berk, W., & Schulz, H. M. (2016). Hydrogen sulfide formation, fate, and behavior in anhydrite-sealed carbonate gas reservoirs: A three-dimensional reactive mass transport modeling approach. *Aapg Bulletin, 100*(5), 843–865. https://doi.org/10.1306/12111514206

Fuller, E. R. (1979). *An evaluation of double-torsion testing—Analysis* (pp. 3–18). ASTM International.

Fulljames, J. R., Zijerveld, L. J. J., & Franssen, R. C. M. W. (1997). Fault seal processes: systematic analysis of fault seals over geological and production time scales. In *Norwegian petroleum society special publications* (Vol. 7, pp. 51–59). Elsevier. https://doi.org/10.1016/S0928-8937(97)80006-9

Gale, J. F., Laubach, S. E., Olson, J. E., Eichhubl, P., & Fall, A. (2014). Natural fractures in shale: A review and new observations. *AAPG Bulletin, 98*(11), 2165–2216. https://doi.org/10.1306/08121413151

Guo, T. (2018). Hydrocarbon accumulation conditions and key technologies for exploration and development of Yuanba gas field. *Petroleum Research, 3*(4), 293–305. https://doi.org/10.1016/j.ptlrs.2018.11.004

Handin, J., & Hager, R. V., Jr. (1958). Experimental deformation of sedimentary rocks under confining pressure: Tests at high temperature. *AAPG Bulletin, 42*(12), 2892–2934. https://doi.org/10.1306/0BDA5C27-16BD-11D7-8645000102C1865D

Handin, J., Hager, R. V., Jr., Friedman, M., & Feather, J. N. (1963). Experimental deformation of sedimentary rocks under confining pressure: Pore pressure tests. *Aapg Bulletin, 47*(5), 717–755.

Handin, J., Heard, H. A., & Magouirk, J. N. (1967). Effects of the intermediate principal stress on the failure of limestone, dolomite, and glass at different temperatures and strain rates. *Journal of Geophysical Research, 72*(2), 611–640. https://doi.org/10.1029/JZ072i002p00611

Hangx, S. J., Spiers, C. J., & Peach, C. J. (2010). Mechanical behavior of anhydrite caprock and implications for CO_2 sealing capacity. *Journal of Geophysical Research: Solid Earth, 115*(B7). https://doi.org/10.1029/2009JB006954

Hao, F., Zhu, W., Zou, H., & Li, P. (2015). Factors controlling petroleum accumulation and leakage in overpressured reservoirs. *AAPG Bulletin, 99*(5), 831–858. https://doi.org/10.1306/010215 14145

Holder, J., Olson, J. E., & Philip, Z. (2001). Experimental determination of subcritical crack growth parameters in sedimentary rock. *Geophysical Research Letters, 28*(4), 599–602. https://doi.org/ 10.1029/2000GL011918

Ingram, G. M., & Urai, J. L. (1999). Top-seal leakage through faults and fractures: The role of mudrock properties. *Geological Society, London, Special Publications, 158*(1), 125–135. https:// doi.org/10.1144/GSL.SP.1999.158.01.10

ISRM Testing Commission. (1988). Suggested methods for determining the fracture toughness of rock. *International Journal of Rock Mechanics and Mining & Geomechanics Abstracts, 25*(2), 71–96.

Jin, X., Shah, S. N., Roegiers, J. C., & Zhang, B. (2014, February). Fracability evaluation in shale reservoirs-an integrated petrophysics and geomechanics approach. In *SPE hydraulic fracturing technology conference and exhibition* (pp. SPE-168589). SPE. https://doi.org/10.2118/ 168589-MS

Jin, Z. (2012). Formation and accumulation of oil and gas in marine carbonate sequences in Chinese sedimentary basins. *Science China Earth Sciences, 55*, 368–385. https://doi.org/10.1007/s11 430-011-4264-4

Jin, Z., Yuan, Y., Sun, D., Liu, Q., & Li, S. (2014b). Models for dynamic evaluation of mudstone/ shale cap rocks and their applications in the Lower Paleozoic sequences, Sichuan Basin, SW China. *Marine and Petroleum Geology, 49*, 121–128. https://doi.org/10.1016/j.marpetgeo.2013. 10.001

Klimchouk, A. (1996). The dissolution and conversion of gypsum and anhydrite. *International Journal of Speleology, 25*(3), 2. https://doi.org/10.5038/1827-806X.25.3.2

Kuruppu, M. D., Obara, Y., Ayatollahi, M. R., Chong, K. P., & Funatsu, T. (2014). ISRM-suggested method for determining the mode I static fracture toughness using semi-circular bend specimen. *Rock Mechanics and Rock Engineering, 47*, 267–274. https://doi.org/10.1007/s00603-013-0422-7

Lee, H. P., Olson, J. E., Holder, J., Gale, J. F., & Myers, R. D. (2015). The interaction of propagating opening mode fractures with preexisting discontinuities in shale. *Journal of Geophysical Research: Solid Earth, 120*(1), 169–181. https://doi.org/10.1002/2014JB011358

Li, P., Hao, F., Guo, X., Zou, H., Yu, X., & Wang, G. (2015). Processes involved in the origin and accumulation of hydrocarbon gases in the Yuanba gas field, Sichuan Basin, southwest China. *Marine and Petroleum Geology, 59*, 150–165. https://doi.org/10.1016/j.marpetgeo.2014.08.003

Ma, Y., Guo, X., Guo, T., Huang, R., Cai, X., & Li, G. (2007). The Puguang gas field: New giant discovery in the mature Sichuan Basin, southwest China. *AAPG Bulletin, 91*(5), 627–643. https:// doi.org/10.1306/11030606062

Ma, Y., Zhang, S., Guo, T., Zhu, G., Cai, X., & Li, M. (2008). Petroleum geology of the Puguang sour gas field in the Sichuan Basin, SW China. *Marine and Petroleum Geology, 25*(4–5), 357–370. https://doi.org/10.1016/j.marpetgeo.2008.01.010

Machel, H. G. (2001). Bacterial and thermochemical sulfate reduction in diagenetic settings—old and new insights. *Sedimentary Geology, 140*(1–2), 143–175. https://doi.org/10.1016/S0037-073 8(00)00176-7

Major, J. R., Eichhubl, P., Dewers, T. A., & Olson, J. E. (2018). Effect of CO_2–brine–rock interaction on fracture mechanical properties of CO_2 reservoirs and seals. *Earth and Planetary Science Letters, 499*, 37–47. https://doi.org/10.1016/j.epsl.2018.07.013

Meléndez-Martínez, J., & Schmitt, D. R. (2016). A comparative study of the anisotropic dynamic and static elastic moduli of unconventional reservoir shales: Implication for geomechanical investigations. *Geophysics, 81*(3), D245–D261. https://doi.org/10.1190/geo2015-0427.1

Meng, T., Hu, Y., Fang, R., Kok, J., Fu, Q., & Feng, G. (2015). Study of fracture toughness and weakening mechanisms in gypsum interlayers in corrosive environments. *Journal of Natural Gas Science and Engineering, 26*, 356–366. https://doi.org/10.1016/j.jngse.2015.06.027

Meredith, P. G., & Atkinson, B. K. (1985). Fracture toughness and subcritical crack growth during high-temperature tensile deformation of Westerly granite and Black gabbro. *Physics of the Earth and Planetary Interiors, 39*(1), 33–51. https://doi.org/10.1016/0031-9201(85)90113-X

Millero, F. J., & Roy, R. N. (1997). A chemical equilibrium model for the carbonate system in natural waters. *Croatica Chemica Acta, 70*(1), 1–38. https://hrcak.srce.hr/134777

Mucci, A. (1983). The solubility of calcite and aragonite in seawater at various salinities, temperatures, and one atmosphere total pressure. *American Journal of Science, 283*(7), 780–799.

Nara, Y., Morimoto, K., Hiroyoshi, N., Yoneda, T., Kaneko, K., & Benson, P. M. (2012). Influence of relative humidity on fracture toughness of rock: Implications for subcritical crack growth. *International Journal of Solids and Structures, 49*(18), 2471–2481. https://doi.org/10.1016/j.ijsolstr.2012.05.009

Naviaux, J. D., Subhas, A. V., Rollings, N. E., Dong, S., Berelson, W. M., & Adkins, J. F. (2019). Temperature dependence of calcite dissolution kinetics in seawater. *Geochimica Et Cosmochimica Acta, 246*, 363–384. https://doi.org/10.1016/j.gca.2018.11.037

Olson, J. E. (1993). Joint pattern development: Effects of subcritical crack growth and mechanical crack interaction. *Journal of Geophysical Research: Solid Earth, 98*(B7), 12251–12265. https://doi.org/10.1029/93JB00779

Olson, J. E., Laubach, S. E., & Lander, R. H. (2009). Natural fracture characterization in tight gas sandstones: Integrating mechanics and diagenesis. *AAPG Bulletin, 93*(11), 1535–1549. https://doi.org/10.1306/08110909100

Piane, C. D., Clennell, M. B., Keller, J. V. A., Giwelli, A., & Luzin, V. (2017). Carbonate hosted fault rocks: A review of structural and microstructural characteristic with implications for seismicity in the upper crust. *Journal of Structural Geology, 103*, 17–36. https://doi.org/10.1016/j.jsg.2017.09.003

Reches, Z., & Lockner, D. A. (1994). Nucleation and growth of faults in brittle rocks. *Journal of Geophysical Research, 99*(B9), 18159–18173. https://doi.org/10.1029/94JB00115

Renner, J., & Rummel, F. (1996). The effect of experimental and microstructural parameters on the transition from brittle failure to Cataclastic flow of carbonate rock. *Tectonophysics, 258*, 151–169. https://doi.org/10.1016/0040-1951(95)00192-1

Rostom, F., Røyne, A., Dysthe, D. K., & Renard, F. (2013). Effect of fluid salinity on subcritical crack propagation in calcite. *Tectonophysics, 583*, 68–75. https://doi.org/10.1016/j.tecto.2012.10.023

Shimamoto, T. (1986). Transition between frictional slip and ductile flow for halite shear zones at room temperature. *Science, 231*, 711–714. https://doi.org/10.1126/science.231.4739.711

Smith, D. A. (1966). Theoretical considerations of sealing and non-sealing faults. *AAPG Bulletin, 50*(2), 363–374. https://doi.org/10.1306/5D25B48F-16C1-11D7-8645000102C1865D

Sone, H., & Zoback, M. D. (2013). Mechanical properties of shale-gas reservoir rocks-Part 2: Ductile creep, brittle strength, and their relation to the elastic modulus. *Geophysics, 78*(5), 393–402. https://doi.org/10.1190/geo2013-0051.1

Swanson, P. L. (1984). Subcritical crack growth and other time-and environment-dependent behavior in crustal rocks. *Journal of Geophysical Research: Solid Earth, 89*(B6), 4137–4152. https://doi.org/10.1029/JB089iB06p04137

Tingay, M. R. P., Hillis, R. R., Swarbrick, R. E., Morley, C. K., & Damit, A. R. (2009). Origin of overpressure and pore-pressure prediction in the Baram province, Brunei. *AAPG Bulletin, 93*(1), 51–74. https://doi.org/10.1306/08080808016

Tromans, D., & Meech, J. A. (2002). Fracture toughness and surface energies of minerals: Theoretical estimates for oxides, sulphides, silicates and halides. *Minerals Engineering, 15*(12), 1027–1041. https://doi.org/10.1016/S0892-6875(02)00213-3

Urai, J. L., Spiers, C. J., Zwart, H. J., & Lister, G. S. (1986). Weakening of rock salt by water during long-term creep. *Nature, 324*(6097), 554–557. https://doi.org/10.1038/324554a0

Watts, N. L. (1987). Theoretical aspects of cap-rock and fault seals for single-and two-phase hydrocarbon columns. *Marine and Petroleum Geology, 4*(4), 274–307. https://doi.org/10.1016/0264-8172(87)90008-0

Wei, S., Feng, Q., Feng, Y., & Yuan, S (2011) Prediction of Triassic gypsum cap rocks in Tongnanba region of Northeast Sichuan Basin. *Petroleum Geology and Experiment, 33*(1), 81–86. (In Chinese) https://doi.org/10.11781/sysydz201101081

Wiederhorn, S. M., & Johnson, H. (1973). Effect of electrolyte pH on crack propagation in glass. *Journal of the American Ceramic Society, 56*(4), 192–197. https://doi.org/10.1111/j.1151-2916.1973.tb12454.x

Williams, D. P., & Evans, A. G. (1973). A simple method for studying slow crack growth. *Journal of Testing and Evaluation, 1*(4), 264–270. https://doi.org/10.1520/JTE10015J

Yielding, G., Freeman, B., & Needham, D. T. (1997). Quantitative fault seal prediction. *AAPG Bulletin, 81*(6), 897–917. https://doi.org/10.1306/522B498D-1727-11D7-8645000102C1865D

Zhang, X., Lu, Y., Tang, J., Zhou, Z., & Liao, Y. (2017). Experimental study on fracture initiation and propagation in shale using supercritical carbon dioxide fracturing. *Fuel, 190*, 370–378. https://doi.org/10.1016/j.fuel.2016.10.120

Zhang, Y., Lu, H., & Hu, H. (2012). Distribution of gypsum-salt cap rocks in Yuanba Area of Northeastern Sichuan Basin. *Xingjiang Petroleum Geology, 33*(4), 450–452. (In Chinese).

Zoback, M. (2010). *Reservoir geomechanics*. Cambridge University Press.

Chapter 8
Fault-Fold System

Abstract The formation of bed folding along weakened zones often leads to the emergence of subsidiary faults, thereby creating intricate fault-fold systems with fractures. These systems are of paramount importance for the exploration of methane. While existing literature on fault-fold systems predominantly relies on field observations, there has been scant research focusing on their subsurface structures. Utilizing three-dimensional seismic data, this study endeavors to identify and quantify the principal fault-fold systems within the No. 3 coal seam of the Shanxi Formation and the No. 15 coal seam of the Taiyuan Formation, situated in the Zhengzhuang block of the southern Qinshui Basin, China. The findings reveal: (1) The presence of five principal fault-fold systems, with four exhibiting a North–South orientation and one aligning East–West; (2) The establishment of an asymmetric model for the fault-fold system through a quantitative analysis of seismic attributes, particularly focusing on the high-angle normal faults along the folds' limbs; (3) The identification of a normal fault on the left side of the No. 3 fault-fold system, with an estimated fault damage zone width of about 200 m, contrasting with the right side, where the fault damage zone width ranges from 50 to 500 m. These initial assessments propose a geological model based on seismic attributes, aiming to enhance the exploration and exploitation of deep methane and petroleum within fractured reservoirs.

8.1 Introduction

The effective extraction and management of coalbed methane (CBM), a critical clean energy resource, are challenged by the complex geophysical and geological conditions of coal seams. These challenges are exacerbated in regions like the southern Qinshui Basin, China, where the intricate interplay of structural and diagenetic fractures within the Paleozoic coal seams significantly influences methane accumulation and recovery (Wang et al., 1996).

Recent studies have advanced our understanding of these phenomena through seismic attribute analysis, highlighting the role of structural fractures, including small faults and intra-layer shear fractures, in controlling subsurface flow and accumulation

Z. Liao, *Applied Seismic Geostructures*, https://doi.org/10.1007/978-981-97-7079-3_8

(Shi et al., 2020). Seismic attributes have been applied to collapse chimneys and fault zones within shale formations (Liao et al., 2020). Marfurt (2010) utilized seismic volume attributes to detect the heterogeneous properties of seismic reflections. Guo et al. (2010) studied the faults and fracture properties within the Woodford shale in the Arkoma Basin using seismic attributes. Arroyal (2009) discovered the properties of fractures in the Woodford shale through a multi-scale fracture study and classified them, while also identifying similarities and differences between potential fault zones and observed fault zones on outcrops. However, despite substantial progress, current methodologies often fall short of accurately predicting the spatial distribution of these fractures and their impact on methane recovery, indicating a persistent gap in our understanding of subsurface dynamics (Guo et al., 2010; Marfurt, 2010).

This study posits that a refined seismic attribute analysis, integrated with geological models, can offer a more precise understanding of fracture networks within coal seams, potentially overcoming existing limitations in CBM recovery strategies.

Aimed at addressing this critical gap, our research seeks to develop an enhanced model of the fault-fold fracture systems in the southern Qinshui Basin's coal seams, integrating seismic attribute analysis with geological and geophysical data. Moreover, using three-dimensional seismic data, important seismic profiles in the study area are selected for stratigraphic tracing and structural interpretation, analyzing the characteristics of faults in both planar and longitudinal directions. Through qualitative and quantitative studies, the fault-fold structural characteristics of the study area are revealed, which can provide insights into subsurface fault-fold morphology and the development of coalbed methane in southern Qinshui Basin.

8.2 Geological Background

The Qinshui Basin, situated in the southeastern part of Shanxi Province, China, forms a critical component of the North China Craton's structural framework (Fig. 8.1). This basin represents a Mesozoic geological feature, predominantly characterized by its NNE orientation, which delineates a second-order structural belt of the extensive Shanxi Uplift. Recent studies have underscored its significance as a tertiary tectonic unit, shaped by multifaceted tectonic episodes spanning the Haisi, Indosinian, Yanshan, and Himalayan periods (Cai et al., 2011; Su et al., 2005). The study area is located in the southern part of the Qinshui Basin, adjacent to the basin's belly in the north. The shape of the slope is horseshoe-shaped, with gentle and open strata, with angles ranging from 2 to 7°, averaging about 4° (Tao et al., 2012; Teng et al., 2015). The working area covers an area of about 225 km^2.

The structural evolution of the Qinshui Basin is marked by its transition from a cratonic basin to a complex anticline structure, a transformation catalyzed by the cumulative impact of these tectonic movements. The southern part of the basin, in particular, has garnered attention due to its unique sedimentary and structural characteristics, which are instrumental in coalbed methane (CBM) accumulation. The Carboniferous Taiyuan Formation and the Permian Shanxi Formation, notable for

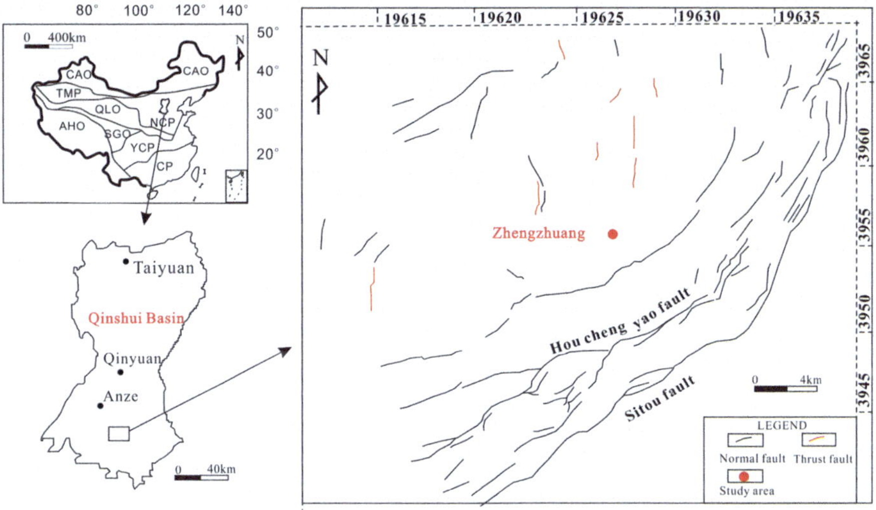

Fig. 8.1 Geological map of Qinshui basin, China

their coal-bearing strata, consist predominantly of sandstone, limestone, mudstone, and coal seams. These formations play a pivotal role in the region's CBM potential, with the Taiyuan Formation presenting a thickness of 83–100 m and the Shanxi Formation ranging from 37 to 72 m (Fig. 8.2; Peng et al., 2017; Zhang et al., 2015).

The sedimentary environment, coupled with the basin's structural features, has facilitated the development of significant coal resources. The stratigraphy reveals a fluvial deltaic sedimentary system, further complicating the geological setting and influencing the distribution and accessibility of CBM resources. Field surveys have indicated the presence of synclines accompanied by normal faults, illustrating the basin's dynamic geological history and the resultant structural complexities that impact CBM extraction (Fig. 8.3; Tao et al., 2012; Teng et al., 2015).

8.3 Methods

This study adopts a multifaceted approach to quantifying fault-fold structures within the southern Qinshui Basin's coal-bearing strata, utilizing seismic attribute analysis to elucidate the complex subsurface geological features. Our methodology is grounded in the application of curvature and variance attributes of seismic data, which are critical for assessing the deformation degree and detecting seismic discontinuities associated with geological structures such as folds, faulting, and fracturing.

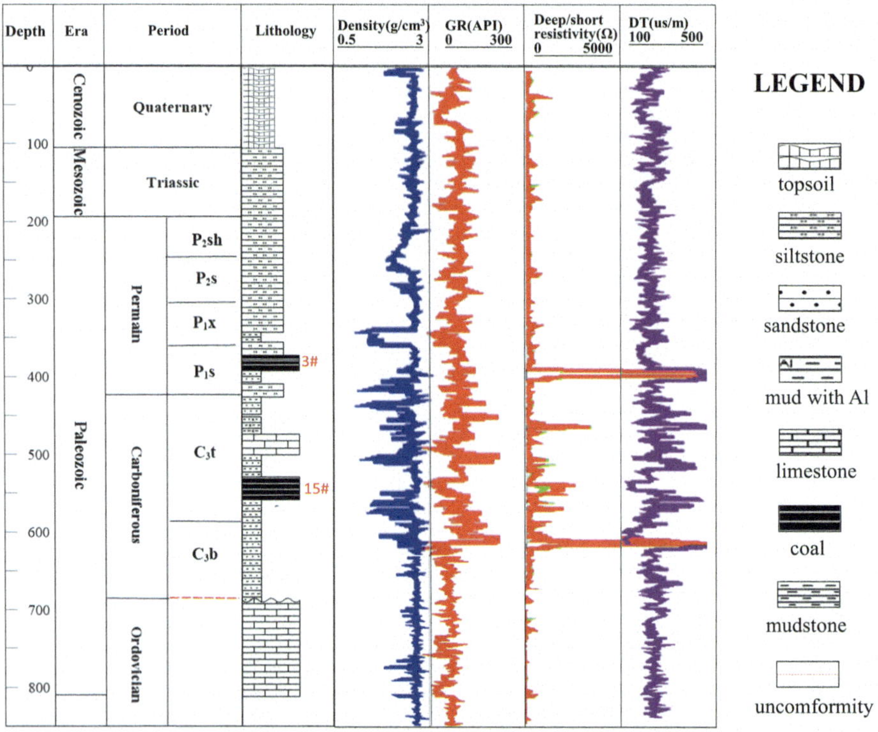

Fig. 8.2 Comprehensive stratigraphic column of southern Qinshui Basin. 3# is Shanxi Coalbed and 15# is Taiyuan Coalbed

Fig. 8.3 **a** A fault-fold system outcrop in field and **b** a schematic model of the fault-fold system

The curvature attribute serves as a pivotal tool for calculating the deformation degree of a point, reflecting synchronous changes with the curvature and demonstrating a high sensitivity to fracture responses. This attribute has been instrumental in the investigation of geological structures including folding, salt domes, faulting, and fracturing. Following the foundational work of Lisle (1994), who established the correlation between fractures and Gaussian curvature, and Roberts (2001), who applied two-dimensional curvature to structural characterization and developed mathematical models for various curvature types, we employ three-dimensional volumetric curvature as delineated by Al-Dossary and Marfurt (2006). This application of three-dimensional curvature endows practical significance to curvature attributes, enhancing our ability to visualize and interpret complex subsurface geological formations.

The variance attribute technique is employed to detect seismic discontinuity along coherent axes, leveraging the calculation of probability variance to identify faults (Marfurt, 2010). This technique is especially valuable in areas with significant tectonic deformation, as it highlights waveform differences between adjacent seismic traces, with variance values approaching 1 indicating the presence of geological discontinuities. The variance attribute's capacity to distinguish features that deviate from surrounding seismic traces provides a robust mechanism for mapping fault and fracture zones within the target strata.

Building on these theoretical foundations, our study introduces an approach for quantifying fault-folds using seismic attribute values. This quantification method involves the following steps: (1), election of a baseline adjacent to the fault-fold, aligned parallel to the extension direction of the area under investigation; (2), Drawing of multiple crosslines perpendicular to both the baseline and the fault zone, ensuring a manageable distance between crosslines; (3), Identification of specific intervals on each crossline for the placement of measurement points; (4), Construction of variation curves for each crossline, using the distance from the measurement point to the baseline as the horizontal axis and the variance value as the vertical axis.

This approach allows for a detailed quantitative analysis of the fault-fold structures, leveraging the accuracy and objectivity of seismic data to provide insights into the spatial distribution and characteristics of underground rock layers. Through the integration of curvature and variance attributes, complemented by time-depth analysis, our methodology offers a comprehensive framework for investigating the subsurface geological features critical to understanding the dynamics of coalbed methane within the basin.

8.4 Results and Analysis

8.4.1 Formation and Structural Characteristics Analysis of Fault-Fold Systems

The geological evolution and structural configuration of coal-bearing strata in the southern Qinshui Basin have been profoundly shaped by the dynamics of plate tectonics through geological time, from the Carboniferous-Permian era to the present day. The tectonic stress field in the Zhengzhuang area of the southern Qinshui Basin has transitioned through four distinct periods, each characterized by unique stress orientations and geological outcomes: (1) Indosinian Orogeny: This period exerted a relatively minor influence on the basin, primarily introducing an SN-directed compressive stress field. The resultant geological impact was localized uplift within the southern segments of the study area. (2) Yanshan Orogeny: Marking a pivotal phase in the basin's deformation, this period was defined by robust NW–SE and nearly horizontal compressive stress fields. This tectonic regime fostered extensive compression, uplift, and folding, giving rise to broad, gentle folds. These folds, primarily NE–NNE-oriented linear synclines and anticlines, signify the substantial restructuring of the basin's architecture. (3) Early Himalayan Phase: Characterized by a NW–SE extensional stress field, this era was instrumental in the genesis of a faulted basin traversing the north–south axis of Shanxi Province. It facilitated the emergence of high-angle normal faults and a myriad of joint structures, especially pronounced in the Changzhi and Jincheng sectors, though the development of folds was less notable during this period. (4) Late Himalayan to Modern Phase: This period witnessed a gradual shift in the tectonic stress field from NNE–SSW compression to a dominant EW compression orientation, transitioning to NE–SW compression. This shift prompted significant crustal uplift and weathering erosion, reshaping the landscape and geological features of the study area.

The structural examination reveals that the Qinshui Basin is predominantly a large-scale complex syncline oriented NE–NNE, displaying asymmetrical fold limbs. The fold structures diverge in orientation, with SW trends in the basin's northern part and NE trends on the southern slope, occasionally exhibiting nearly EW and arcuate patterns. The synclines within the working area exhibit amplitudes under 50 m and axial lengths between 5 and 10 km, classified as long axial, linearly correlated folds.

Faults within this geological setting are relatively sparse, with few instances exhibiting displacements exceeding 20 m, indicative of two-block sliding mechanisms. The majority of fault development is concentrated in the basin's southwest, where high-angle normal faults, primarily trending NE, NNE, and NEE, are prevalent. Notably, the Sitou Fault and Houchengyao Fault, alongside folds-associated faults and a series of nearly EW and NE–SW trending arcuate fault zones, represent the primary fault structures within the area.

8.4.2 Identification of Folds and Faults Based on Three-Dimensional Seismic Technology

To elucidate the development of folds and faults within the Permian Shanxi Formation's No. 3 coal seam and the Carboniferous Taiyuan Formation's No. 15 coal seam, it is imperative to undertake a detailed extraction of target layer attributes. This includes the analysis of time-depth, curvature, and variance attributes. Specifically, within the designated study area, time-depth attributes extracted along the No. 3 coal seam of the Shanxi Formation (as depicted in Fig. 8.4e) highlight the seam's terrain characteristics—elevated in the east and descending towards the west.

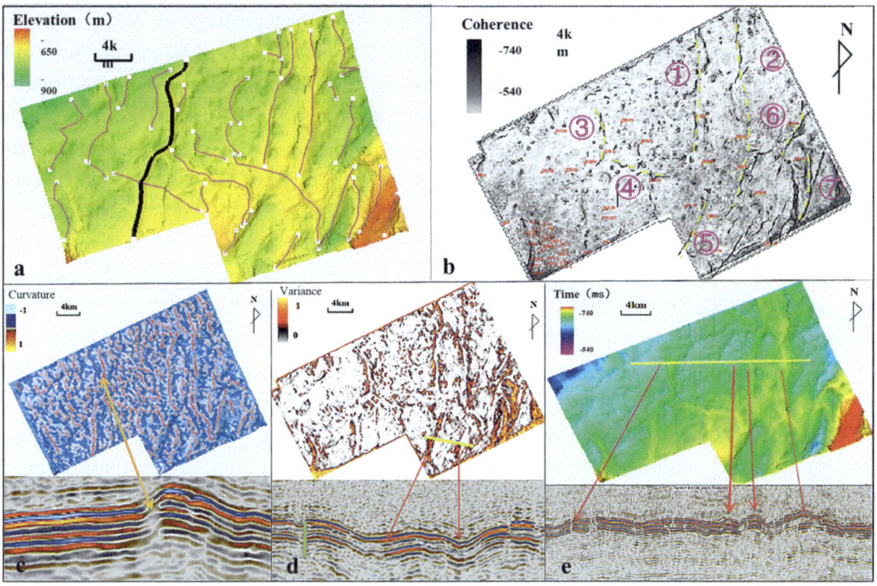

Fig. 8.4 **a** The map of elevation depth on top of Coal 3#. **b** The attribute map of coherence on top of Coal 3#. **c** The attribute map of curvature on top of Coal 3#. **d** The attribute map of variance on top of Coal 3#. **e** The time map on top of Coal 3#

Subsequently, curvature attributes analyzed along the same coal seam (refer to Fig. 8.4c) provide insights into fault development within the target layer. The illustration reveals that areas of positive curvature, indicative of anticlinal or uplifted structures, are delineated in red, whereas areas of negative curvature, denoting synclinal or depressed formations, are marked in blue. This pattern of alternating curvature values aids in the precise identification of fault zones, with the positive curvature regions signaling the upthrown blocks of faults and the negative curvature areas marking the downthrown blocks. This underscores the curvature attribute's critical role in detailing folding degrees in subterranean rock layers and distinguishing between upthrown and downthrown fault blocks. Notably, a profile at the base of Fig. 8.4c showcases a fold located in the northwest sector of the study area, accompanied by a fracture on its left limb.

Further, the extraction of variance attributes along the No. 3 coal seam (illustrated in Fig. 8.4d) enables a comprehensive evaluation of fault distribution within the target layer. Here, regions of higher variance are colored in red, signifying stratigraphic discontinuities that may point to the presence of folds or faults. Additionally, coherence attributes, detailed along the No. 3 coal seam in Fig. 8.4b, assist in assessing fault distribution, with darker shades marking the sites of geological structures like fractures or faults.

Integrating these seismic attributes has led to the identification of 30 significant folds within the study area, as captured in Fig. 8.4a. These folds, predominantly linear with orientations majorly SN and nearly EW, are components of an extensive fold system. The southern region of the Qinshui Basin is characterized by a synclinal structure extending NE–SW, with its core elevated in the east and tapering towards the west. This area exhibits gentle limb dips, an axial length approximating 18 km, and a short axis around 4 km. Moreover, a succession of nearly EW-trending folds and faults, situated above the principal NE-oriented structures, has been discovered. The fold development in the floor of the No. 15 coal seam mirrors that observed in the roof of the No. 3 coal seam.

This investigation also discerned seven primary fracture zones (shown in Fig. 8.4b), primarily linked to faults with minimal displacements. These faults are categorized based on their lateral reach: primary faults (F1, F2), spanning nearly the entire study area, and secondary faults (F3, F4, F5, F6, F7) with more confined lateral extents. The delineation between primary and secondary faults, as indicated in the attribute maps, is solely based on their lateral spans, thereby not necessarily reflecting a difference in fault scale.

8.4.3 Quantification of Fault-Fold Belts Based on Attribute Values

Leveraging the delineation of seven significant faults and 30 folds, our analysis identified five distinct fault-fold belts within the study area, as demonstrated in Fig. 8.5. Focusing on one fault-fold belt (indicated by the green arrow in Fig. 8.5), as exemplified in Fig. 8.6a, we meticulously quantified curvature, variance, and time-depth attributes. The analysis of curvature attributes reveals alternating positive and negative values, suggesting potential sites for fault or fracture development. Variance data fluctuations signal the emergence of folds or faults within the coal seam layers, while time-depth maps detail the subsurface morphological features beneath the seismic lines. This tripartite attribute analysis aids in pinpointing zones of faults or fractures.

Illustrating this methodology with the variance map of the No. 3 coal seam, we consider seismic lines C1–C5, each separated by approximately 500 m and extending over a length of about 3 km. We sampled 50 points along each line at 60-m intervals, starting from the point nearest to the established baseline. This setup allowed for the plotting of a graph, using the distance from the baseline to each measurement point as the x-axis (abscissa) and the corresponding attribute value as the y-axis (ordinate), as depicted in Fig. 8.6b. This graph reveals notable variability in the width of fault fragmentation across different plane locations.

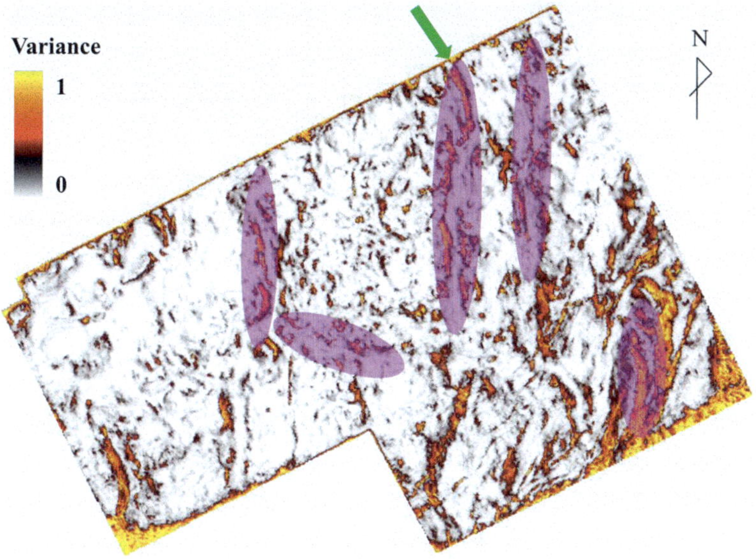

Fig. 8.5 Interpreted 5 major fault-fold systems in variance attribute of Coal 3#

The analysis underscores that faults or fractures are present on both sides of the folds, yet specific characteristics such as morphology, width, and variations remain undetermined. To address this, a comparative study between the No. 3 and No. 15 coal seams was conducted, as shown in Fig. 8.7. Integrating planar attribute maps, cross-sectional maps, and combined quantified data for these seams, we observe a shaded area on the left side of Fig. 8.8a, marking the fault at the inflection point of the fold's left limb. Conversely, a shaded area on the right side signifies a fault on the fold's right limb. This observation points to an asymmetric distribution of folds within the fault-fold belt, with a normal fault forming on the left side and a variable-width fault on the right side, ranging from 50 to 500 m and then narrowing to 200 m from the first to the third seismic line.

8.5 Discussions

8.5.1 Asymmetry in Fault-Fold Systems

Our comparative analysis between the No. 3 and No. 15 coal seams underscores the asymmetrical distribution of fault-fold systems, a characteristic that challenges traditional models of subsurface geology. This observation aligns with the theoretical models proposed by Cawood and Bond (2020), suggesting a reevaluation of existing geological paradigms. Furthermore, the development of a three-dimensional model of

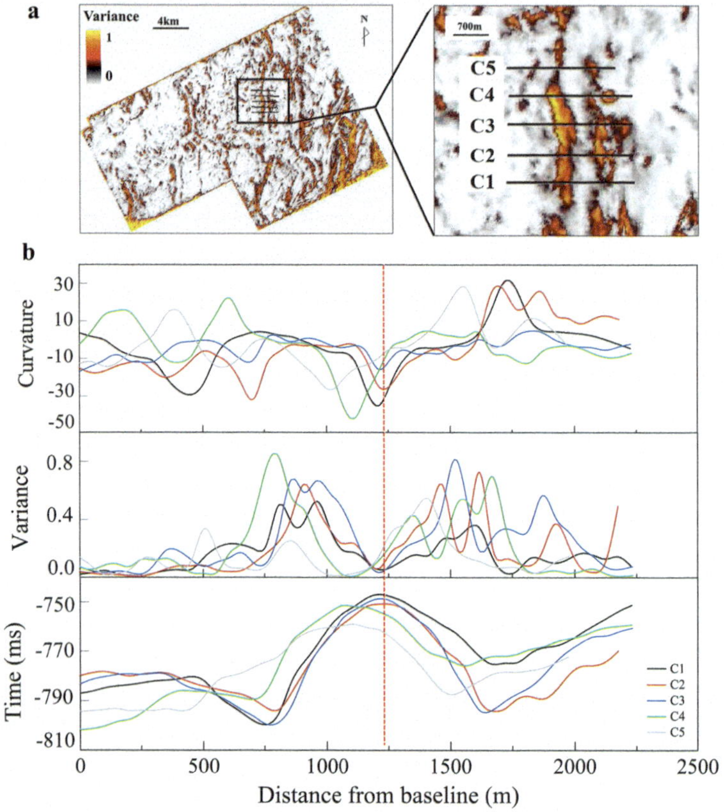

Fig. 8.6 **a** Attribute map of variances on top of Coal 3#. The zoom-in figure shows the fault-fold of interest with 5 perpendicular profiles (baseline) across the fault-fold system. **b** Profiles of attributes for the fault-fold system

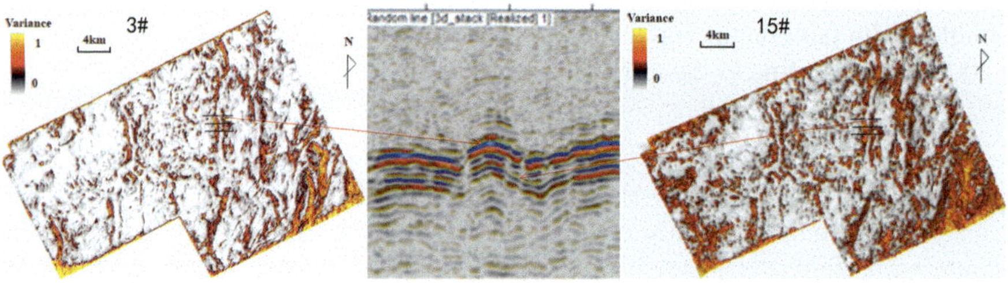

Fig. 8.7 Profile of amplitude (center) and two attribute maps of variances on top of Coal 3# (left) and 15# (right), respectively

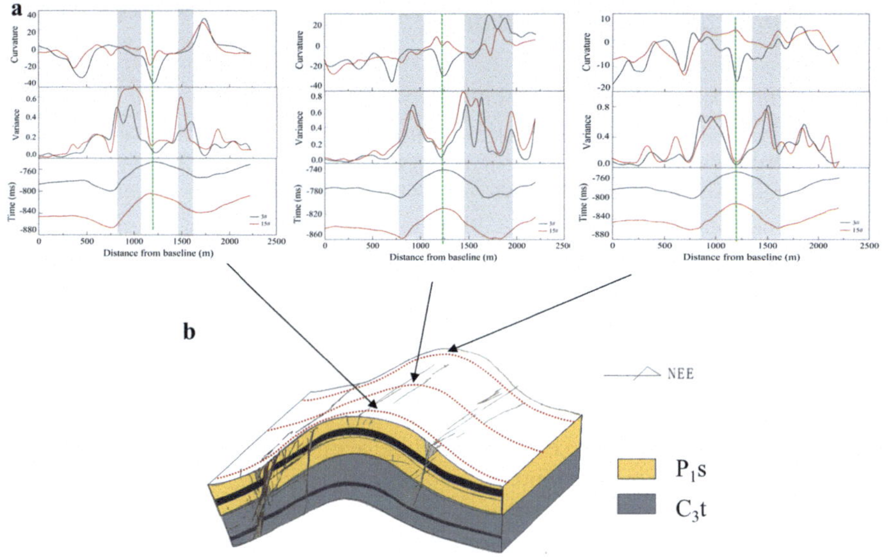

Fig. 8.8 **a** Fault-fold attribute profiles on top of Coal 3# and 15# and **b** schematic of the fault-fold model

the Zhengzhuang fault-fold belt provides a comprehensive framework for assessing the geomechanical properties of the region. Building on this comprehensive data analysis, we synthesized a three-dimensional model of the Zhengzhuang fault-fold belt within the southern sector of the Qinshui Basin, encapsulated in Fig. 8.8b. This model offers an enriched understanding of the complex interplay between geological structures in the region, advancing our understanding of the spatial relationship between faults and folds.

The observed asymmetry in the distribution of fault-fold systems resonates with the work of Paolo et al. (2022), who emphasized the role of regional tectonic regimes in influencing the asymmetrical nature of fault-fold systems. The controlling mechanisms behind this observed asymmetry can be attributed to the variable intensities of the compressive stress fields, particularly during the Yanshan orogeny and the early stages of the Himalayan orogeny. The differential application of these stresses likely contributed to the formation of high-angle normal faults with varying displacement magnitudes. The insights offered by Yang et al. (2020) support this interpretation, suggesting that the directional influence of compressive and extensional tectonics plays a critical role in defining the architecture of fault-fold systems.

8.5.2 *Implications and Limitations*

The asymmetrical characteristics of these geological structures have significant implications for coalbed methane (CBM) exploitation strategies. Understanding the spatial distribution and orientation of these asymmetrical fault-fold systems can enhance predictive models for CBM reservoirs, facilitating more effective exploration and extraction methods. The correlation between fault-fold asymmetry and CBM accumulation patterns, as discussed by Li et al. (2023), underscores the importance of integrating asymmetrical analysis into CBM resource assessments.

However, our study has limitations. While the identified seismic attributes significantly enhance the detection of geological structures, the resolution of seismic data and the interpretative nature of seismic attribute analysis may introduce uncertainties in characterizing fault and fold systems. Future research should aim at integrating high-resolution seismic data and machine learning algorithms to improve the accuracy of geological interpretations (Arrowsmith et al., 2022).

To refine the understanding of these asymmetrical structures and their controlling mechanisms further, future research should aim to integrate multidisciplinary approaches, combining high-resolution seismic imaging with advanced computational models. The potential integration of machine learning algorithms, as suggested by Shahri et al. (2021), could offer new insights into the predictive modeling of geological structures, enabling a more nuanced interpretation of subsurface data.

8.6 Conclusion

The exploration and analysis of the Qinshui Basin's geological structures have yielded significant insights into its complex subsurface morphology, contributing valuable data to the fields of geology and geophysics. Through detailed seismic attribute analysis, this study has delineated the basin's current structural composition, identified extensive fold systems, and classified major fracture zones, thereby enhancing our understanding of the region's geological dynamics. Major findings include:

(1) The Qinshui Basin is characterized by a large, complex syncline structure with a general NE–NNE trend. The structural symmetry between the eastern and western limbs, along with the dominantly NE and SW trending folds on the northern and southern slopes, respectively, highlights the basin's intricate geological fabric. Despite the presence of high-angle normal faults, the fault structures within the basin are relatively underdeveloped, indicating a predominance of normal faults of relatively small scale.

(2) The identification of thirty major folds within the basin reveals a predominantly linear fold system with SN and nearly EW orientations. This system underpins the synclinal structure distributed across the NE–SW axis of the basin, with the core exhibiting a pronounced elevation gradient from east to west. This structural nuance, coupled with the gentle limb dips and the presence of nearly EW-trending folds and faults above the main NE-trending structures, underscores the dynamic geological processes at play.
(3) The delineation of seven major fracture zones, primarily oriented north–south with minor displacements, alongside the classification of faults into primary and secondary categories, provides a nuanced understanding of the basin's fault dynamics. Moreover, the quantitative analysis of the fault-fold belt No. 3 reveals an asymmetric distribution, indicative of the historical tectonic stress fields influenced by the Yanshan and Himalayan orogenies. This asymmetry, characterized by a consistent normal fault width on the left side and a variable fracture width on the right, elucidates the basin's response to complex tectonic stresses.

In summary, the findings from this study not only shed light on the Qinshui Basin's structural characteristics but also provide a foundational framework for future geological and geophysical explorations. The insights into the basin's syncline structure, fold system, fracture zones, and fault-fold belts enhance our comprehension of its geological evolution, influenced by significant orogenic events. These results contribute to the broader geological knowledge, offering potential implications for resource exploration within the Qinshui Basin and similar geological settings.

8.7 Collapse Column

Abstract

Seismic survey data collected for coal gas exploration show that there are many collapse columns distributed in the subsurface of Qinshui Basin, China. The interesting features of the collapse columns are observed by the seismic attributes, including the circular discontinuous patches on the horizon of the Shanxi Formation and multiple parallel discontinuities in vertical profiles of amplitudes. We speculate that the wide presence of these collapse columns are point constraints for the migration and accumulation of coal gas on a large scale.

Summary

The collapse column is an interesting morphology of collapse in the coal basins (Zou et al., 2019). Collapse columns are demonstrated using seismic surveys within Paleozoic stratum, including the Permian Shanxi Formation (3# coal) and Carboniferous Taiyuan Formation (15# coal), in Qinshui Basin, Shanxi Province, China (Wang et al., 2016; Fig. 8.9). The seismic amplitude and attribute of coherence from the seismic

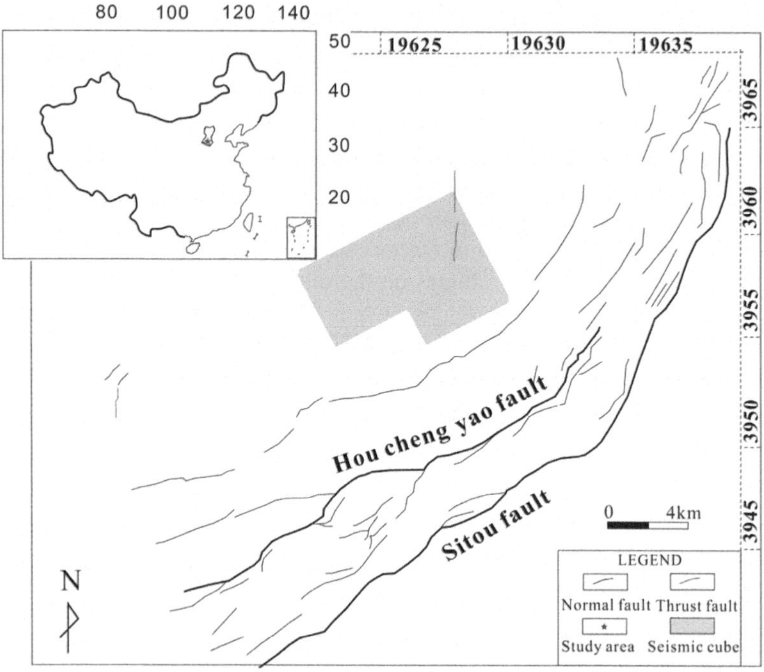

Fig. 8.9 Structural setting of the study area in Qinshui Basin, China, showing the seismic survey area (*shaded area*), and the Houchengyao Fault and Sitou Fault striking northeast. Inset the location of the study area in China

survey (prestack time migrated) have been used to characterize the morphology of collapse columns. The main observations include: (1) there are circular/oval patches of discontinuities on the horizon of Shanxi Formation (Fig. 8.10); (2) the vertical profiles of amplitude show cylindrical/funnel-shaped discontinuities, and (3) collapse columns are widely distributed on the same horizon of Shanxi Formation in this area (Fig. 8.11). These collapse columns formed as a result of the collapse of overlying strata (Permian and Carboniferous Formations, 3# and 15# coals) in the Ordovician karst cave. Compared to the fault damage zone, the collapse column has some similar characteristics in the vertical profile of amplitude (Botter et al., 2014; Liao et al., 2019). Another alternative interpretation for structures with similar seismic features could be the velocity pulldown due to the overburden.

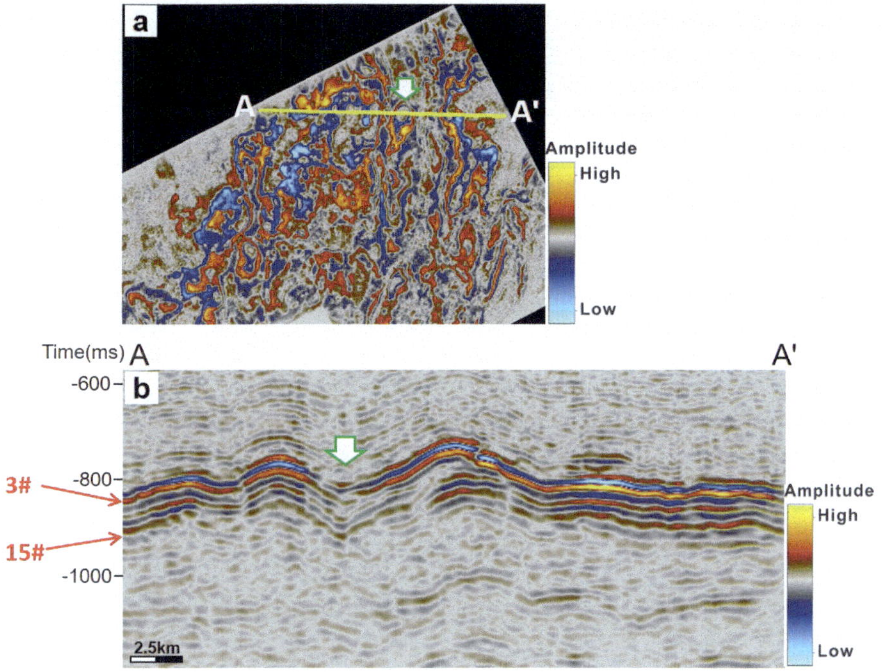

Fig. 8.10 Collapse column detection by **a** time section of ~−850 ms and **b** a vertical slice through seismic amplitude along the yellow line. 3# and 15# are two main coal bedding within Permian Taiyuan Formation and Carboniferous Shanxi Formation, respectively. Note lateral changes of the waveform in **b** crossing though both 3# and 15#. These are interpreted as collapse columns (*green arrow*) in coal bedding, though the time slice shows circular/oval pattern elusively

In this study area, the collapse columns are point structures with frequent recurrences. The amplitude and coherence mapping reveal collapse columns with diameters ranging 50 to 250 m, and up to 600 m of height. There are cylindrical damage-zones with dense fractures developed around these collapse columns (Fig. 8.12), which are difficult to quantify due to seismic scattering noises. Nonetheless, the geometric parameters may be overestimated due to the low resolution of the seismic data. The application of seismic amplitude and coherence to characterize the collapse columns improve the in-depth understanding of the geomorphology and formation processes of collapse columns, which is significant for hazard prediction and mitigation in coal mining.

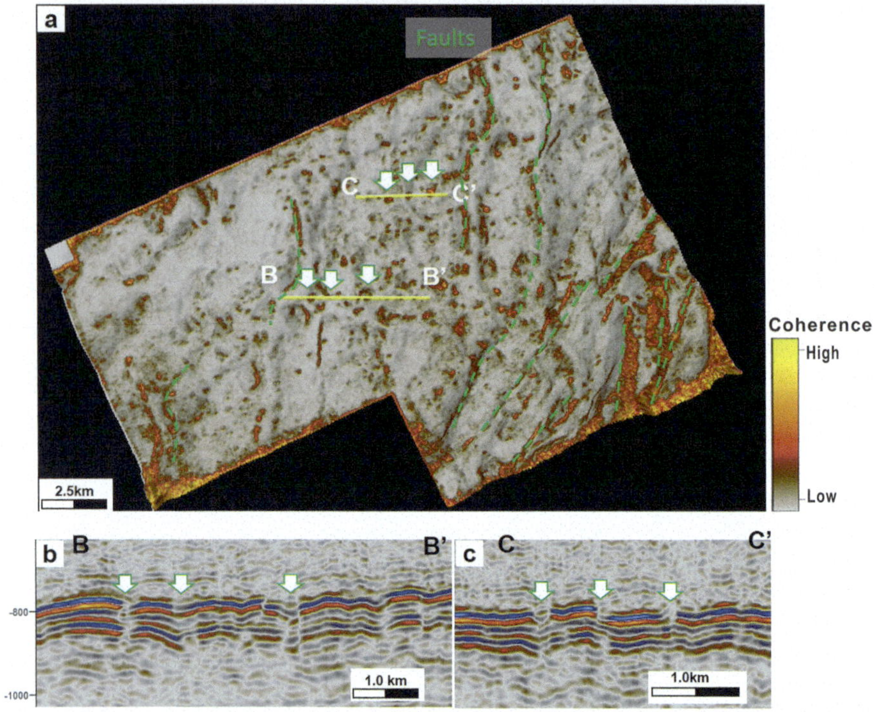

Fig. 8.11 Collapse column detection by **a** seismic attribute of coherence on horizon of Taiyuan Formation (3#) and **b**, **c** vertical slices through seismic amplitude along the yellow lines of BB' and CC', respectively. Two beddings of coal, 3# and 15#, are marked. The small-size circular/oval patches of low coherence values (bright yellow in coherence attribute) are collapse columns scattering in the study area, while the greyish area is of high coherence values that indicate the continuous coal bedding. Lateral changes of the waveform in **b**, **c** demonstrate the collapse columns with green arrows. The lineaments are faults (*green dashed lines*), particularly in the southeast part of the study area

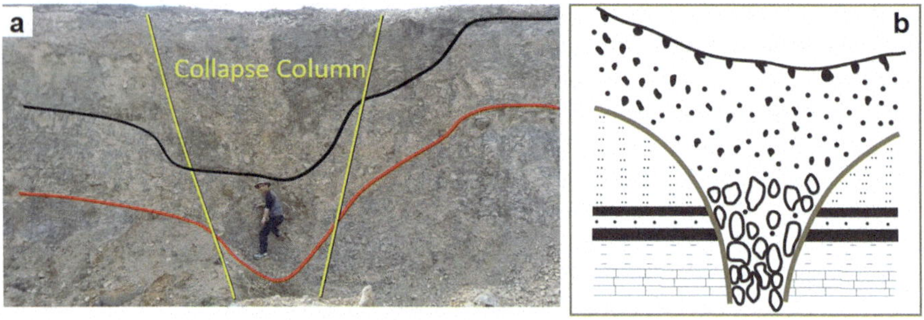

Fig. 8.12 **a** Outcrop observation of the collapse column in the Qinshui Basin. The yellow lines show the general funnel-shaped collapse column, whereas black and red lines indicate the beddings. **b** The schematic model showing the morphology of the collapse column in the coal reservoir (after Zuo et al., 2019)

References

Al-Dossary, S., & Marfurt, K. J. (2006). 3D volumetric multispectral estimates of reflector curvature and rotation. *Geophysics, 71*(5), P41–P51.

Arrowsmith, S. J., Trugman, D. T., MacCarthy, J., Bergen, K. J., Lumley, D., & Magnani, M. B. (2022). Big data seismology. *Reviews of Geophysics, 60*(2), e2021RG000769. https://doi.org/10.1029/2021RG000769

Arroyal, R. M. P. (2009). *Characterization and origin of fracture patterns in the Woodford Shale in southeastern Oklahoma for application to exploration and development*. Doctoral dissertation, University of Oklahoma

Botter, C., Cardozo, N., Hardy, S., Lecomte, I., & Escalona, A. (2014). From mechanical modeling to seismic imaging of faults: A synthetic workflow to study the impact of faults on seismic. *Marine and Petroleum Geology, 57*, 187–207. https://doi.org/10.1016/j.marpetgeo.2014.05.013

Cai, Y., Liu, D., Yao, Y., Li, J., & Qiu, Y (2011) Geological controls on prediction of coalbed methane of No. 3 coal seam in Southern Qinshui Basin, North China. *International Journal of Coal Geology, 88*(2–3), 101–112. https://doi.org/10.1016/j.coal.2011.08.009

Cawood, A. J., & Bond, C. E. (2020). Broadhaven revisited: A new look at models of fault–fold interaction. *Geological Society, London, Special Publications, 487*(1), 105–126. https://doi.org/10.1144/SP487.11

Guo, Y., Zhang, K., & Marfurt, K. J (2010) Seismic attribute illumination of Woodford Shale faults and fractures, Arkoma Basin, OK. In *SEG Technical Program Expanded Abstracts 2010* (pp. 1372–1376). Society of Exploration Geophysicists. https://doi.org/10.1190/1.3513097

Li, S., Qin, Y., Tang, D., Shen, J., Wang, J., & Chen, S. (2023). A comprehensive review of deep coalbed methane and recent developments in China. *International Journal of Coal Geology, 279*, 104369. https://doi.org/10.1016/j.coal.2023.104369

Liao, Z., Liu, H., Carpenter, B. M., Marfurt, K. J., & Reches, Z. (2019). Analysis of fault damage-zones by using 3D seismic coherence in Anadarko Basin, Oklahoma. *AAPG Bulletin, 103*(8), 1771–1785. https://doi.org/10.1306/1219181413417207

Liao, Z., Zhang, L., Wen, L., & Zeng, L. (2020). Collapse columns in Permian and carboniferous formations of coal, Qinshui Basin, China. *Interpretation, 8*(4), SR33–SR35. https://doi.org/10.1190/INT-2020-0031.1

Lisle, R. J. (1994). Detection of zones of abnormal strains in structures using Gaussian curvature analysis. *AAPG Bulletin, 78*(12), 1811–1819.

Marfurt, K. J. (2010). *The shape of seismic interpretation.*

Paolo, P., Calamita, F., & Tavarnelli, E. (2022). Along-strike variation of fault-related inversion folds within curved thrust systems: The case of the Central-Northern Apennines of Italy. *Marine and Petroleum Geology, 142*, 105731. https://doi.org/10.1016/j.marpetgeo.2022.105731

Peng, C., Zou, C., Zhou, T., Li, K., Yang, Y., Zhang, G., & Wang, W. (2017). Factors affecting coalbed methane (CBM) well productivity in the Shizhuangnan block of southern Qinshui basin, North China: Investigation by geophysical log, experiment and production data. *Fuel, 191*, 427–441. https://doi.org/10.1016/j.fuel.2016.11.071

Roberts, A. (2001). Curvature attributes and their application to 3D interpreted horizons. *First Break, 19*(2), 85–100.

Shahri, A. A., Shan, C., Zäll, E., & Larsson, S. (2021). Spatial distribution modeling of subsurface bedrock using a developed automated intelligence deep learning procedure: A case study in Sweden. *Journal of Rock Mechanics and Geotechnical Engineering, 13*(6), 1300–1310. https://doi.org/10.1016/j.jrmge.2021.07.006

Shi, J., Zeng, L., Zhao, X., Zhang, Y., & Wang, J. (2020). Characteristics of natural fractures in the upper Paleozoic coal bearing strata in the southern Qinshui Basin, China: Implications for coalbed methane (CBM) development. *Marine and Petroleum Geology, 113*, 104152. https://doi.org/10.1016/j.marpetgeo.2019.104152

Su, X., Lin, X., Liu, S., Zhao, M., & Song, Y. (2005). Geology of coalbed methane reservoirs in the Southeast Qinshui Basin of China. *International Journal of Coal Geology, 62*(4), 197–210. https://doi.org/10.1016/j.coal.2005.01.002

Tao, S., Wang, Y., Tang, D., Xu, H., Lv, Y., He, W., & Li, Y. (2012). Dynamic variation effects of coal permeability during the coalbed methane development process in the Qinshui Basin, China. *International Journal of Coal Geology, 93*, 16–22. https://doi.org/10.1016/j.coal.2012.01.006

Teng, J., Yao, Y., Liu, D., & Cai, Y. (2015). Evaluation of coal texture distributions in the southern Qinshui basin, North China: Investigation by a multiple geophysical logging method. *International Journal of Coal Geology, 140*, 9–22. https://doi.org/10.1016/j.coal.2014.12.014

Wang, J., Cui, Y., Shi, J., et al. (2016). Logging response and parameters reconstruction of coal reservoir fractures in the southern Qinshui Basin. *Natural Gas Geosciences, 27*(11), 2086–2092. (in Chinese) https://doi.org/10.1016/j.jngse.2016.11.001

Wang, S., Zhang, M., & Zhuang, X. (1996). The mechanism of microfractures and cleats of coal reservoirs and their implication for methane recovery. *Earth Science: Journal of University of Geosciences, 21*(6), 637–640. (in Chinese).

Yang, X., Peel, F. J., McNeill, L. C., & Sanderson, D. J. (2020). Comparison of fold-thrust belts driven by plate convergence and gravitational failure. *Earth-Science Reviews, 203*, 103136. https://doi.org/10.1016/j.earscirev.2020.103136

Zhang, Z., Qin, Y., Fu, X., Yang, Z., & Guo, C. (2015). Multi-layer superposed coalbed methane system in southern Qinshui Basin, Shanxi Province, China. *Journal of Earth Science, 26*, 391–398. https://doi.org/10.1007/s12583-015-0541-z

Zou, J., Hong, Z., Peng, S., Shi, Y., Song, H., Li, M., & Zhang, Z. (2019). Investigation on failure behavior of collapse column in China's coal mine based on discontinuous deformation numerical method. *PLoS ONE, 14*(8), e0219733. https://doi.org/10.1371/journal.pone.0219733

Part III
New Technologies and New Applications

Chapter 9
Characterizing Fault Damage Zones Using Forward Modeling Within Superdeep Carbonates in Northwest China

Abstract By using forward modeling and associated seismic attributes of variance, four fault models of idealized damage-zones are characterized, and the frequency effect is evaluated on the width estimation of fault damage-zones in the subsurface. Main results indicate: (1) The general geometrical pattern of damage-zones could be identified by using simulated amplitude and seismic variance with main frequencies of 10, 25, and 40 Hz; (2) The estimated widths of damage-zones at a low frequency of 10 Hz are larger (up to two-fold) than those at frequencies of 25 and 40 Hz; For large damage-zones (>400 m), the width is best estimated by a frequency of 25 Hz; (3) Scattering noise and diffraction around the fault are found in data at a high frequency of 40 Hz, which results in width overestimation of the damage-zones by ~17%. The internal structures are difficult to distinguish as scattering noise and chaotic reflections dominate seismic signals. More factors that may influence the accuracy of damage-zone width estimation via seismic attributes, include the bedding thickness, fracture density, and velocity.

9.1 Introduction

Fault damage-zones in the subsurface may display complex geometrical shapes, with multiple interacted fault cores, and surrounding damage-zones, as observed in exposed structures (Fig. 9.1; Aydin & Johnson, 1978; Faulkner et al., 2003; Liao et al., 2020). These damage-zones form as a result of continued slip along the existing fault and ongoing fragmentation of the protolith. These damage-zones usually occur at a later stage of fault evolution (Cartwright et al., 1995; Krantz, 1988). It is challenging to estimate the width of such complex damage-zones in the subsurface due to lack of detailed data, which is critical to understand the internal structures and mechanical behavior, and to inform hydrological models (e.g., Busetti et al., 2012; Caine et al., 1996; Faulkner et al., 2010). The main objective here is to investigate the geometrical features of damage-zones using an approach of seismic simulation based on idealized fault models.

Z. Liao, *Applied Seismic Geostructures*, https://doi.org/10.1007/978-981-97-7079-3_9

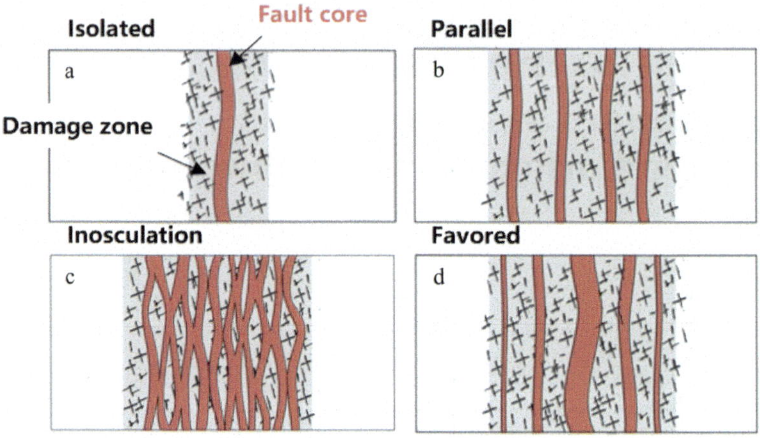

Fig. 9.1 Schematic models of the damage-zones adapted from Liao et al. (2020). **a** Isolated model: Damage-zone of single fault-core. **b** Parallel model: Composite damage-zone with approximately equal-size fault-cores parallel/sub-parallel. **c** Inosculation model: Composite damage-zone with multiple, anastomosing fault-cores. **d** Favored model: Composite damage-zone with one dominating fault-core and a few secondary ones. Size not to the scale

We recently developed four idealized fault models of damage-zones (Liao et al., 2020) by using multi-attribute analyses from a fine-resolution three-dimensional seismic survey. The internal architecture and fracture distribution of the damage-zones in the subsurface are set up for tight sandstone reservoirs in northeast Sichuan, China. The damage-zones are mapped with widths in a range of 200–1000 m along faults ranging 3–15 km in length, and up to 1000 m of cumulative slip. We defined four idealized fault models based on the identification of numerous fault cores and associated damage zones, including: isolated, parallel, inosculation, and favored. Each of the models is attributed to proper faulting processes of the mechanical contrasts and fault-core distortion. Thus, we concluded that the derived models are realistic examples of complex damage-zones in the subsurface (details in Liao et al., 2020).

The seismic attribute, variance, is defined as the coherent part of the seismic reflector divided by the average acoustic energy of the input seismic traces (Chopra & Marfurt, 2007). Trace-to-trace amplitude variation could be detected by variance attribute. Considering our interest is the feature of subsurface architecture, the seismic variance should be the appropriate attribute for this research.

In this study, we explore the effects of the seismic attribute, variance (the opposite of coherency), on the estimation of the damage-zone morphology in the subsurface, focusing on the effect of the main frequency on the width of damage-zones. We applied forward simulations based on idealized fault models. Reverse time migration (RTM) is a useful tool for seismic imaging in complex structures, but it will produce low-frequency artifacts and distort seismic images with cross-correlation imaging conditions (Huang et al., 2015; Liu & Liu, 2018; Rickett, 2003). Considering that reflectivity varies greatly in a damage-zone, RTM was determined to not be an appropriate fit for simulation. Least-squares migration (LSM) is a linearized inversion method (Nemeth et al., 1999) and can give the true amplitude images that migration resolution is improved and migration artifacts are suppressed (Duquet et al., 2000; Fomel et al., 2002; Nemeth et al., 1999). However, LSM is a low efficiency method. LSRTM, combining the LSM and RTM operators, can suppress low-frequency artifacts within RTM and has improved computation efficiency compared to LSM (Dai & Schuster, 2013; Huang et al., 2015; Liu & Liu, 2018). Thus, it is the chosen method for damage-zone simulation and provides reliable seismic attributes. We applied the least-squares reverse time migration (LSRTM) to simulation of the idealized damage-zone models.

Seismic attributes are measures of seismic data that are used to visually enhance or quantify structures of interest in seismic images (Chopra & Marfurt, 2007). Seismic attributes, such as dip-azimuth (Guo et al., 2008; Marfurt & Kirlin, 2000), curvature (Al-Dossary & Marfurt, 2006; Roberts, 2001), coherence (Marfurt et al., 1998; Liao et al., 2019), seismic variance (Liao et al., 2019; Marfurt & Rich, 2010), could efficiently be used to characterize appropriate fault zone architecture and properties. Seismic variance has demonstrated the ability to identify general dimensions and shapes of damage-zones, however, with limitations related to seismic noise and unknown complex features. It is evident that the seismic attribute contains noise that should be eliminated and does not always provide the best insight into the structures (Chopra & Marfurt, 2007; Li & Lu, 2014). How much noise contained within the seismic variance attribute becomes critical for determining the structure of the damage-zones. With the expectation to quantify the effect from the seismic variance frequency on width estimation, for the idealized fault models of damage-zones we implement multiple forward simulations with various main frequencies and calculated attributes of variance correspondingly. The application of idealized fault models avoids background geological noise. We propose that these simulations can be applied to the estimation of widths of the damage-zones and also discuss the effect of main frequencies on the width error.

9.2 Model Setup and Approach

We analyze damage-zones using forward modeling, and the approaches are shown below:

9.2.1 Geological Models of Damage-Zones

Fault damage-zones are commonly described to consist of a fault core and surrounding damage-zone (Chester & Logan, 1986; Savage & Brodsky, 2011). Liao et al. (2020) proposed four idealized fault models of damage-zones (Fig. 9.1) using a fine-resolution three-dimensional seismic survey and associated seismic attributes, including:

(a) Isolated model for a typical damage-zone of an independent fault, which is commonly discussed in outcrop research (Fig. 9.1a);
(b) Parallel model (Fig. 9.1b) for a large composite damage-zone with three large fault-cores that are separated by corresponding damage-zones. It is suggested that the parallel model is formed by the superposition of several sub-parallel faults of an approximately similar amount of slip that is spaced at equal distance from each other;
(c) Inosculation model (Fig. 9.1c) for a composite damage-zone with multiple anastomosing fault-cores and damage-zones without individual fault-cores. It evolves by the coalescence of multiple, closely-spaced fault segments that formed in earlier phases;
(d) Favored model (Fig. 9.1d) for a damage-zone with multiple fault-cores where the central fault-core is the largest one that accommodates most of the deformation. It may start as a parallel model with multiple subparallel faults, and one fault core became weaker to accommodate additional deformation via localized slip in later stages. The central fault-core is thus a favored slip surface.

9.2.2 *Velocity Models for Damage-Zones*

The damage-zones likely consist of fragmented and fractured rock body with one or more fault-cores of localized slip. The fragmented and fractured rock is known to influence both travel times and amplitudes of seismic waves (Anderson et al., 1974; Boadu & Long, 1996). Seismic waves will be attenuated when passing through the fractured medium, resulting in amplitude and amplitude-related seismic attributes attenuation (Maultzsch et al., 2007; Thompson et al., 2010). We applied a two-dimensional seismic forward model, considering velocity models as analogs for realistic fault damage-zones (Botter et al., 2014). We assumed the protolith is homogeneous and all faults are set with a dip angle of 81° and slip displacement of 50 m according to the field conditions. Considering fractures are usually below the resolution of seismograma, the fractures in our models are represented as attenuated seismic velocities (Table 9.1):

$$\begin{cases} v_{pf} &= v_{pro} \times 89\% \\ v_{sf} &= v_{pro} \times 95\% \\ v_{dam} &= v_{fc} + v' F(r) \end{cases} \tag{9.1}$$

where v_{pf}, v_{sf}, v_{pro}, v_{dam}, v_{fc} is the assumed velocities of primary fractures near the fault-core, secondary fractures in the damage-zones, protolith, damage-zones, and fault-core, respectively. v' is the velocity gradient between fault-core and the protolith. $F(r)$ is a random function that describes the velocity increase gradually from fault-core to the protolith (Faulkner et al., 2010; Maultzsch et al., 2007;

Table 9.1 Parameters of fault geometries

Modeling parameters			Velocity parameters (m/s)			
Model size (m)		5000 × 3000		Protolith (v_{pro})	Primary fractures (v_{pf})	Secondary fractures (v_{sf})
Grid size (m)		5 × 5	Bedding 1	3700	3293	3515
Sampling interval (ms)		5	Bedding 2	3900	3471	3705
Source	Number	20	Bedding 3	4400	3916	4180
	Interval (m)	250	Bedding 4	4900	4361	4655
Fracture width (m)	Primary	2	Fault-core (v_{fc})	3000		
	Secondary	1	Damage-zone (v_{dam})	$v_{dam} = v_{fc} + v'F(r)$		

Thompson et al., 2010). The geometrical parameters are shown in Table 9.1 (Faulkner et al., 2010). Damage-zones are idealized as about 1–2 orders wider than the fault-cores (e.g., Liao et al., 2019; Sagy et al., 2001). Primary fractures are set near the fault-core and the secondary-fracture are set in the damage-zones with a random pattern. In order to study the impact of model setting, we designed four layers from 1000 to 2500 m for comparison (Fig. 9.2), two thin beds in the shallow area and two thick beds in the deep area. Fracture density of the top bed and the bottom bed are set less than the middle two beds. These are idealized models to the problem, probably inaccurate, but it provides a reasonable approximation to model the damage-zone in the sense of seismic attenuation (Anderson et al., 1974; Boadu & Long, 1996; Chopra & Marfurt, 2007; Li & Lu, 2014).

9.2.3 Seismic Simulation and Attribute of Variance

We use Least Square Reverse Time Migration Theory (LSRTM) to implement seismic imaging (Huang et al., 2015). LSRTM simulator is based on wave equations using time extrapolation, which makes the simulation suitable for various structures with inclination angles and variation of transverse velocities.

The inputs for the LSRTM simulator include the velocity model, fracture setting (Eq. 9.1), grid size, observation system, and main frequency (Table 9.1). The velocity model is significant in determining the reflectivity with the contrast of velocities usually causing stronger seismic reflectivity. It is known that the wave velocity decreases approximately 50% within the fault-core as estimated from in-situ measurements or measurements on core samples in the lab (Nes et al., 2000). We assume the velocity of fault-core is 3000 m/s and beds with a velocity gradient as in Table 9.1. The velocity is set to increase from the fault-core to the margin of the damage-zone in the function of $F(r)$. The fault-core is set independently from the beds to avoid the interaction of reflections. The spatial size of the models is set

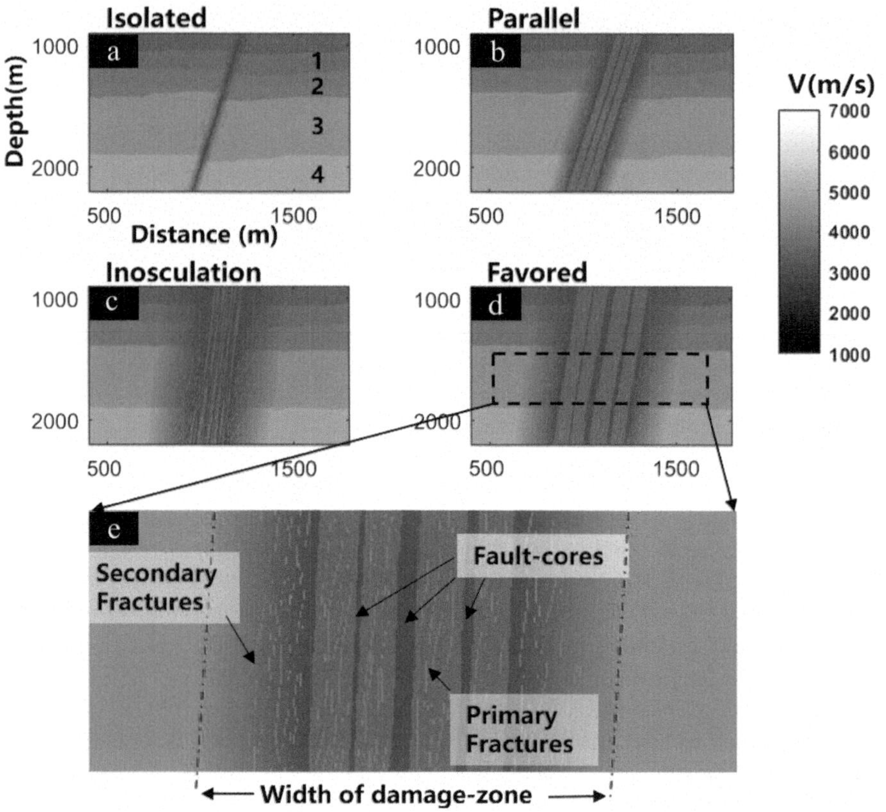

Fig. 9.2 Idealized velocity models for forward modeling, **a–d** corresponding to the damage-zones in Fig. 9.1. **e** Schematic velocity setting for internal structure and fractures within the favored model of damage-zone

as 5000 m × 3000 m, the grid size is 5 m × 5 m, and the seismic sampling interval is 5 ms. 20 sources are evenly distributed at 1m depth with a spacing distance of 250 m. The other parameters are shown in Table 9.1.

To detect the discontinuities, faults, in the subsurface, we used the attribute of seismic variance, which integrates the cumulative seismic response to the structural deformation (Chopra & Marfurt, 2007; Iacopini & Butler, 2011; Liao et al., 2019). Variance can be described as the variation of amplitude:

$$\mathrm{V_{ij}} = \left(\mathrm{A_{ij}} - \mathrm{A_{ave}}\right)^2 \tag{9.2}$$

in which V_{ij} is the variance of ij th grid node, A_{ij} is the amplitude of ij th grid node, A_{ave} is the average amplitude. In a seismogram, the amplitude is a convolution of the wavelet and impedance. Wavelength, $\lambda = v/f$, is a function of frequency in our model. Variance, presented as Eq. (9.2), is obviously sensitively to the variation of amplitude from Eq. (9.1). The trace-to-trace variability is detected over a sample interval, and the large variance coefficients are used for delineation of the faults, while the low values are for the protolith. With fine-resolution, it could be used for distinguishing the internal structure of the damage-zones on a large scale. We thus expect the high variance zone as the fault-core and damage-zones and low values for the protolith. This approach is usually applied using seismic reflection surveys, which has been validated to conform with field observations (e.g., Mitchell & Faulkner, 2009; Sagy et al., 2001; Savage & Brodsky, 2011). We adopt this approach for our forward modeling in this study.

9.3 Results

9.3.1 Fault Damage-Zones

We applied LSRTM and the above approaches to obtain twelve simulations for four models of damage-zones with main frequencies of 10, 25, and 40 Hz. The simulation results are illustrated via seismic amplitude with zero-phase wavelet (Fig. 9.3) and the seismic attribute of variance (Fig. 9.4). Figure 9.3 displays seismic amplitude corresponding to vertical slices through the four models of damage-zones with different main frequencies. We analyze the characteristics of damage-zones based on the simulations.

For the isolated model, the fault can be identified from the disturbance of amplitude waveforms for all three main frequencies (Fig. 9.3a–c). Figure 9.3a shows that dark and thick bands of waveforms appear both horizontally for bed interfaces and vertically for the fault. These dark and thick bands turn much thinner at the main frequency of 40 Hz. It is difficult to identify the internal structure of damage-zones of the isolated fault, particularly at 1800 m (the 3 beds, Fig. 9.3a–c), which could be attributed to the low resolution of the data and relatively small size of the fault. The signal of waveform discontinuity for the fault is mixed with increasing noise from shallow to the deep.

For the other three models, the general geometrical shape of the damage-zones could be recognized from the amplitude maps (Fig. 9.3d–l). The internal structures are unclear at the frequency of 10 Hz, but they could be delineated at higher frequencies of 25 and 40 Hz. Structural details, e.g., parallel fault cores, could be partially identified in Fig. 9.3d–e and the shallower area of Fig. 9.3f. In a deep area of Fig. 9.3f, noise occurred similar to what is observed for the isolated fault in Fig. 9.3c. The structural details in Fig. 9.3g–l are not easily distinguished from the amplitude results for the rest models due to their structural complexities. Only the geometrical patterns of large damage-zones are shown, while the internal structures are not visible and dominated by chaotic reflections in deep areas.

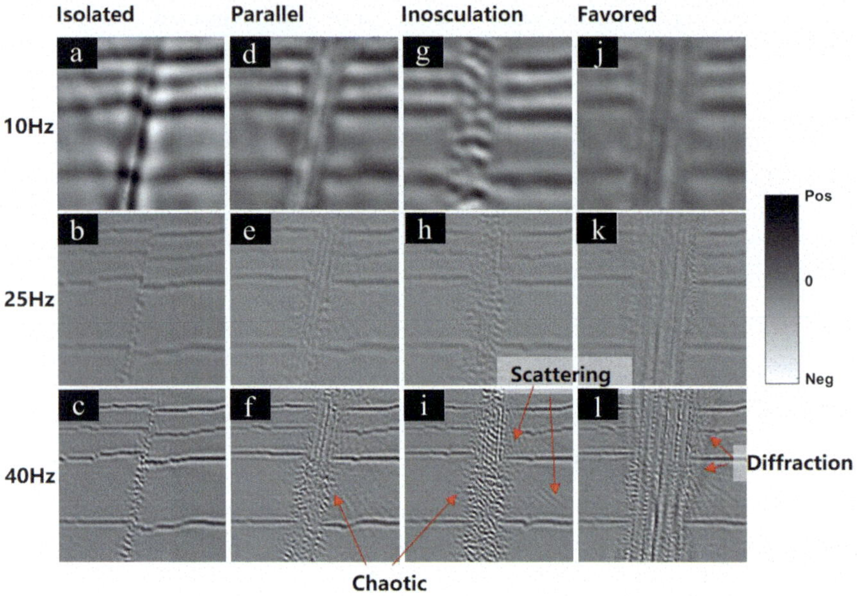

Fig. 9.3 Fault detection by vertical slices through seismic amplitude. Results with main frequencies of 10, 25, and 40 Hz for **a–c** isolated model, **d–f** parallel model, **g–i** inosculation model, and **j–l** favored model, respectively. Note poor resolution in low frequencies (e.g., **a, d, j, g**) and scattering, chaotic noises, and diffraction in (**f, i, l**)

9.3.2 Widths of Damage-Zones

We further investigate the seismic attribute of variance, which is a measurement of the discontinuity of a seismic section. Figure 9.4 presents the corresponding vertical profiles of seismic variance at the three main frequencies. The vertical zones of red-yellow bands indicate the reverse faults. The high main frequencies of 25 and 40 Hz delineate the main damage-zones of all the damage-zone models. Compared to sections of amplitude, the noise in the deep area is filtered in seismic variance. The discontinuities in red-yellow color vertically illustrate geometries of the damage-zones. It is still difficult to elucidate the internal differences between the three models by seismic variance. Moreover, the low frequency of 10 Hz produces significant noise around the damage-zones. E.g., large bright patches near the faults that would lead to misinterpretation as discontinuities (Fig. 9.4a). Figure 9.4d shows smaller damage-zones in shallow areas for the parallel model and, in contrast, exaggerated damage-zones in the deep area. Figure 9.4g illustrates the bifurcation of faults in shallow areas for the inosculation model, which is not the case in our geology model. These are examples that caused by poor resolutions due to low frequencies.

To further explore the effect of frequency on characterizing the damage-zones, we followed the approach (Liao et al., 2019) to apply seismic variance to estimate the widths of the damage-zones. We prepared 84 profiles of seismic variance (Figs. 9.5,

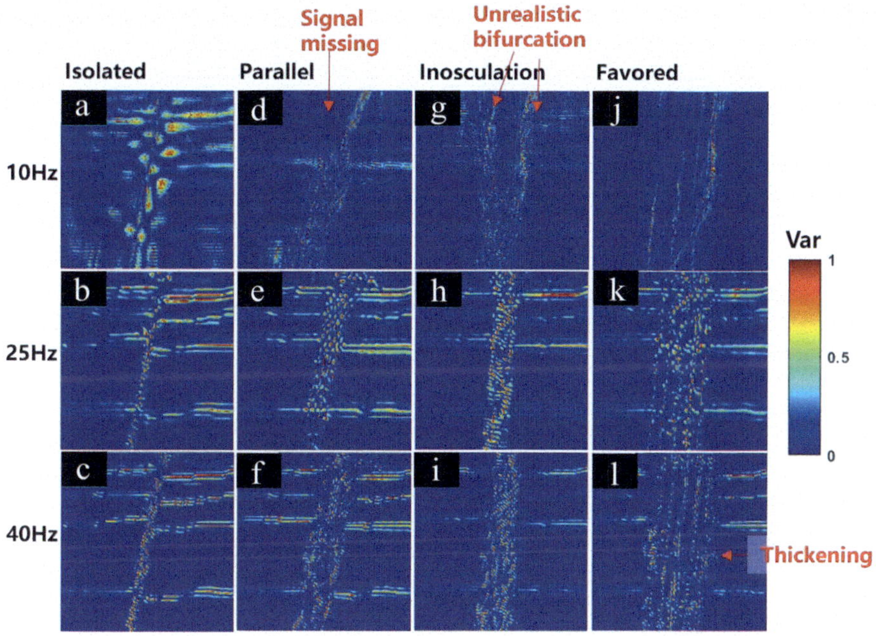

Fig. 9.4 Fault detection by vertical slices through seismic attributes of variance. Results of attribute variance with main frequencies of 10, 25, and 40 Hz for **a–c** isolated model, **d–f** parallel model, **g–i** inosculation model, and **j–l** favored model, respectively. Note scattering patches (**a**), unrealistic fault bifurcation (**g**), and thickening (**l**)

and 9.6) crossing the faults in 1500–1800 m. The seven profiles in each section display distribution of the seismic variance for one main frequency, which we generalized as the characteristics of one of the damage-zones. A typical damage-zone, e.g., an isolated model, could be represented by a zone of high variance, 0.1–0.9 (grey in Fig. 9.5a–c). We argue a small fraction of the highest variance, >0.8, could be interpreted as the fault core (Liao et al., 2019). The width of the damage-zone is estimated to be 400 m using main frequencies of 10 Hz, and decreases to 250 m and 200 m at the main frequencies of 25 Hz and 40 Hz, respectively. The distribution of seismic variance in the frequency of 10 Hz (Fig. 9.5a) obviously does not comply with the damage-zones in Fig. 9.5b, c in two aspects: (1) the widths are different; (2) the structure is distorted at low frequency.

For the other three models, we also prepared the profiles of distributions of seismic variance. The widths are estimated, and structures are evaluated based on the profiles from the 10 Hz, 25 Hz, and 40 Hz main frequencies, respectively. As shown for the parallel model in Fig. 9.5d–l, the widths of damage-zones are varied with few differences for two reasons: (1) the profiles are selected from the deep area that avoids the aforementioned exaggeration (the isolated model) in shallow area; (2) these damage-zones are set with a relatively large scale (>400 m) compared to the isolated fault, which is now distinguishable at low frequency. Another essential feature is the

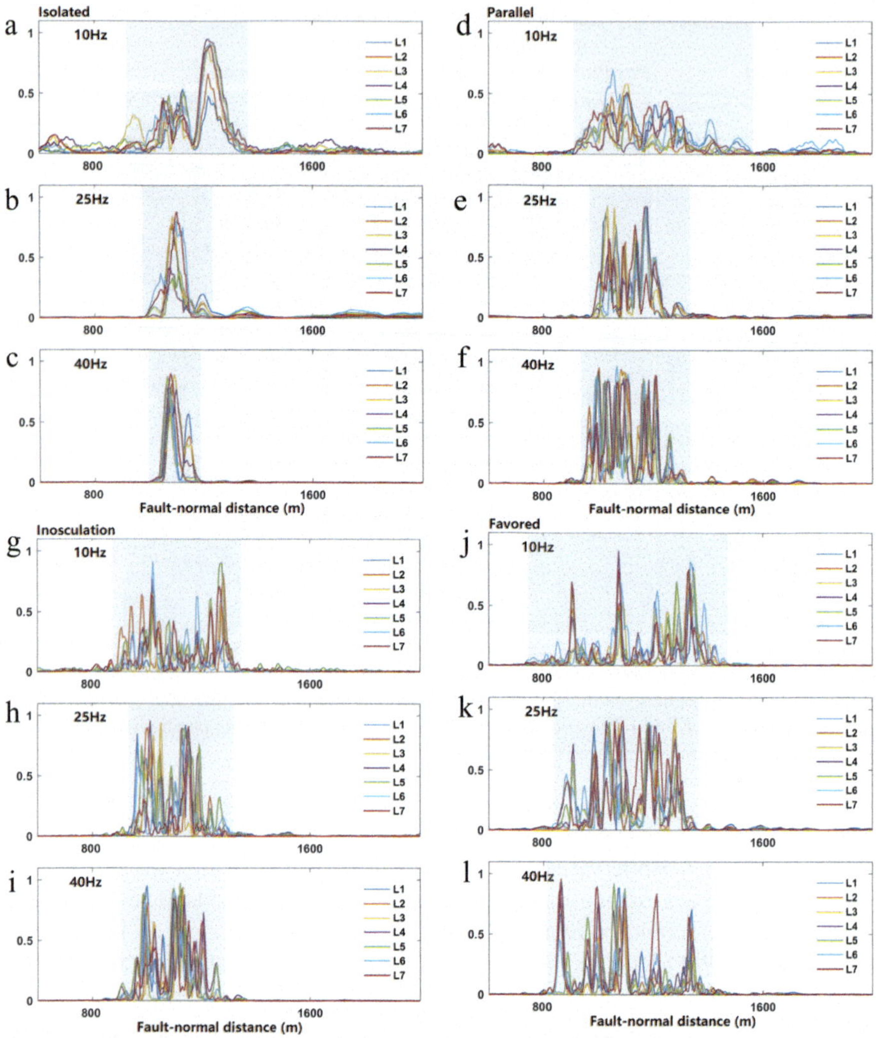

Fig. 9.5 Profiles of the seismic variance values across the damage-zones. Profiles of attribute variance with main frequencies of 10, 25, and 40 Hz for **a–c** isolated model, **d–f** parallel model, **g–i** inosculation model, and **j–l** favored model, respectively. Profiles locations in Fig. 9.2e; variance values above background are interpreted as damage-zones (*grey-zone*)

profiles of seismic variance show small values (~0) for protolith, which is due to the idealized background setting that is not realistic in nature. Similar observations are found for the inosculation and favored models.

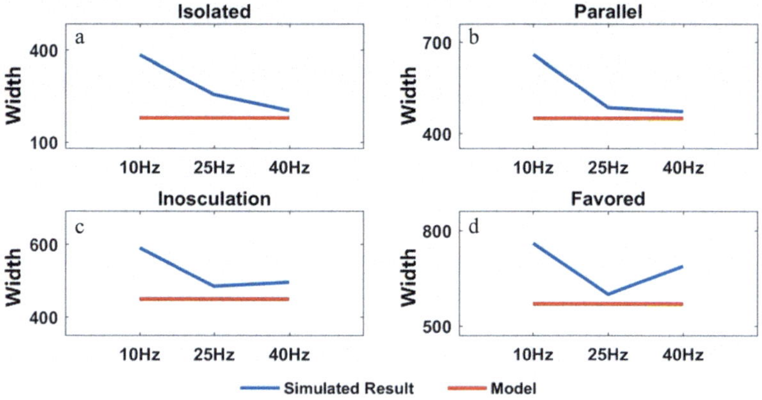

Fig. 9.6 Width estimation from the seismic variances (Fig. 9.6) versus frequencies for **a** isolated model, **b** parallel model, **c** inosculation model, and **d** favored model. The red line illustrates the actual widths from the models

9.4 Discussions

LSRTM was used to simulate the characterizations of four idealized fault model of damage-zones. If we could build a damage-zone velocity model with main features fitting to nature, and the simulation setting (e.g., grid size) is suitable, the simulation would provide us a feasible way to evaluate the effect of frequency on the estimation of damage zone width.

9.4.1 Bed Thickness, Fracture Density, and Velocity Gradient

We first discuss how the velocity models affect the quality of our simulation results. In the case of a frequency of 40 Hz, the relationship between fracture density and imaging quality is clearly presented (Figs. 9.3 and 9.4). In the third bed that are set thick (500 m) with high fracture density, the reflection of damage-zones is detected with more noise due to chaotic reflection than in the fourth bed with less fracture density. The chaotic noise widens the width of damage-zones and decreases the visibility of the structure details (e.g., Fig. 9.3f). For the second bed that is set thin (200 m) with high fracture intensity in the shallow area, the structure details are better preserved by the seismic signal and illustrated (e.g., Fig. 9.3c, f, i). Compared to the third bed that is thick, it illustrates that there is less chaotic reflection but more seismic scattering in the shallow areas with lower seismic velocity. The results show that the bed-controlled fracture density and velocity gradient could cause different types of noise, significantly affecting the interpretation results using amplitude and seismic variance.

9.4.2 Spectral Effect on the Estimation of Widths

LSRTM is a relatively accurate migration method based on data simulated by the finite difference method (Huang et al., 2015). Besides the grid size limited by computing power, frequency is the main factor to affect simulated results (Figs. 9.3 and 9.4). We set main frequencies as 10, 25, and 40 Hz to analyze how the fault damage-zones can be imaged. In general, the higher the wave frequency is, the smaller the size of geological bodies can be resolved by seismic wavelet. Figure 9.6 shows the main frequencies in relation to the widths of damage-zones (Table 9.2). The red lines are for the actual width of damage-zones by the model setting; the blue lines are widths estimated from the simulation results. For the isolated model with damage-zone of 180 m in width, the higher the frequency, the more accurate the estimation; for the parallel model and inosculation model with large damage-zones (>400 m), there is not much difference between frequencies of 25 and 40 Hz. As for the favored model, only a frequency of 25 Hz best fits the actual width; both the results from the low frequency of 10 Hz and a high value of 40 Hz suffer from the influence of noise on the results. With regards to the high frequency in the example of 40 Hz for the favored model, diffraction around the fault is found. Mixed with scattering noise, it will lead to width overestimation of the damage-zones, e.g., about 17% for the favored model. The Fresnel zone (Kozlov et al., 2014) could result in the overestimation of the width of damage-zone in low frequency (e.g. isolated model of 10 Hz) in small fault damage zones, while the overestimation by Fresnel zone could hardly be seen in large fault damage zones (e.g. composited damage-zones).

Our results illustrate that the structural details of the three composite models of damage-zones are challenging to detect, which is possibly limited by the resolution of the seismic data. The accumulated, distorted signal could also be attributed to the noise of chaotic reflection in the deep subsurface. However, it is feasible to apply the seismic attribute of variance to measure the widths of damage-zones (>200 m) with acceptable overestimations. We failed to qualitatively define all the styles of fault damage-zones because of the scattering noise from the fault zones. It strongly depends on the size of the structures that deserves investigation in the future. For wide damage-zones (>400 m), the distribution of damage-zones and internal structures can be interpreted by seismic attributes integrated with geological modeling and outcrop observations.

Table 9.2 Damage-zones width and resolution (λ)

	Model	10 Hz		25 Hz		40 Hz	
	Width (m)	Width (m)	λ (m)	Width (m)	λ (m)	Width (m)	λ (m)
Isolated	180	385	300–490	256	120–196	204	75–123
Parallel	450	660		485		472	
Inosculation	465	590		485		496	
Favored	570	760		600		668	

9.5 Summary

In this study, we displayed an approach to apply seismic simulation to characterize the structure of damage-zones and evaluate the frequency effect on the estimation of damage zone widths. Though the internal structures of the damage-zones are hard to distinguish, the widths of damage-zones could be calculated with some overestimation using seismic attributes at different main frequencies of 10, 25, and 40 Hz. Such overestimation appeared in previous attribute analyses of seismic surveys, could be complemented by using higher frequency, e.g., 25 and 40 Hz, to improve the signal–noise ratio. It is therefore been shown that seismic variance can be applied to characterize the widths of fault damage-zones, which will further provide insights for evaluating the internal structures of faults.

References

Al-Dossary, S., & Marfurt, K. J. (2006). 3D volumetric multispectral estimates of reflector curvature and rotation. *Geophysics, 71*, 41–51. https://doi.org/10.1190/1.2242449

Anderson, D. L., Minster, B., & Cole, D. (1974). Effect of oriented cracks on seismic velocities. *Journal of Geophysical Research, 79*(26), 4011–4015. https://doi.org/10.1029/JB079i026p04011

Aydin, A., & Johnson, A. M. (1978). Development of faults as zones of deformation bands and as slip surfaces in sandstone. *Pure and Applied Geophysics, 116*(4–5), 931–942. https://doi.org/10.1007/BF00876547

Boadu, F. K., & Long, L. T. (1996). Effects of fractures on seismic-wave velocity and attenuation. *Geophysical Journal International, 127*(1), 86–110. https://doi.org/10.1111/j.1365-246X.1996.tb01537.x

Botter, C., Cardozo, N., Hardy, S., Lecomte, I., & Escalona, A. (2014). From mechanical modeling to seismic imaging of faults: A synthetic workflow to study the impact of faults on seismic. *Marine and Petroleum Geology, 57*, 187–207. https://doi.org/10.1016/j.marpetgeo.2014.05.013

Busetti, S., Mish, K., Hennings, P., & Reches, Z. (2012). Damage and plastic deformation of reservoir rocks: Part 2 Propagation of a hydraulic fracture. *AAPG Bulletin, 96*, 1711–1732. https://doi.org/10.1306/02011211011

Caine, J. S., Evans, J. P., & Forster, C. B. (1996). Fault zone architecture and permeability structure. *Geology, 24*, 1025–1028. https://doi.org/10.1130/0091-7613(1996)024%3c1025:FZAAPS%3e2.3.CO;2

Cartwright, J. A., Trudgill, B. D., & Mansfield, C. S. (1995). Fault growth by segment linkage: An explanation for scatter in maximum displacement and trace length data from the Canyonlands Grabens of SE Utah. *Journal of Structural Geology, 17*, 1319–1326. https://doi.org/10.1016/0191-8141(95)00033-A

Chester, F. M., & Logan, J. M. (1986). Implications for mechanical properties of brittle faults from observations of the Punchbowl Fault Zone, California. *Pure and Applied Geophysics, 124*, 79–106. https://doi.org/10.1007/BF00875720

Chopra, S., & Marfurt, K. J. (2007). *Seismic attributes for prospect identification and reservoir characterization*. Society of Exploration Geophysicists and European Association of Geoscientists and Engineers.

Dai, W., & Schuster, G. T. (2013). Plane-wave least-squares reverse-time migration. *Geophysics, 78*(4), S165–S177. https://doi.org/10.1190/geo2012-0377.1

Duquet, B., Marfurt, K. J., & Dellinger, J. (2000). Kirchhoff modeling, inversion for reflectivity, and subsurface illumination. *Geophysics, 65*, 1195–1209. https://doi.org/10.1190/1.1444812

Faulkner, D. R., Jackson, C. A. L., Lunn, R. J., Schlische, R. W., Shipton, Z. K., Wibberley, C. A. J., & Withjack, M. O. (2010). A review of recent developments concerning the structure, mechanics and fluid flow properties of fault zones. *Journal of Structural Geology, 32*, 1557–1575. https://doi.org/10.1016/j.jsg.2010.06.009

Faulkner, D. R., Lewis, A. C., & Rutter, E. H. (2003). On the internal structure and mechanics of large strike-slip fault zones: Field observations of the Carboneras fault in southeastern Spain. *Tectonophysics, 116*(3–4), 235–251. https://doi.org/10.1016/S0040-1951(03)00134-3

Fomel, S., Berryman, J., Clapp, R., & Prucha, M. (2002). Iterative resolution estimation in least-squares Kirchhoff migration. *Geophysical Prospecting, 50*, 577–588. https://doi.org/10.1046/j.1365-2478.2002.00341.x

Guo, H., Lewis, S., & Marfurt, K. J. (2008). Mapping multiple attributes to three- and four-component color models—A tutorial. *Geophysics, 73*, W7–W19. https://doi.org/10.1190/1.2903819

Huang, J., Li, C., Wang, R., & Li, Q. (2015). Plane-wave least-time reverse time migration for rugged topography. *Journal of Earth Science, 26*(4), 471–480. https://doi.org/10.1007/s12583-015-0556-5

Iacopini, D., & Butler, R. W. H. (2011). Imaging deformation in submarine thrust belts using seismic attributes. *Earth and Planetary Science Letters, 302*, 414–422. https://doi.org/10.1016/j.epsl.2010.12.041

Kozlov, I., Romanenkov, D., Zimin, A., & Chapron, B. (2014). SAR observing large-scale nonlinear internal waves in the White Sea. *Remote sensing of environment, 147*, 99–107. https://doi.org/10.1016/j.rse.2014.02.017

Krantz, R. W. (1988). Multiple fault sets and three-dimensional strain: Theory and application. *Journal of Structural Geology, 10*, 225–237. https://doi.org/10.1016/0191-8141(88)90056-9

Li, F., & Lu, W. (2014). Coherence attribute at different spectral scales. *Interpretation, 2*(1), SA99–SA106. https://doi.org/10.1190/INT-2013-0089.1

Liao, Z., Li, W., Zou, H., Hao, F., Marfurt, K. J., & Reches, Z. (2020). Composite damage zone in the subsurface. *Geophysical Journal of International (submitted)*. https://doi.org/10.1093/gji/ggaa158

Liao, Z., Liu, H., Carpenter, B. M., Marfurt, K. J., & Reches, Z. (2019). Analysis of fault damage-zones by using 3D seismic coherence in Anadarko Basin, Oklahoma. *AAPG Bulletin, 103*(8), 1771–1785. https://doi.org/10.1306/1219181413417207

Liu, X., & Liu, Y. (2018). Plane-wave domain least-squares reverse time migration with free-surface multiples. *Geophysics, 83*, 477–487. https://doi.org/10.1190/geo2017-0570.1

Marfurt, K. J., & Kirlin, R. L. (2000). 3-D broad-band estimates of reflector dip and amplitude. *Geophysics, 65*, 304–320. https://doi.org/10.1190/1.1444721

Marfurt, K. J., Kirlin, R. L., Farmer, S. L., & Bahorich, M. S. (1998). 3-D seismic attributes using a semblance-based coherency algorithm. *Geophysics, 63*, 1150–1165. https://doi.org/10.1190/1.1444415

Marfurt, K. J., & Rich, J. (2010). Beyond curvature—Volumetric estimates of reflector rotation and convergence. *SEG Technical Program Expanded Abstracts, 2010*, 1467–1472. https://doi.org/10.1190/1.3513118

Maultzsch, S., Chapman, M., Liu, E., & Li, X. (2007). Modelling and analysis of attenuation anisotropy in multi-azimuth VSP data. *Geophysical Prospecting, 55*(5), 627–642. https://doi.org/10.1111/j.1365-2478.2007.00645.x

Mitchell, T. M., & Faulkner, D. R. (2009). The nature and origin of off fault damage surrounding strike-slip fault zones with a wide range of displacements: A field study from the Atacama fault system, northern Chile. *Journal of Structural Geology, 31*, 802–816. https://doi.org/10.1016/j.jsg.2009.05.002

Nemeth, T., Wu, C., & Schuster, G. (1999). Least-squares migration of incomplete reflection data. *Geophysics, 64*, 208–221. https://doi.org/10.1190/1.1444517

Nes, O. M., Holt, R. M., & Fjær, E. (2000). The reliability of core data as input to seismic reservoir monitoring studies. *Society of Petroleum Engineers, 5*, 79–86. https://doi.org/10.2118/65180-MS

Rickett, J. (2003). Illumination-based normalization for wave-equation depth migration. *Geophysics, 68*, 1371–1379. https://doi.org/10.1190/1.1598130

Roberts, A. (2001). Curvature attributes and their application to 3D interpreted horizons. *First Break, 19*, 85–100.

Sagy, A., Reches, Z., & Roman, I. (2001). Dynamic fracturing: Field and experimental observations. *Journal of Structural Geology, 23*, 1223–1239. https://doi.org/10.1016/S0191-8141(00)00190-5

Savage, H. M., & Brodsky, E. E. (2011). Collateral damage: Evolution with displacement of fracture distribution and secondary fault strands in fault damage-zones. *Journal of Geophysical Research, 116*(B03405), 428–452. https://doi.org/10.1029/2010JB007665

Thompson, M. D., Grunow, A. M., & Ramezani, J. (2010). Cambro-Ordovician paleogeography of the Southeastern New England Avalon Zone: Implications for Gondwana breakup. *Geological Society of America Bulletin, 122*(1–2), 76–88. https://doi.org/10.1130/B26581.1

Chapter 10
Elastic Characteristics of Fault Damage Zones Within Superdeep Carbonates in Northwest China

Abstract This chapter investigates the internal architecture of fault damage zones using seismic attributes of curvatures based on 3D seismic reflection data from superdeep carbonate reservoirs in northwest China. Our results show that there are three typical fault damage models and their geometrical attributes including main directions and fracture density are characterized using statistical methods. We study seismic parameters (e.g., anisotropy and attenuation) relevant to elastic scaling down fault damage models under different frequencies using finite-element modeling. The equivalent stiffness tensors of the three models are significantly affected by frequency. Comparing the three models at the same seismic frequency, the stiffness coefficients are mainly controlled by their densities. We also discuss frequency-dependent seismic attenuation of the three fault damage models and found out that their maximum attenuation appeared at the same frequency, while the different values are affected by directions orientations of the impermeable fault cores. This work quantitatively appraises interpret the complex fault damage zones and highlights the importance and usefulness of combining the methods of seismic, rock physics, and geology to characterize their attributes.

10.1 Introduction

Fault zones usually consist of isolated or several fault cores surrounded by a fractured damage zone (Faulkner et al., 2003, 2006; Liao et al., 2019; Savage & Brodsky, 2011; Shipton & Cowie, 2003). It has three primary components, country rock, fault core, and damage zone (Caine et al., 1996; Sagy et al., 2001; Savag & Brodsky, 2011). Ordovician carbonate reservoirs over 7 km deep (~22,965 ft) are governed by small-displacement strike-slip faults in the Shuntuoguole low uplift of Tarim Basin, Northwest China (Qi, 2021). Multistage tectonic activities, associated with numerous karst corrosions and collapses owing to atmospheric freshwater seepage during the Ordovician period, generate strike-slip fault damage zones with numerous sub-faults, fractures, and small-scale caves (Deng et al., 2019; Pu et al., 2018; Qi, 2021). Such fault damage zones can act as hydrocarbon migration channels and fracture-cave

Z. Liao, *Applied Seismic Geostructures*, https://doi.org/10.1007/978-981-97-7079-3_10

reservoir spaces, which are limited from outcrop observations and have to resort to seismic attributes for characterization (Bravo & Aldana, 2010; Chopra & Marfurt, 2007, 2013; Marfurt & Rich, 2010).

Iacopini et al. (2012) apply seismic mapping for the recognition of reduced signals through seismic disturbance volumes to detect structural deformations. Particularly, seismic curvatures (e.g., Al-Dossary & Marfurt, 2006; Bravo & Aldana, 2010) should be efficiently calculated to define a suitable fault zone architecture. Botter et al., (2016, 2017) applied seismic attributes to characterize fault damage zones. Based on laboratory ultrasonic measurements, Ba et al. (2018) inversed P-wave attenuation for carbonate reservoirs, which provides insights for deep reservoirs prediction. Liao et al., (2019, 2020) used three-dimensional (3D) seismic attributes to map damage zones relevant to subsurface strike-slip faults in different areas, with a detailed analysis of damaged and large-scale splays. However, little effort has been made to characterize the elastic properties (e.g., velocity anisotropy and seismic attenuation) of fault damage zones.

Several theoretical models of fractured media (e.g., Chapman, 2003; Cheng, 1993; Hudson, 1990) have been developed to analytically calculate the equivalent elastic parameters of fractured rocks. However, most of the theoretical or semi-analytical models are confined to ideal conditions with an oversimplification of fracture properties. As an effective alternative, numerical simulations have been extensively applied for studying the elastic properties of complex rocks (e.g., Vlastos et al., 2003; Quintal et al., 2011; Zhao et al., 2015; Guo et al., 2017; Ba et al., 2017; Wei et al., 2021). Ba et al. (2017) proposed a double double-porosity model for rock an elasticity based on saturation and fabric heterogeneities, which affect wave propagation. Unlike conventional numerical methods, Rubino et al., (2009, 2016) proposed a two-dimensional (2D) numerical upscaling method with least-square optimization for frequency-dependent effective stiffness tensors. This method has been applied to fractured rocks (Cao et al., 2019) and extended to 3D cases (Jian et al., 2021). Here, we apply this approach to calculate the velocity anisotropy and seismic attenuation of fault zones.

On the basis of the stereotype of fault damage zones (Chen et al., 2020; Liao et al., 2020), three typical fault-zone patterns can be identified in the study area, including paratactic multiple cores, paratactic interactive cores, and one primary-four subsidiary cores. We implement an integrated study for these fault damage zones using seismic curvature attributes, velocity anisotropies, and seismic attenuations. In this study, seismic curvature attributes are used to map fault-zone structures in this study. Finally, seismic properties (velocity anisotropy and seismic attenuation) are numerically simulated and are applied for the elastic characteristics of fault zones.

10.2 Geological Setting

The Tarim Basin is a representative superimposed basin (Jia & Wei, 2002) in Northwest China. Shunbei (SHB) oil and gas field lies in the northwestern Shuntuoguole low uplift that acts as a transition zone between the Manjiaer and Awati depressions (Fig. 10.1). Its north and south are bounded by Shaya and Katak uplifts, respectively. The Shuntuoguole low uplift, an important oil and gas exploration area, is a comparatively stable paleotectonic unit in the Tarim Basin due to its weak fold deformation (Jiao, 2018; Qi, 2021). From Cambrian to mid-Ordovician, a north–south tectonic low-uplift zone has formed in the Shuntuoguole area, associated with strike-slip faults (Fig. 10.1) caused by the joint action of stress in mid-Caledonian. These strike-slip faults are steep and straight, accompanied by numerous sub-fractures with different scales and orientations. They connect source rocks and fault-karst carbonate reservoirs at the bottom of Cambrian (Deng et al., 2019; Jiao, 2018; Qi, 2021), and provide favorable channels for oil and gas migration and accumulation. Figure 10.2 shows that the mudstone and argillaceous limestone in the Upper Ordovician Sangtamu Formation acts as regional and local seals in the SHB reservoirs (Jiao, 2018; Qi, 2021), which allows the low uplift to have favorable geological conditions for great oil and gas fields.

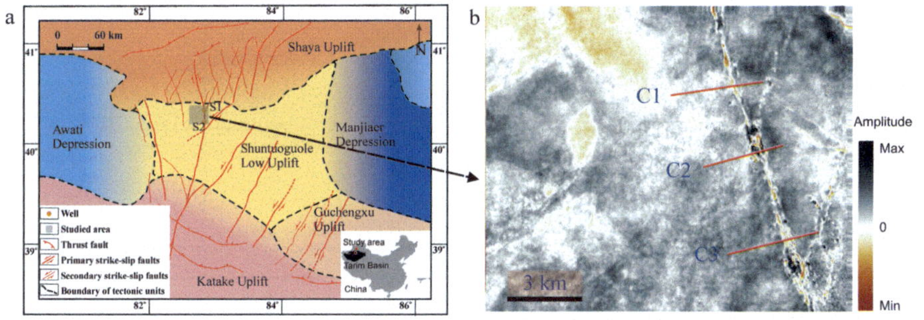

Fig. 10.1 Location of the SHB oil and gas field in Tarim Basin (left panel, modified from Jiao, 2018) with the background colors showing different tectonic units (uplifts, low uplift, and depressions) where numerous strike-slip faults are distributed. The study area is indicated by a gray square with an enlarged map of seismic amplitude slice (*right panel*) along the top of the mid-Ordovician Yijianfang Formation, where C1, C2, and C3 denote three seismic survey in-lines across the major strike-slip fault

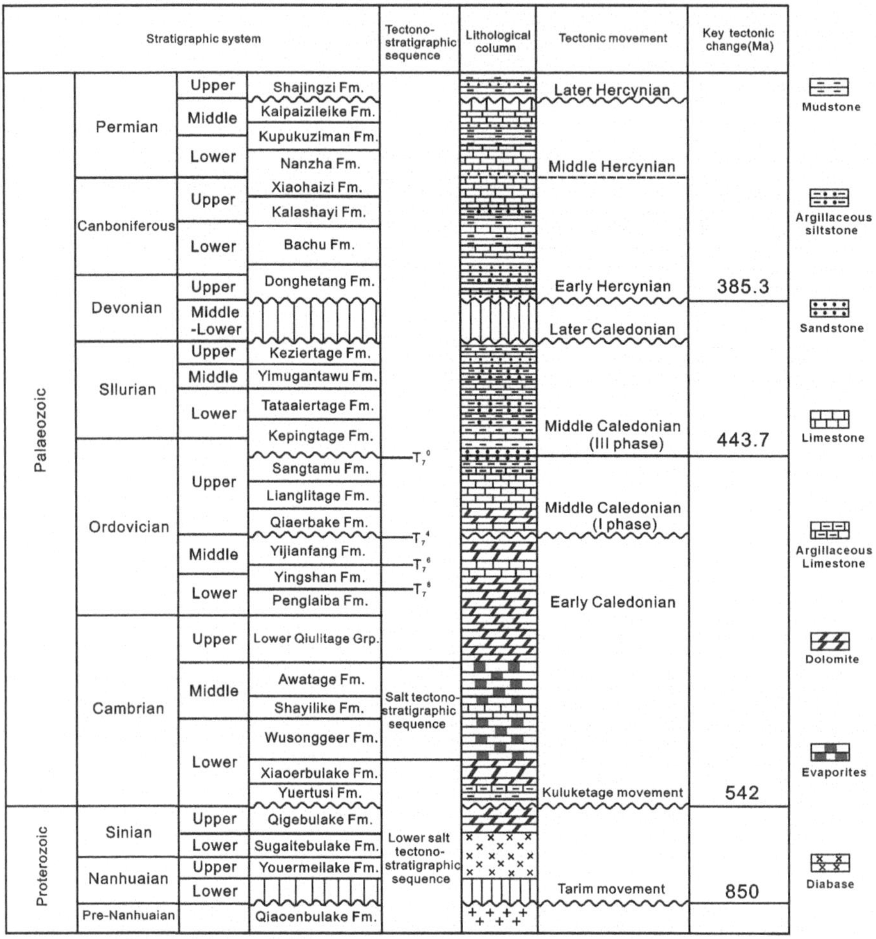

Fig. 10.2 Simplified stratigraphic section of Tarim Basin stratigraphy and the regional tectonic movement time (slightly changed from Tang et al., 2012)

10.3 Fault-Zone Patterns

10.3.1 Seismic Curvature Attributes for Mapping Fault Zones

3D seismic reflection data are available in the SHB area (gray square in Fig. 10.1). The data sampling rate is 1 ms and the common reflection element is 25 m × 25 m. The frequency ranges from 10 to 45 Hz. We calculated 3D seismic curvature attributes to detect faults and damage zones using seismic curvature methods (e.g., Al-Dossary &Marfurt, 2006; Rich, 2008; Bravo & Aldana, 2010; Roberts, 2001). Curvature in 2D is defined by the radius of a circle tangent to a curve (Roberts, 2001). In 3D, we must fit two circles tangent to a surface. Based on Robert's notation (2001), the fitting formula of the trend surface is written as follows:

$$z(x, y) = Ax^2 + By^2 + Cxy + Dx + Ey + G, \tag{10.1}$$

where A, B, C, D, E, G, characterizing the shape of surface, are the fitting coefficients. $z(x, y)$ is the height of the surface. Based on 3D seismic data, the fitting coefficients are calculated using the dip attribute. The seismic data can be regarded as a function of amplitude for spatial location and time sampling points. Thus, its gradient is written as follows (Al-Dossary &Marfurt, 2006):

$$\mathrm{div}(\mathbf{u}) = \frac{\partial u}{\partial x}\boldsymbol{i} + \frac{\partial u}{\partial y}\boldsymbol{j} + \frac{\partial u}{\partial t}\boldsymbol{k} = p\boldsymbol{i} + q\boldsymbol{j} + r\boldsymbol{k}, \tag{10.2}$$

where $\mathbf{u} = u(x, y, t)$ is the amplitude, p, q, and r are the components of dip and where $\boldsymbol{i}, \boldsymbol{j}$, and $\boldsymbol{k}$ are unit vectors along the x, y, and t directions, respectively.

The first derivative of Eq. (10.1) with respect to x and y is the gradient of Eq. (10.2). The fitting coefficients are given by,

$$\begin{aligned} A &= \frac{1}{2}\frac{\partial p}{\partial x}, B = \frac{1}{2}\frac{\partial q}{\partial y}, \\ C &= \frac{1}{2}\frac{\partial p}{\partial y} + \frac{1}{2}\frac{\partial q}{\partial x}, \\ G &= 0, D = p, E = q. \end{aligned} \tag{10.3}$$

Once we obtain the fitting coefficients, all kinds of curvature attributes are computed based on the Eq. (10.3). The calculation formula of the most-positive curvature K_{pos} and most-negative curvature K_{neg} are:

$$\begin{aligned} K_{pos} &= (A + B) + \sqrt{(A - B)^2 + C^2} \\ K_{neg} &= (A + B) - \sqrt{(A - B)^2 + C^2}, \end{aligned} \tag{10.4}$$

The shape index s which identifies small faults, is calculated using the following formula:

$$s = \frac{2}{\pi} * \tan^{-1}[(K_1 + K_2)/(K_1 - K_2)]. \tag{10.5}$$

where K_1 and K_2 are the maximum curvature and minimum curvature, respectively.

Figure 10.3 shows the most-positive and most-negative curvature anomalies around a geologically interesting fault, which provides an understanding of the delineation of faults by curvature attributes. The left panel in Fig. 10.4 shows seismic data profiles for three in-lines C1, C2, and C3 at seismic survey shown in Fig. 10.1. On the seismic profiles, the interfaces $T_7{}^4$ (red line) and $T_7{}^6$ (blue line) denote the

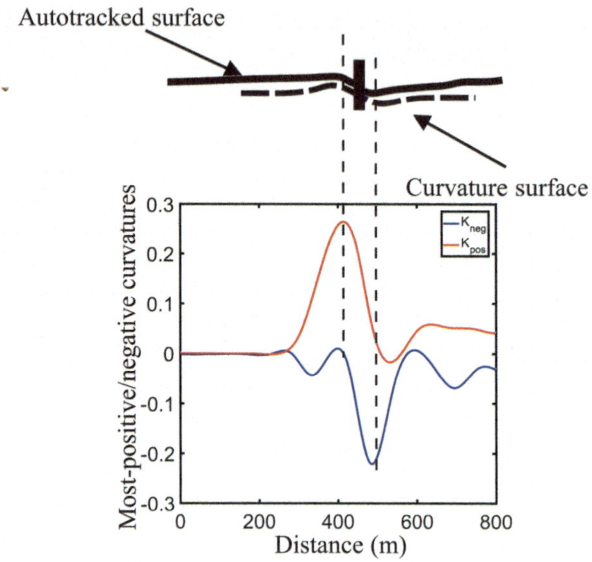

Fig. 10.3 Schematic of a fault delineation using the most-positive (*red line*) and most-negative (*blue line*) curvatures

Middle Ordovician Yijianfang Formation top and Yingshan Formation top, respectively. We observe that a major strike-slip fault (the right panel of Fig. 10.1) penetrates these two target layers. An obvious in-phase axis distortion and dislocation at the strike-slip fault is observed, and the amplitudes of the reservoir are weaker than that of the non-reservoir. The widths of fault damage zone are different, and the time thickness of target layers is ~128–158 ms. The right panel in Fig. 10.4 shows the corresponding overlays of seismic data, curvature attributes, and shape indexes. The pink and blue areas covered the picture in the right panel are the most-negative and most-positive curvature attributes after threshold segmentation, respectively. The most-negative and most-positive curvature attributes illustrate the major strike-slip fault with numerous fractures (marked by white dotted rectangles), which matches well with the corresponding discontinuities in the seismic profiles.

10.3.2 Models of Fault Damage Zones

Liao et al., (2019, 2020) defined the high curvature values as a fault core, the intermediate-level values as the damage zone, and the low curvature zone as the protolith. The most-positive and most-negative curvatures enable seismic signals to identify the discontinuities of sequence strata. Based on the 3D seismic curvatures, we follow Fig. 10.4 to draw the most-negative and most-positive curvature values across the major strike-slip fault of the target layers (T_7^4 and T_7^6) with an attempt to identify some fault-zone patterns. The curvature values above the background (green-dotted line) and below the background (celadon-dotted line) are regarded as

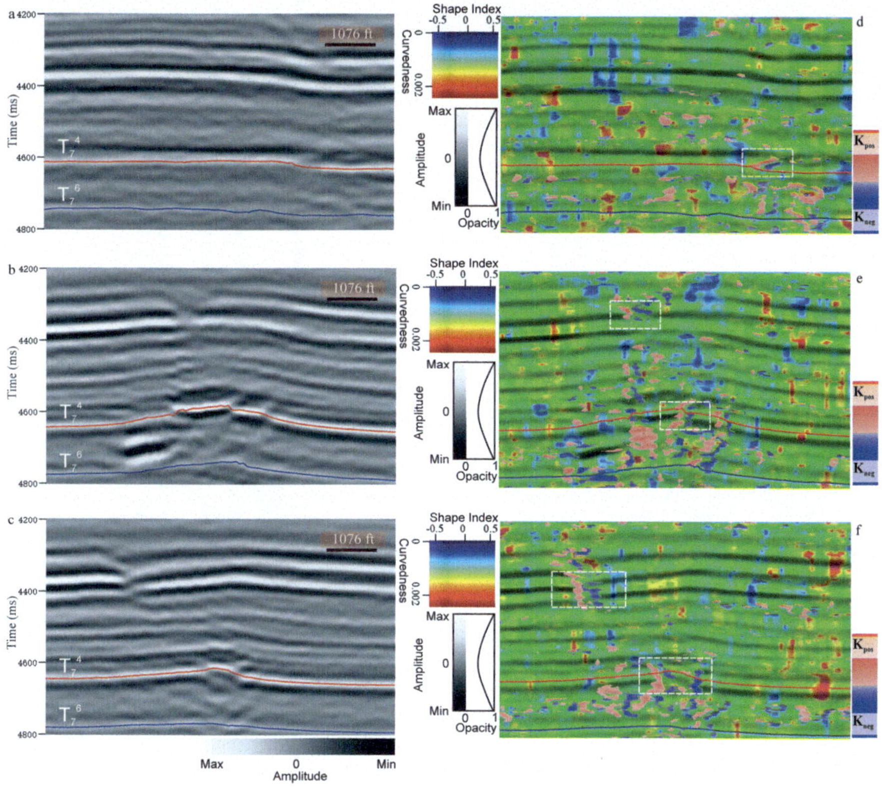

Fig. 10.4 Seismic data profiles (*left panel*) for three seismic survey in-lines C1, C2, and C3 shown in Fig. 10.1, and the corresponding overlays (*right panel*) of seismic data, curvature attributes, and shape indexes. The interfaces $T_7{}^4$ and $T_7{}^6$ denote the Middle Ordovician Yijianfang Formation top and Yingshan Formation top, respectively. The cherry-red and bright-blue areas indicate fault damage zones with numerous fractures (*marked by white dotted rectangles*) matching to the discontinuities in the seismic data profiles

fault damage zones in Fig. 10.5. We investigate various fault damage zones through the orientation of the major strike-slip fault as shown in Fig. 10.1. Three representative fault-zone patterns are summarized as shown in Fig. 10.5. (1) Paratactic multiple fault cores are shown in Fig. 10.5a for a fault damage zone of ~2100-m width, with several fault cores (large curvature values) parallel to each other. Such multiple fault cores can be simplified as an idealized fault-zone model (labeled Model A) shown in Fig. 10.5b. (2) Interactive fault cores are shown in Fig. 10.5c for a fault damage zone of ~1000-m width, with its idealized fault-zone model (labeled Model B) shown in Fig. 10.5d. (3) One primary fault core accompanied by several subsidiary cores (decreasing curvature values) is shown in Fig. 10.5e for a damage zone of ~1100-m width, with its idealized model (labeled Model C) shown in Fig. 10.5f. The background colors of the models on the right panel in the Fig. 10.5 qualitatively denote the damage degree of the fault damage zone.

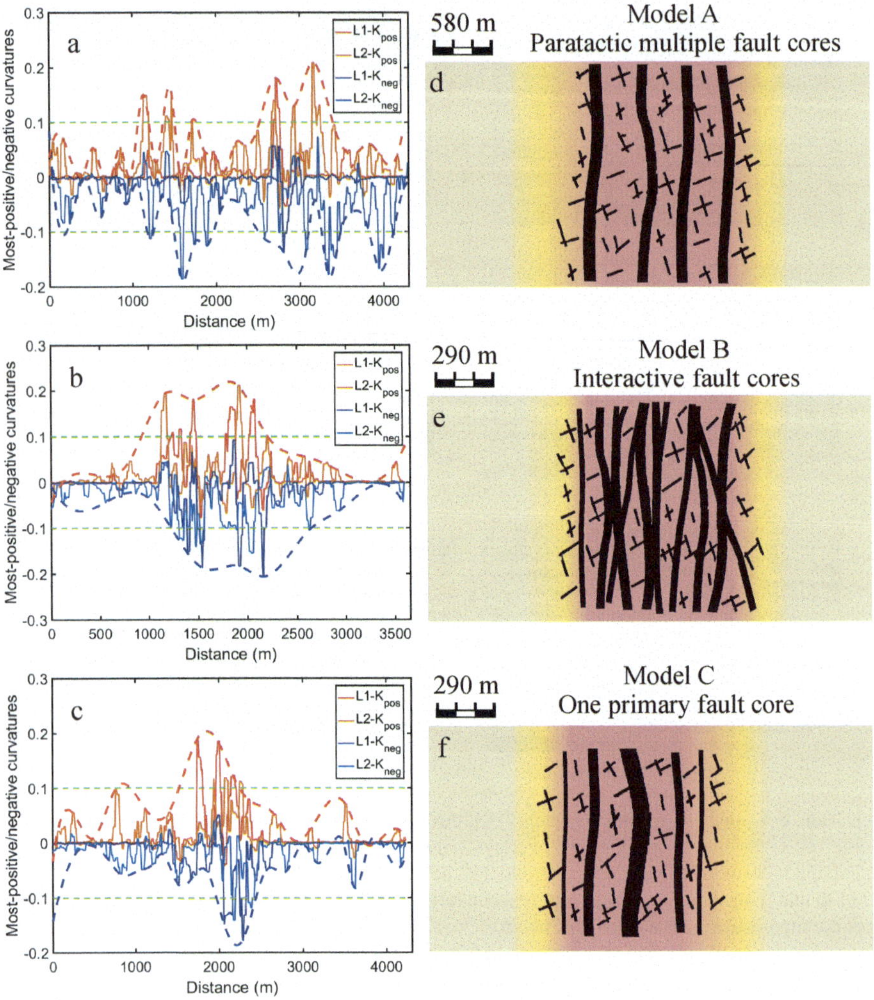

Fig. 10.5 The distribution of most-positive and most-negative curvature values across the major strike-slip fault of the target layers (L1 and L2, namely, T_7^4 and T_7^6 in Fig. 10.4). Curvature values above background (*green-dotted line*) and below background (*celadon-dotted line*) are regarded as fault damage zones. Three representative fault-zone patterns are summarized as **a** paratactic multiple cores with the idealized fault-zone model **b**, **c** paratactic interactive cores with the idealized fault-zone model (**d**), and **e** one primary core with several subsidiary cores with the idealized fault-zone model (**f**). Dashed lines (*red and blue*) in the distribution of curvature values indicate the generalized damage intensity curves

10.3.3 Geometrical Characteristics of the Fault Damage Models

We investigated geometrical attributes (density and dip-angle of the fractures) of the fault damage models built in Sect. 10.3.2. We applied the method proposed by Mauldon et al. (2001) and Healy et al. (2017) to characterize the fracture density, which uses a circular scanning window to estimate the number of total fractures by counting the number of endpoints of visible complete fractures within the window. Furthermore, we used an equal area rose diagram instead of the histogram of fracture length since it provides indirect orientation distribution (see Nemec (1988) for more details).

Figure 10.6 shows the geometric properties of three superdeep fault damage models with obvious differences in densities and orientations. We see that the fracture exhibits high angles in superdeep reservoirs. Figure 10.6a–c depict fracture density estimated of the models, where the red and yellow areas in the maps indicate more fractures than others, which agrees with the setting of the models. Figure 10.6d–f display the equal-area rose diagram of fracture length, indirectly providing orientation distribution, from which the estimated dips (nearly vertical) are consistent with those in the fractured models.

10.4 Elastic Characteristics of Fault Damage Zones

The size and orientation of fractures are generally multiscale and multidirectional, leading to complex distributions that have to resort to numerical methods for effective elastic properties. We study the velocity anisotropy and seismic attenuation of superdeep fault damage zones under different frequencies by applying a finite-element numerical upscaling procedure. For reducing computational cost, the width and length of a fault zone are downscaled equally to 0.143 m and 0.0726 m for satisfying the simulation parameters while keeping geometric parameters unchanged.

10.4.1 Biot's Equation for Fractured Rocks

Biot's consolidation system of equations in the space-frequency domain (Biot, 1962) is provided by.

$$\nabla \cdot \boldsymbol{\sigma} = 0, \tag{10.6}$$

$$\nabla P_f = -i\omega \frac{\eta}{\kappa} \mathbf{W}, \tag{10.7}$$

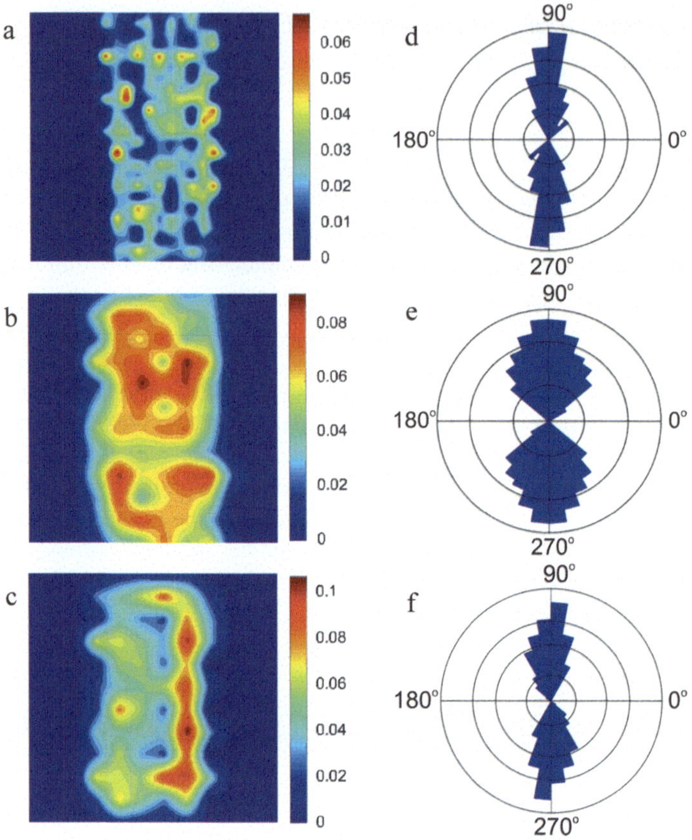

Fig. 10.6 Geometrical characteristics of three superdeep fault damage zone models in Fig. 10.5d–f. The fracture density distributions of the models are shown in **a**, **b**, and **c**, where the red and yellow areas in the maps indicate more fractures than others. The equal-area rose diagrams of fracture length, which provide dip distribution, are displayed in (**d**–**f**)

where ∇ depicts the Nabla operator for spatial derivatives, $\boldsymbol{\sigma} \equiv (\sigma_{ij})$ represents the total stress vector. P_f denotes the fluid pressure, κ is the permeability, and ω is the angular frequency. $\mathbf{W}$ denotes the average relative fluid–solid displacement tensor. η indicates the viscosity of the pore fluid. Thus, stress–strain relationships can be rewritten as

$$\sigma_{ij} = 2\mu e_{ij} + \delta_{ij}(\lambda_u \nabla \cdot \mathbf{U} - \alpha N \zeta), \tag{10.8}$$

$$P_f = -\alpha N \nabla \cdot \mathbf{U} + N\zeta, \tag{10.9}$$

$$e_{ij} = \frac{1}{2}\left(\frac{\partial u_i}{\partial j} + \frac{\partial u_j}{\partial i}\right) \tag{10.10}$$

where δ_{ij} denotes the Kronecker delta, and $\zeta = -\nabla \cdot \mathbf{W}$ depicts a measurement of local variations in the fluid content. $\mathbf{U}$ the solid displacement vector, e_{ij} denotes strain tensor and μ represents the shear modulus of the rock with saturated water, which is equivalent to that of the dry frame μ_m. Furthermore, the Biot constant α, fluid storage coefficient N, and Lamé parameter λ_u are expressed as

$$\lambda_u = K_m + \alpha^2 N - \frac{2}{3}\mu_m, \tag{10.11}$$

$$N = \left(\frac{\alpha - \phi}{K_s} + \frac{\phi}{K_f}\right)^{-1}, \tag{10.12}$$

$$\alpha = 1 - \frac{K_m}{K_s}, \tag{10.13}$$

where ϕ is the porosity. K_s denotes the bulk moduli of solid grains, K_m is the bulk moduli of dry frame, and K_f denotes the bulk moduli of fluid phase.

Substituting Eqs. (10.11) to (10.8) and taking Eq. (10.9), we obtain a new form of 2D total stress tensor in porous fractured medium, which is written as follows:

$$\sigma_{ij} = 2\mu e_{ij} + \delta_{ij}\left(\lambda \nabla \cdot \mathbf{U} - \alpha P_f\right), \tag{10.14}$$

where $\lambda = K_m - \frac{2}{3}\mu_m$. When $i, j = x, y$, it is the 2D case. Substituting $\zeta = -\nabla \cdot \mathbf{W}$ into Eq. (10.9), we have

$$P_f = -\alpha N \nabla \cdot \mathbf{U} - N \nabla \cdot \mathbf{W}, \tag{10.15}$$

Taking Eqs. (10.15) and (10.7), the P_f is given by

$$P_f = -\alpha N \nabla \cdot \mathbf{U} - N \nabla \cdot \left(\frac{ik}{\eta\omega} \nabla P_f\right). \tag{10.16}$$

In order to obtain the solid displacement fields, we do relaxation tests and solve Eqs. (10.6) and (10.16) with the corresponding boundary conditions applying finite-element method (FEM). After computing the displacement field, the stress and strain components of each element are computed using Eqs. (10.10) and (10.14). The corresponding average over the model's volume can be calculated using these parameters

$$\overline{e_{ij}^s} = \frac{1}{S} \int_D e_{ij}^s dS, \tag{10.17}$$

$$\overline{\sigma_{ij}^s} = \frac{1}{S} \int_D \sigma_{ij}^s dS, \tag{10.18}$$

where $s = 1$–3 denotes the sth simulation test described in appendix A, and S is the surface of the domain D.

Based on the prerequisite that an equivalent homogeneous anisotropic elastic solid can stand for average response of the interest model, the average stress and strain components can be related through the Voigt stiffness matrix $\mathbf{M} \equiv M_{ij}$. For general anisotropy and under plane strain conditions, this relation is given by

$$\begin{pmatrix} \overline{\sigma_{xx}^s} \\ \overline{\sigma_{yy}^s} \\ \overline{\sigma_{xy}^s} \end{pmatrix} \equiv \begin{pmatrix} M_{11} & M_{12} & M_{16} \\ M_{12} & M_{22} & M_{26} \\ M_{16} & M_{26} & M_{66} \end{pmatrix} \begin{pmatrix} \overline{e_{xx}^s} \\ \overline{e_{yy}^s} \\ 2\overline{e_{xy}^s} \end{pmatrix}, \tag{10.19}$$

where $s = 1$–3 depicts the sth simulation test, $\overline{\sigma_{ij}^s}$ and $\overline{e_{ij}^s}$ are cells of the averaging stress and strain vector, respectively. Note that the matrix $\mathbf{M}$ considers the greatest extent of anisotropy.

The relationship established based on Eq. (10.19) holds for the three simulation tests presented in appendix A. Thus, the six unknown components of the stiffness matrix possibly are calculated using a least-squares procedure, therefore, we consider the following cost function $F(\mathbf{M})$ according to Rubino et al. (2016).

$$\begin{aligned} F(\mathbf{M}) = \sum_{s=1}^{3} \Bigg(& \left\| \overline{\sigma_{xx}^s} - M_{11}\overline{e_{xx}^s} - M_{12}\overline{e_{yy}^s} - 2M_{16}\overline{e_{xy}^s} \right\|^2 \\ & + \left\| \overline{\sigma_{yy}^s} - M_{12}\overline{e_{xx}^s} - M_{22}\overline{e_{yy}^s} - 2M_{26}\overline{e_{xy}^s} \right\|^2 \\ & + \left\| \overline{\sigma_{xy}^s} - M_{16}\overline{e_{xx}^s} - M_{26}\overline{e_{yy}^s} - 2M_{66}\overline{e_{xy}^s} \right\|^2 \Bigg), \end{aligned} \tag{10.20}$$

where $\overline{\sigma_{ij}^s}$ and $\overline{e_{ij}^s}$ are the average stress and strain, respectively. The $s = 1$–3 depicts the sth simulation test. The components of the equivalent stiffness matrix M_{ij} can be obtained by minimizing the cost function $F(\mathbf{M})$ as follows

$$\frac{\partial F(\mathbf{M})}{\partial M_{ij}} = 0. \tag{10.21}$$

We can solve the linear system of six equations yielded by Eq. (10.21) to get the equivalent stiffness tensor M_{ij}. Please see appendix B for more details.

10.4.2 Velocity Dispersion and Seismic Attenuation of Fault Damage Zones

Equations (10.6) and (10.17) are solved based on the relaxation simulation tests using the FEM. We need to mesh a fault damage zone saturated with brine and discretize

Table 10.1 Calculation parameters for the 2D cracked models (slightly changed from Rubino et al., 2016)

Parameters	Notation	Matrix	2D fractures
Frame			
Shear modulus, (Pa)	μ or μ_m	31.0e9	0.02e9
Permeability, (m^2)	κ	1e−15	1e−10
Bulk modulus, (Pa)	K_m	26.0e9	0.04e9
Porosity	ϕ	0.1	0.8
Grain			
Density, (g/cm^3)	ρ_s	2.75	2.65
Bulk modulus, (Pa)	K_s	37.0e9	37.0e9
Brine			
Bulk moduli, (Pa)	K_f	2.25e9	
Density, (g/cm^3)	ρ_f	1.09	
Viscosity, (Pa.s)	η	0.01	

the relevant matrix equation. Table 10.1 shows the physical properties of the fluid parameters, fractures with saturated brine, and the embedding background matrix.

With the complex-valued coefficients M_{ij} computed by the FEM, we can obtain the attenuation and velocity of P waves as functions of frequency by the plane-wave analysis (Rubino et al., 2016) as follows,

$$\left(FR - T^2\right)k^4 - \omega^2\rho(F + R)k^2 + \omega^4\rho^2 = 0, \tag{10.22}$$

where

$$F = M_{11}\sin^2\theta + 2M_{16}\sin\theta\cos\theta + M_{66}\cos^2\theta, \tag{10.23}$$

$$R = 2M_{26}\sin\theta\cos\theta + M_{66}\sin^2\theta + M_{22}\cos^2\theta, \tag{10.24}$$

$$T = (M_{66} + M_{12})\sin\theta\cos\theta + M_{16}\sin^2\theta + M_{26}\cos^2\theta, \tag{10.25}$$

with k as the wavenumber and θ as the incidence angle.

We obtain the real solution with the negative imaginary part $k(\omega, \theta)$. Then, the corresponding P-wave phase velocities and inverse quality factors are provided as

$$V_p = \frac{\omega}{\Re(k(\omega, \theta))}, \tag{10.26}$$

$$Q_p^{-1} = \frac{\Im\left(k(\omega, \theta)^2\right)}{\Re\left(k(\omega, \theta)^2\right)}. \tag{10.27}$$

Table 10.2 The real parts of M_{11} and M_{12} for different frequencies resulting from three typical fault-zone patterns

Frequency (Hz)		0.01	1	10	100	1000
Re (M_{11}) (Pa)	Model A	1.13e10	1.20e10	1.37e10	1.53e10	1.55e10
	Model B	1.50e10	1.54e10	1.62e10	1.79e10	1.82e10
	Model C	1.71e10	1.73e10	1.78e10	1.95e10	1.96e10
Re (M_{12}) (Pa)	Model A	2.59e9	2.81e9	2.87e9	4.07e9	4.20e9
	Model B	4.49e9	4.61e9	4.82e9	5.61e9	5.83e9
	Model C	5.72e9	5.75e9	5.76e9	5.86e9	5.91e9

In this study, we set $\theta = -\pi/12$ because of vertically dipping fractures considered with strike-slip faults. We use the numerical procedure to simulate three typical fault-zone patterns shown in Fig. 10.5. The real parts of effective stiffness coefficient M_{11} for different frequencies, including 0.01 Hz, 1 Hz, 10 Hz, 100 Hz, and 1000 Hz, are shown in Table 10.2.

Based on these equivalent stiffness tensors, we can calculate the velocity and attenuation of P waves that could be employed to describe the elastic characteristics of fault damage zones. As we observe in Fig. 10.7a, the P-wave velocities increase with increasing frequencies, showing a similar tendency for these three fault-zone patterns. Strong variations occur in the seismic frequencies (0–100 Hz) because the difference between protolith and faults becomes smaller due to the stiffening of the fluid as the frequency increases (Milani et al., 2016). The P-wave velocities for the multiple-core model are lower than those of the other two models because of different fracture numbers occurring in these models. Figure 10.7b displays inverse P-wave quality factors of these three models for different frequencies. We see that the maximum attenuation occurs at 100 Hz for all models. The multiple-core model has a higher attenuation than that of the other two models because it has the largest percentage of brine-saturated fractures. When the incident angle of wave is fixed, Fig. 10.7 shows the obvious velocity dispersion and seismic attenuation of models, which are different. Moreover, Eqs. (10.22)–(10.25) include the elastic parameters, so they directly affect the velocity and seismic attenuation, which also is affected by the model. Specifically, the velocity and seismic attenuation is affected by the density, length and orientation of the cracks shown in the Fig. 10.6.

10.4.3 Anisotropic Velocity of Fault Damage Zones

We can analyze the fracture-induced anisotropy using these equivalent stiffness tensors. Based on Eq. (10.22), the incident-angle P-wave phase velocity is calculated by setting $\theta = 0°\sim360°$, $\omega = 10$ Hz. Figure 10.8 shows the strong anisotropic characteristic of P-wave phase velocity with incident angles of three fault-zone models. When the frequency and elastic parameters are determined, the velocity variation

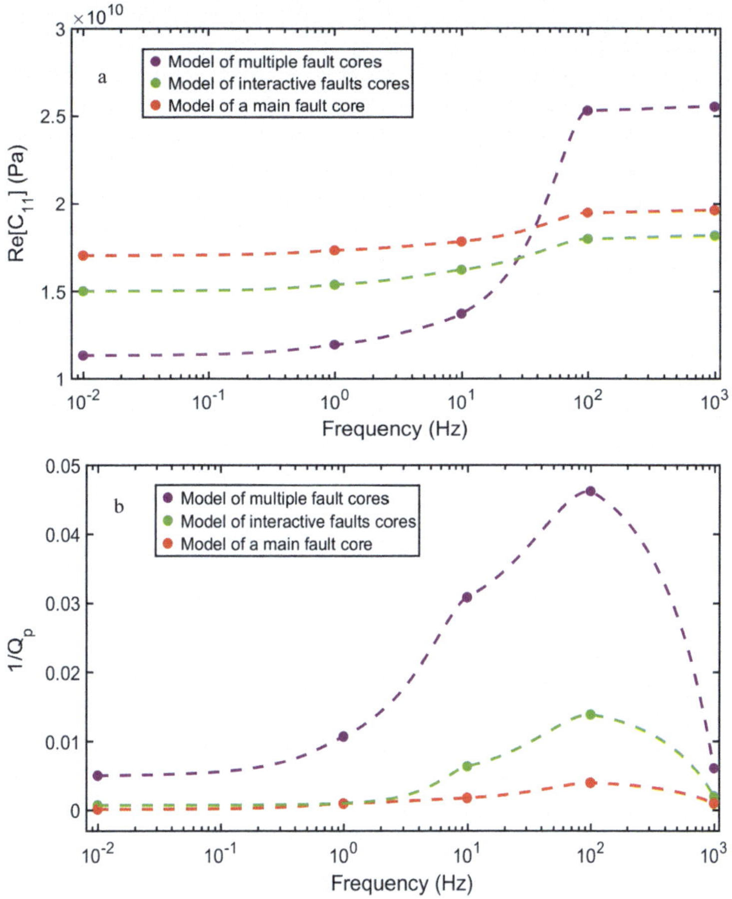

Fig. 10.7 Velocity dispersion and seismic attenuation of three fault damage zone models in Fig. 10.5. **a** Depicts the P-wave velocity as a function of frequency, and **b** displays the dependent-frequency seismic attenuation

shows a certain regularity based on Eq. (10.26), and it has a smooth curve. We observe that the maximum velocities of the three models appear at incident angles of ~40°, 95°, and 20°, respectively. The phase velocity for incident angle is symmetric, which minor and long symmetry axis shifts with different models. Figure 10.8 shows that the Model B has the highest anisotropy and density. The anisotropic characteristic of P-wave phase velocity is affected by the joint incidence of waves, fracture density and distribution. For better verification results, further investigation on elastic characteristics of the fault damage zone should also consider a comparison using laboratory measurements. Additionally, we need to consider that the maximum size of the model satisfies the simulation conditions if the model is very complex.

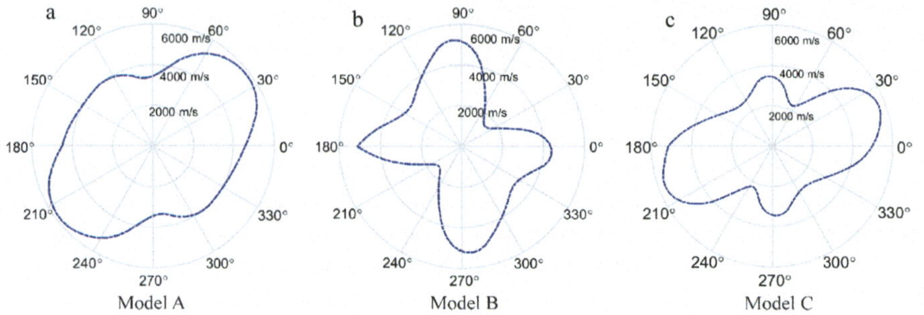

Fig. 10.8 Anisotropic characteristic of incident-angle P-wave phase velocity of three fault-zone models

10.5 Conclusions

The seismic attributes of curvatures of 3D seismic reflection data from superdeep carbonate reservoirs are considered useful for characterizing the internal architecture of fault damage zones in the Tarim Basin, Northwest China. We revealed three typical fault damage models, including paratactic multiple fault cores, interactive fault cores, and one primary-several subsidiary fault cores, which are investigated for their dip-angles and fracture density. To satisfy the simulation parameters, the width and length of models are equally downscaled, keeping geometric parameters unchanged.

We used the FEM to solve space-frequency Biot's equations for investigating seismic parameters (velocity anisotropy and attenuation) relevant to the elastic of fault damage models under different frequencies. The equivalent stiffness tensors of different fault damage models, used to characterize the fracture-caused anisotropy, are considerably affected by frequency. The P-wave velocities for the paratactic multiple cores model are lower than those of the other two models because fracture numbers are different for these models. Additionally, we study their frequency-dependent seismic attenuation. We observe the maximum attenuation of three fault damage models, which appeared at the same frequency, while the different values are affected by the distribution of fault cores. In addition, our result shows the strong anisotropic characteristic of P-wave phase velocities with the incident angle of three fault-zone models, which can be used to appraise these reservoirs with respect to the migration, filling, and accumulation of hydrocarbons. We have a better understanding of the elastic parameters of fault damage zones. Furthermore, we also conjectured that comprehensive analysis is an exceptional tool for superdeep carbonate reservoirs.

References

Al-Dossary, S., & Marfurt, K. J. (2006). 3D volumetric multispectral estimates of reflector curvature and rotation. *Geophysics, 71*, P41–P51. https://doi.org/10.1190/1.2242449

Ba, J., Xu, W., Fu, L. Y., Carcione, J. M., & Zhang, L. (2017). Rock anelasticity due to patchy saturation and fabric heterogeneity: A double double-porosity model of wave propagation. *Journal of Geophysical Research: Solid Earth, 122*(3), 1949–1976. https://doi.org/10.1002/2016JB013882

Ba, J., Zhang, L., Wang, D., Yuan, Z., Cheng, W., Ma, R., & Wu, C. (2018). Experimental analysis on P-wave attenuation in carbonate rocks and reservoir identification. *Journal of Seismic Exploration, 27*(4), 371–402.

Biot, M. A. (1962). Mechanics of deformation and acoustic propagation in porous media. *Journal of Applied Physics, 33*(4), 1482–1498. https://doi.org/10.1063/1.1728759

Botter, C., Cardozo, N., Qu, D., Tveranger, J., & Kolyukhin, D. (2017). Seismic characterization of fault facies models. *Interpretation, 5*(4), SP9–SP26. https://doi.org/10.1190/INT-2016-0226.1

Botter, C., Cardozo, N., Hardy, S., Lecomte, I., Paton, G., & Escalona, A. (2016). Seismic characterisation of fault damage in 3D using mechanical and seismic modelling. *Marine and Petroleum Geology, 77*, 973–990. https://doi.org/10.1016/j.marpetgeo.2016.08.002

Bravo, L., & Aldana, M. (2010). Volume curvature attributes to identify subtle faults and fractures in carbonate reservoirs: Cimarrona Formation, Middle Magdalena Valley Basin, Colombia. In *SEG Technical Program Expanded*

Caine, J. S., Evans, J. P., & Forster, C. B. (1996). Fault zone architecture and permeability structure. *Geology, 24*(11), 1025–1028.

Cao, C., Fu, L. Y., Ba, J., & Zhang, Y. (2019). Frequency-and incident-angle-dependent P-wave properties influenced by dynamic stress interactions in fractured porous media. *Geophysics, 84*(5), MR173–MR184. https://doi.org/10.1190/geo2018-0103.1

Chapman, M. (2003). Frequency-dependent anisotropy due to meso-scale fractures in the presence of equant porosity. *Geophysical Prospecting, 51*(5), 369–379. https://doi.org/10.1046/j.1365-2478.2003.00384.x

Chen, Y., Liao, Z., Fu, L. Y., Zhou, G., Xu, L., Marfurt, K. J., Mu, X., & Zou, H. (2020). Effect of main frequencies on characterizing fault damage zones using forward modeling and attribute of variance. *Interpretation, 8*(4), SP157–SP165. https://doi.org/10.1190/INT-2020-0017.1

Cheng, C. H. (1993). Crack models for a transversely isotropic medium. *Journal of Geophysical Research: Solid Earth, 98*(B1), 675–684. https://doi.org/10.1029/92JB02118

Chopra, S., & Marfurt, K. J. (2007). *Seismic attributes for prospect identification and reservoir characterization.* Society of Exploration Geophysicists and European Association of Geoscientists and Engineers

Chopra, S., & Marfurt, K. J (2013) Volumetric estimates of seismic reflector rotation and convergence—Tools for map** rotation about faults and seismic stratigraphy. *The Leading Edge, 32*(4), 402–408. https://doi.org/10.1190/tle32040402.1

Deng, S., Li, H., Zhang, Z., Zhang, J., & Yang, X. (2019). Structural characterization of intracratonic strike-slip faults in the central Tarim Basin. *Aapg Bulletin, 103*(1), 109–137. https://doi.org/10.1306/06071817354

Faulkner, D. R., Lewis, A. C., & Rutter, E. H. (2003). On the internal structure and mechanics of large strike-slip fault zones: Field observations of the Carboneras fault in southeastern Spain. *Tectonophysics, 367*(3–4), 235–251.

Faulkner, D. R., Mitchell, T. M., Healy, D., & Heap, M. J. (2006). Slip on'weak'faults by the rotation of regional stress in the fracture damage zone. *Nature, 444*(7121), 922–925.

Guo, J., Rubino, J. G., Glubokovskikh, S., & Gurevich, B. (2017). Effects of fracture intersections on seismic dispersion: theoretical predictions versus numerical simulations. *Geophysical Prospecting, 65*(5), 1264–1276. https://doi.org/10.1111/1365-2478.12474

Healy, D., Rizzo, R. E., Cornwell, D. G., Farrell, N. J., Watkins, H., Timms, N. E., Gomez-Rivas, E., & Smith, M. (2017). FracPaQ: A MATLAB™ toolbox for the quantification of fracture patterns. *Journal of Structural Geology, 95*, 1–16.

Hudson, J. A. (1990). Overall elastic properties of isotropic materials with arbitrary distribution of circular cracks. *Geophysical Journal International, 102*(2), 465–469. https://doi.org/10.1111/j.1365-246X.1990.tb04478.x

Iacopini, D., Butler, R. W. H., & Purves, S. (2012). Seismic imaging of thrust faults and structural damage: A visualization workflow for deepwater thrust belts. *First Break, 30*(5).

Jia, C., & Wei, G. (2002). Structural characteristics and petroliferous features of Tarim Basin. *Chinese Science Bulletin, 47*, 1–11.

Jian, S., Fu, L. Y., Cao, C., Han, T., & Du, Q. (2021). 3D finite-element modeling of effective elastic properties for fracture density and multiscale natural fractures. *Journal of Geophysics and Engineering, 18*(4), 567–582.

Jiao, F. (2018). Significance and prospect of ultra-deep carbonate faultkarst reservoirs in Shunbei area, Tarim Basin. *Oil & Gas Geology, 39*(2): 207–216. In Chinese with English abstract.

Liao, Z. H., Liu, H., Carpenter, B. M., Marfurt, K. J., & Reches, Z. (2019). Analysis of fault damage zones using 3D seismic coherence in the Anadarko Basin, Oklahoma. *AAPG Bulletin, 103*(8), 1771–1785. https://doi.org/10.1306/1219181413417207

Liao, Z., Li, W., Zou, H., Hao, F., Marfurt, K. J., & Reches, Z. E. (2020). Composite damage zones in the subsurface. *Geophysical Journal International, 222*(1), 225–230. https://doi.org/10.1093/gji/ggaa158

Marfurt, K. J., & Rich, J. (2010). Beyond curvature: Volumetric estimates of reflector rotation and convergence. In *80th Annual international meeting, SEG, expanded abstracts* (pp. 1467–1472)

Mauldon, M., Dunne, W. M., & Rohrbaugh, M. B., Jr. (2001). Circular scanlines and circular windows: New tools for characterizing the geometry of fracture traces. *Journal of Structural Geology, 23*(2–3), 247–258.

Milani, M., Rubino, J. G., Müller, T. M., Quintal, B., Caspari, E., & Holliger, K. (2016). Representative elementary volumes for evaluating effective seismic properties of heterogeneous poroelastic media. *Geophysics, 81*(2), D169–D181. https://doi.org/10.1190/geo2015-0173.1

Nemec, W. (1988). The shape of the rose. *Sedimentary Geology, 59*(1–2), 149–152.

Pu, R. H., Li, K. B., Dong, M. C., Cao, Z. C., & Xu, P. Y. (2018). The 3-D seismic characteristics and significance of the strike-slip fault in the Tadong Area (Tarim Basin, China). *Interpretation, 7*, T1–T19.

Qi, L. (2021). Structural characteristics and storage control function of the Shun I fault zone in the Shunbei region, Tarim Basin. *Journal of Petroleum Science and Engineering, 203*, 108653.

Quintal, B., Steeb, H., Frehner, M., & Schmalholz, S. M. (2011). Quasi-static finite element modeling of seismic attenuation and dispersion due to wave-induced fluid flow in poroelastic media. *Journal of Geophysical Research: Solid Earth, 116*(B1). https://doi.org/10.1029/2010JB007475

Rich, J. (2008). Expanding the applicability of curvature attributes through clarification of ambiguities in derivation and terminology. *SEG Technical Program Expanded Abstracts, 27*(1), 3713.

Roberts, A. (2001). Curvature attributes and their application to 3D interpreted horizons. *First Break, 19*(2)

Rubino, J. G., Caspari, E., Müller, T. M., Milani, M., Barbosa, N. D., & Holliger, K. (2016). Numerical upscaling in 2-D heterogeneous poroelastic rocks: Anisotropic attenuation and dispersion of seismic waves. *Journal of Geophysical Research: Solid Earth, 121*(9), 6698–6721. https://doi.org/10.1002/2016JB013165

Rubino, J. G., Ravazzoli, C. L., & Santos, J. E. (2009). Equivalent viscoelastic solids for heterogeneous fluid-saturated porous rocks. *Geophysics, 74*(1), N1–N13. https://doi.org/10.1190/1.3008544

Sagy, A., Reches, Z. E., & Roman, I. (2001). Dynamic fracturing: Field and experimental observations. *Journal of Structural Geology, 23*(8), 1223–1239. https://doi.org/10.1016/S0191-8141(00)00190-5

Savage, H. M., & Brodsky, E. E. (2011). Collateral damage: Evolution with displacement of fracture distribution and secondary fault strands in fault damage zones. *Journal of Geophysical Research: Solid Earth, 116*(B3). https://doi.org/10.1029/2010JB007665

Shipton, Z. K., & Cowie, P. A. (2003). A conceptual model for the origin of fault damage zone structures in high-porosity sandstone. *Journal of Structural Geology, 25*(3), 333–344.

Tang, L., Huang, T., Qiu, H., Qi, L., Yang, Y., Xie, D., Yu, Y., Zhao, Z., & Chen, S. (2012). Salt-related structure and deformation mechanism of the Middle-Lower Cambrian in the middle-west parts of the Central Uplift and adjacent areas of the Tarim Basin. *Science China Earth Sciences, 55*, 1123–1133. https://doi.org/10.1007/s11430-012-4414-3

Vlastos, S., Liu, E., Main, I. G., & Li, X. Y. (2003). Numerical simulation of wave propagation in media with discrete distributions of fractures: effects of fracture sizes and spatial distributions. *Geophysical Journal International, 152*(3), 649–668. https://doi.org/10.1046/j.1365-246X.2003.01876.x

Wei, Y., Ba, J., Carcione, J. M., Fu, L. Y., Pang, M., & Qi, H. (2021). Temperature, differential pressure, and porosity inversion for ultradeep carbonate reservoirs based on 3D rock-physics templates. *Geophysics, 86*, M77–M89.

Zhao, L., Yao, Q., Han, D. H., Yan, F., & Nasser, M. (2015). Characterizing the effect of elastic interactions on the effective elastic properties of porous, cracked rocks. *Geophysical Prospecting, 64*(1), 157–169. https://doi.org/10.1111/1365-2478.12243

Chapter 11
Artificial Intelligence Model for 3D Seismic Fault Analysis

Abstract Horizon extraction and fault detection are essential in seismic interpretation and closely related to each other. Most existing methods tend to deal with these two tasks independently, and may not work well in interpreting seismic images with complex geologic structures. We propose a multi-task learning (MTL) network with two branches to extract all horizons and detect faults simultaneously by estimating a relative geologic time (RGT) map as well as computing a fault map. These two branches share training datasets, feature maps, and network parameters during the training. The RGT estimation branch, constructed with a transformer architecture, is more lightweight compared to previous CNN methods but provides a larger and structure-oriented receptive field to adaptively capture global structural information for estimating a globally optimal RGT map. The fault detection branch is a simple convolutional neural network (CNN) which merges feature maps shared by the transformer and the derivatives of the estimated RGT to compute a fault map. The fault detection branch provides boundary control for the RGT estimation branch while the later provides global constraints for the former to improve its robustness to noise. Note that our RGT estimation by globally fitting all structures in a seismic image is a volumetric horizon interpretation method with which we are able to obtain a whole volume of horizons, all at once, by simply extracting contours of the RGT map. In our method, we further enable convenient human interactions by integrating manually interpreted horizons (or horizon segments) into the network, which imposes expert knowledge on the network to estimate reasonable RGT results from seismic images with complex fault systems, unconformities, and poor data quality. Moreover, when using 3D horizons as constraints, we are able to decompose the computationaly expensive 3D RGT estimation from a seismic volume into independently parallel 2D estimations slice by slice and combine them to obtain a laterally consistent 3D result.

Xinming Wu, School of Earth and Space Sciences, University of Science and Technology of China, Hefei, China.

Z. Liao, *Applied Seismic Geostructures*, https://doi.org/10.1007/978-981-97-7079-3_11

11.1 Introduction

Extracting horizons and detecting faults from a seismic image are two important tasks of seismic interpretation. Horizons can be used for ancient sedimentary environment analysis (Posamentier et al., 2007), structural modeling (Patel & McMechan, 2003), and stratigraphic interpretation (Vail et al., 1977). Faults can be used for structural modeling (Caumon et al., 2009), well location screening (Rivenæs et al., 2005), and tectonic analysis (Kusznir & Karner, 2007). The discontinuities of reflections are typical features of faults in a seismic image. Therefore, different methods are proposed for fault detection to highlight the reflection discontinuities in a seismic image by computing additional attributes such as semblance (Marfurt et al., 1998), coherency (Lou et al., 2019, 2021a, 2021b; Marfurt et al., 1999), variance (Van Bemmel & Pepper, 2000), curvature (Al-Dossary & Marfurt, 2006), and fault likelihood (Hale, 2013; Wu & Hale, 2016). However, these methods based on reflection discontinuities are sensitive to noise and other structures similar to faults in a seismic image. Therefore, some methods are proposed to further enhance fault-related features and suppress the noise in the attribute images, such as ant tracking (Pedersen et al., 2003) and optimal surface voting (Wu & Fomel, 2018a, 2018b). Some other authors (Di et al., 2020) proposed to improve fault detection by using prior horizon information. Fault detection is also important for horizon interpretation. When we interpret the horizons on both sides of faults, the fault detection maps can provide boundary constraints.

Interpreters interpret horizons by tracking reflections in a seismic image. Because manual interpretation is time-consuming, numerous automatic and semi-automatic methods are proposed for improving the interpretation efficiency. These horizon extraction methods are mostly based on seismic instantaneous phase unwrapping (Stark, 2003; Wu & Zhong, 2012), seismic waveform classification (Figueiredo et al., 2007, 2014, 2015), seismic local slopes (Arias, 2016; Bakker et al., 1999; Fomel, 2002; Lou et al., 2021a, 2021b; Marfurt, 2006; Wu & Fomel, 2018a, 2018b; Wu & Janson, 2017; Yu et al., 2013). These methods can extract horizon correctly along continuous reflections in the seismic image, but may not work well in tracking the reflections across faults. To solve this problem, some methods (Luo & Hale, 2013; Wu et al., 2016) remove faults from a seismic image by computing fault locations and fault displacement vectors. Some authors (Wu & Hale, 2015) use manually picked points on both sides of faults to constrain the process of horizon extraction.

Estimating an RGT volume or map from a seismic image provides a volumetric method of extracting all horizons in the seismic image (Qayyum et al., 2017; Stark, 2003, 2004; Wu & Zhong, 2012). In the RGT volume, every point is assigned an RGT value, and RGT values will increase with depth in general. The contours extracted from an RGT volume represent seismic horizons, and faults and unconformities are highlighted by discontinuities in the RGT volume. An RGT map can be also applied in sedimentologic interpretation (Zeng et al., 2012), geologic body detection (Zeng et al., 2012), structural implicit modeling (Grose et al., 2017), and missing well-log data prediction (Bader et al., 2018).

For RGT estimation, a straightforward but labouring method is to track or pick horizons, and interpolate these horizon surfaces to obtain an RGT volume (Bruin et al., 2007; Zeng et al., 1998). Although the estimated RGTs can strictly fit the interpreted horizon surfaces, the RGT surfaces interpolated among these horizons are hardly estimated well. To solve this problem, the method based on phase-unwrapping (Stark, 2003; Wu & Zhong, 2012) are proposed to estimate RGT by unwrapping the seismic instantaneous phase. Fomel (2010) uses a slope-based method to pick horizons and estimate RGT maps by using plane-wave destruction and setting seed points to further follow the local slopes.

In recent years, CNN-based methods have successfully applied in seismic data processing (Yu et al., 2019; Zu et al., 2020), seismic interpretation (Gao et al., 2021; Shi et al., 2019; Wu et al., 2019), and seismic inversion (Geng et al., 2022; Yang & Ma, 2019). Some methods based on CNN are proposed to estimate RGT maps from seismic images, and obtain the structural information (e.g., faults and horizons) from the predicted RGT images (Bi et al., 2021; Geng et al., 2019a, 2019b). Geng et al., (2019a, 2019b) use a 2D end-to-end CNN for RGT estimation from seismic images. From the predicted RGT map, all horizons in the seismic image can be extracted efficiently by extracting RGT contours. Bi et al. (2021) consider RGT estimation and fault detection are related to each other, but deal with these two tasks separately. They use one 3D end-to-end CNN to compute an RGT volume from a 3D seismic volume and use another 3D end-to-end CNN to detect faults from the predicted RGT volume. Although numerous methods have been proposed for both tasks of seismic horizon and fault interpretation, most of the existing methods deal with them as two independent tasks. In addition, most of the existing methods may not be able to robustly estimate reasonable results from seismic images with complex fault systems, unconformities, and poor data quality. Therefore, it is still necessary to incorporate prior structural information into the horizon and fault interpretation methods, especially the deep-learning-based ones, to improve their robustness. Vision transformer (ViT) (Vaswani et al., 2017) is the first work to use self-attention in computer vision, and it achieved state-of-the-art performance in image classification. Unlike the traditional CNN architecture, the transformer architecture excels in capturing global features. Now, transformer architecture is obtaining greater attention as a viable block used in geophysics, such as earthquake detection (Mousavi et al., 2020; Saad et al., 2022), deblending (Zu et al., 2022), and reservoir parameter prediction (Yang et al., 2023).

In this paper, we propose an MLT network (Fig. 11.1) with the constraints of prior structural information to simultaneously compute an RGT map and a fault image from a seismic image as well as interpreted horizons. The MTL network consists of two coupled branches of RGT estimation and fault detection. We construct the first branch of RGT estimation with an encoder-decoder architecture. The encoder uses a transformer backbone to capture global structural information. The decoder uses a CNN backbone to extract local structural features and employs Selective Feature Fusion (SFF) modules to merge both global (concatenated from the encoder) and local information for the RGT estimation. Compared to the previous CNN method (Geng et al., 2019a, 2019b), our RGT estimation branch is more lightweight but provides an adaptively anisotropic receptive field to help our network build RGT maps based on the structural information in seismic images. In this branch, the inputs include a seismic image and a map containing interpreted horizon segments with pre-assigned RGT values, and the output is an estimated RGT map. The interpreted horizon segments with RGT values are used as one of the inputs and as a constraint in loss functions, which helps guide our network to estimate reasonable RGT results, especially in complicated cases. Our method allows to input the prior constraints of incomplete horizons or even sparse horizon points that may be only interpreted near the complex structure areas (e.g., both sides of faults, unconformities and geobody boundaries), without the need of using complete horizons as constraints. In this way, we can build a convenient interaction between interpreters and our network, where interpreters can modify the input horizon segments in real time to update RGT and fault estimations until reasonable results are obtained. Moreover, when using 3D horizons as constraints, we are able to estimate 2D RGT maps slice by slice in a 3D seismic volume and further merge them to obtain a full 3D RGT result with low computational time and memory. The contours extracted from an RGT volume represent seismic horizons, and we can obtain a whole volume of horizons from the result of the RGT estimation branch.

The second branch of fault detection is a simple end-to-end CNN architecture whose inputs consists of the derivatives of the RGT volume and feature maps shared from the encoder of the RGT estimation branch. With shared feature maps and network parameters, these two branches can provide controls for each other. In the process of forward propagation during the training, the first branch can transfer feature maps with sufficient seismic structural information and obvious fault structures to the second branch. These feature maps, shared by the first branch of RGT estimation, help improve the robustness of the fault detection branch to obtain more clear and continuous fault maps. In the process of backward propagation during the

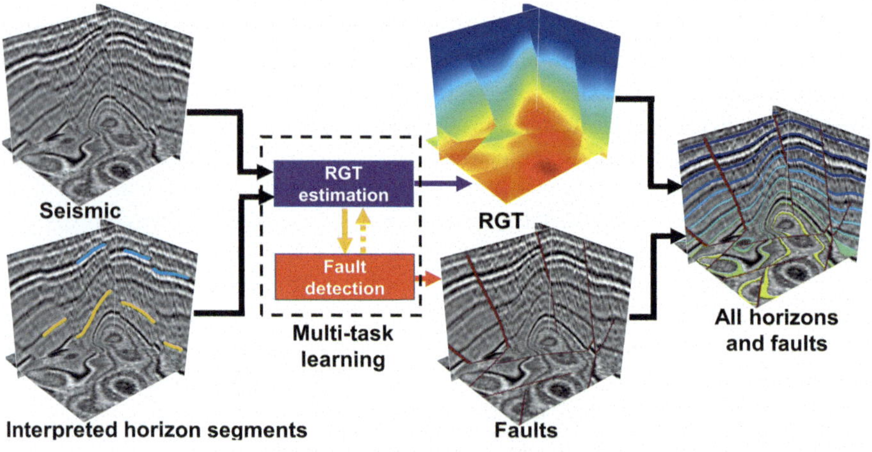

Fig. 11.1 We propose a multi-task learning network to simultaneously extract all horizons and faults from a seismic image and interpreted horizon segments. The network includes RGT estimation and fault detection branches, which are related to each other

training, the loss in the fault detection branch can provide boundary constraints for the first branch of RGT estimation to obtain sharper results near faults.

In summary, our contributions in this paper are as follows:

- We propose an MTL learning method for RGT estimation, horizon extraction, and fault detection, in which we use a lightweight transformer architecture to capture the global structure patterns in a seismic image.
- In addition to seismic data, we propose to add prior horizon constraints to guide the network to obtain reasonable RGT and fault images in complicated cases.
- In order to implement the 3D RGT estimation task accurately and efficiently, we decompose the 3D problem into independently parallel 2D estimations slice by slice where we use 3D horizons as constraints to maintain the lateral consistency of the independent 2D estimations.

11.2 Framework of Multi-task Learning

We use an MTL network (Fig. 11.2) with a transformer backbone and a simple end-to-end CNN to compute RGT and fault maps from a seismic image and horizon segments. To integrate expert knowledge into our network, we use horizon segments with pre-assigned RGT values as the input of our network and the constraints of loss functions in the RGT estimation branch. We use the Dice loss to solve the highly imbalanced distribution of samples for fault detection. Finally, we quantitatively evaluate the estimated results by comparing them with manual interpretation.

11.2.1 RGT Estimation

Figure 11.3 shows more detail of our MTL network architecture. The branch of RGT estimation is modified from a network with transformer blocks (Yang et al., 2023) named Global–Local Path Networks (GLP). The inputs of this branch are a single seismic image and horizon segments, and the output is an RGT map. The inputs are embedded as patches that are fed into the transformer block. The transformer block includes multiple sets of an efficient Self-Attention module, the feedforward network (FFN), and a patch merging layer. The Self-Attention module can capture vertical and horizontal long-range structural information, and we consider this character is beneficial for RGT estimation that requires globally fitting structural features in a seismic image. The vertical direction features in a seismic image represent the relative relations of layers, and the horizontal direction features in a seismic image represent the continuity of each layer. The FFN module provides positional information for the transformer block. Because of the patch merging layer, the encoder part can generate multi-scale features, which are also used in the decoder. Through four transformer blocks, we can obtain a multi-scale feature map with a size of multi-scale feature map with a size of $\frac{H}{2^{i+1}} \times \frac{W}{2^{i+1}} \times C_i (i \in \{1, 2, 3, 4\})$.

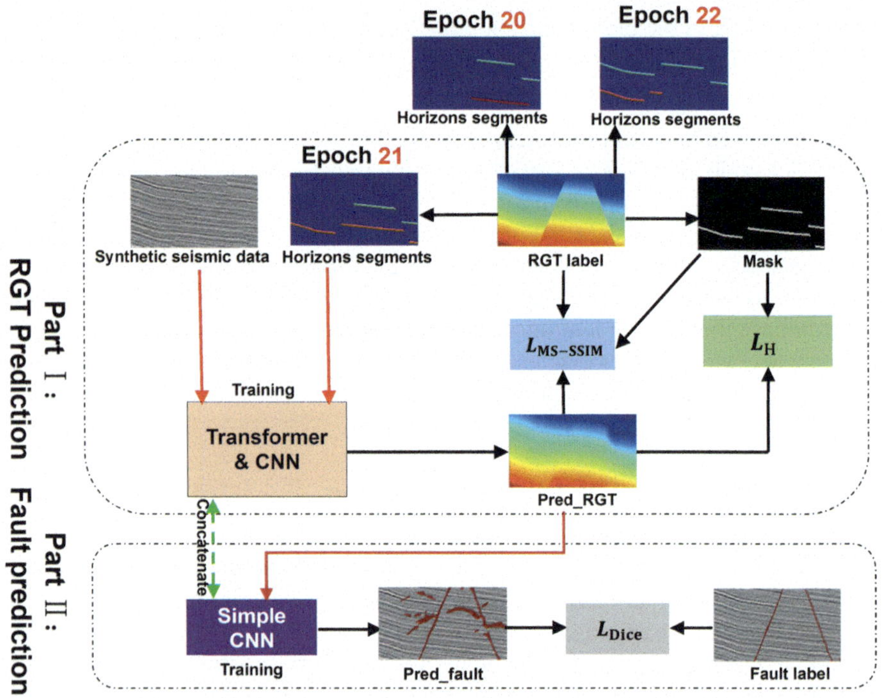

Fig. 11.2 The workflow of training our MTL model. In every epoch, the horizon segments and the corresponding mask are randomly selected from the RGT label, which can improve the diversity of training data. The RGT estimation network includes a transformer backbone and CNN, and the fault detection network is a simple CNN

In the lightweight decoder, we aim to obtain meaningful local information, and integrate it with global information from the encoder with skip connections. The channels of input features from the encoder are reduced into 64 with a 1×1 convolution filter to decrease the computational complexity. Next, we use an upsampling layer to enlarge the size of the feature map from $\frac{H}{32} \times \frac{W}{32} \times 64$ to $\frac{H}{16} \times \frac{W}{16} \times 64$. Then, we use an SFF with an attention mechanism to select and fuse global and local features.

The structure of SFF is shown in Fig. 11.4, where first use 1×1 convolution filter to reduce the dimensions of global features to 64, which is equal to the dimensions of local features. Then, these two-part features are concatenated with the size of $\frac{H}{16} \times \frac{W}{16} \times 128$. Next, two convolution modules are used to integrate information from these features. After the information integration, we obtain two attention maps from a convolution layer and a sigmoid layer. The two maps represent what the local features and global features are focusing on. In the attention map of global features (Fig. 11.4b), the areas of horizons and faults appear as larger values in the attention maps of global and local features, respectively. This is consistent with our expectation that horizons represent global relationships and faults represent detailed information.

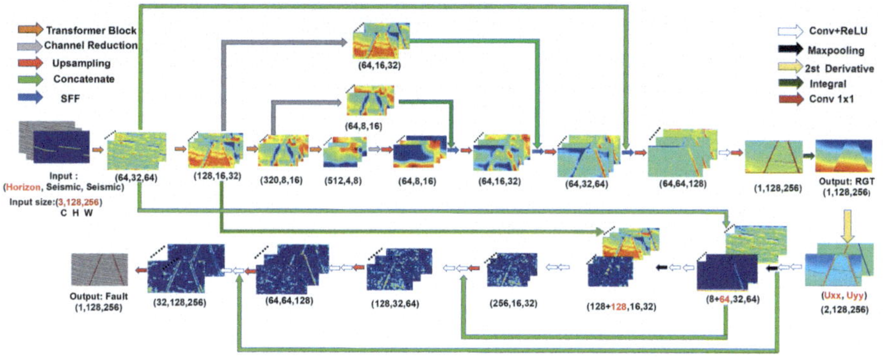

Fig. 11.3 The proposed MTL network architecture consists of two branches of RGT estimation and fault detection. The RGT estimation branch contains a transformer encoder to capture the global structural relationships from the input [Horizon, Seismic, Seismic] and a CNN decoder to compute the local information and fuse both global and local information by the module named SFF. The fault estimation branch is a simple end-to-end CNN architecture, and the inputs of this branch are the 2nd derivatives of the RGT prediction and the feature maps from the transformer encoder of the first branch, which can provide clear fault features and more detailed seismic structural features, respectively

Finally, the two maps are multiplied with original global and local features, and are further added together to obtain a hybrid feature.

After three SFF modules and four upsampling layers, we obtain an $H \times W \times 1$ feature map which estimates the vertical derivative of an RGT map. We further integrate this feature map in the vertical direction to obtain a final RGT map. The reason why we design such an intermediate layer of integration instead of directly

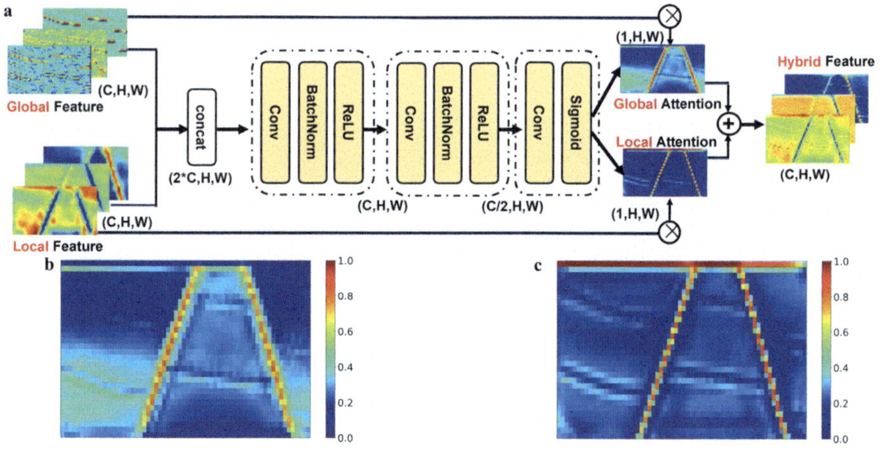

Fig. 11.4 **a** The structure of the SFF module; **b** The attention map of global features; **c** The attention map of local features

output an RGT map in our network is that the integration will accumulate potential prediction errors and amplify the loss misfit during the training, and it can help our network learn RGT features better. In a later section, we use examples to demonstrate the advantage of the integral process.

11.2.2 Fault Detection

We use a simple CNN (button part in Fig. 11.3) to detect faults by the shared features from the RGT estimation branch. First, we observe that fault features are implicitly indicated as discontinuities in the estimated RGT map. Therefore, we calculate the 2nd derivatives (U_{xx} and U_{yy})of the predicted RGT map to obtain features for highlighting faults. In the fault detection network, the two derivative features are used as inputs followed by one convolution block and a max pooling layer with stride 4. The convolution block includes two sets of a 3×3 convolution layer and a ReLU activation. As shown in Fig. 11.3, the outputs of the first convolution and max pooling layers are 8 feature maps containing obvious fault structures. These feature maps computed only from the RGT map show clear fault structures in global scales but may miss detailed features of small-scale faults in the original seismic image. Because the transformer encoder of the RGT estimation branch can extract meaningful features from the seismic image, we merge the output feature maps of the first and second transformer blocks into the encoder of the fault detection branch. Then, we use two convolution blocks and one max pooling layer with stride 2 to obtain the other multiscale feature maps in the encoder of the fault detection. In the decoder part of the fault detection, we use three convolution blocks, three 2×2 upsampling layers, and 2 concatenations with feature maps from the encoder to restore the original resolution as well as fuse fault features between the encoder and decoder. A fault probability map is finally computed, through a 1×1 convolution layer followed by a sigmoid layer.

11.2.3 Loss Function

(1) *RGT loss:* During the training process, the RGT estimation is supervised by two loss functions based on the known horizon segments and RGT labels, respectively. We expect that the estimated RGT values at each known horizon segment should be exactly the same. We impose such constraint on the RGT prediction during the training by defining the following loss $\mathcal{L}_H$:

$$\mathcal{L}_{\mathrm{H}} = \sum_{i=0}^{N} \sum_{j=0}^{M_i} (y_{i,j} - \overline{y}_i)^2, \tag{11.1}$$

where N is the number of horizon segments, M is the number of the points at each horizon. $y_{i,j}$ represents an RGT value estimated at the j-th point of the i-th horizon, and $\overline{y}_i$ is the average of the RGT values estimated at the i-th horizon. In this loss measurement, we do not care about specific RGT values estimated at the known horizon segments, and we only make sure the RGT values estimated at each horizon are identical. With the known positions of the horizon segments, this loss provides a self-supervision during the training.

Another loss function supervising the RGT estimation is based on RGT labels and we define it with the Structural Similarity (SSIM) metric (Wang et al., 2003). SSIM has been is widely used in the regression problem, such as super resolution (Li et al., 2021), depth estimation (Vasiljevic et al., 2019), and image restoration (Zhao et al., 2016). Compared with pixel-wise measurements such as MSE or MAE, SSIM focus more on the structural information. Our RGT map is a combination of geological structures such as horizons and faults. Therefore, we use the criterion based on Multi-Scale Structural Similarity (MS-SSIM) as the second loss function in RGT estimation branch, and MS-SSIM can be written as follows:

$$\text{MS-SSIM}(x, y) = \left[l_M(x, y)\right]^{\alpha M} \cdot \prod_{j=1}^{M}\left[c_j(x, y)\right]^{\beta_j}\left[s_j(x, y)\right]^{\gamma_j}, \tag{11.2}$$

x and y are two images that we want to evaluate. With M scales of one image, this method can measure image features at different resolutions. This measurement mimic the process of seismic structural interpretation by geophysicists or geologists who typically zoom in and zoom out the seismic image to observe the structures in different scales. In RGT estimation problem, *l, c, and s* represent the measurements of RGT values, sharp change in an RGT map, and relative relationships among the contours (corresponding to horizons in a seismic image) in an RGT map, respectively. We calculate the SSIM in one part of an image, and we slide an 11×11 Gaussian window pixel by pixel to compute the mean SSIM of all local images. In addition, we replace the predicted RGT values with ground truth values at the horizons segments before calculating this MS-SSIM loss. These replaced values, if consistent with the predicted ones, will affect the surrounding pixels by applying the Gaussian window which makes the network more sensitive to the input horizon segments. The range of SSIM is from 0 to 1, and the loss function of MS-SSIM can be written as follows:

$$\mathcal{L}_{\text{MS-SSIM}} = 1 - \text{MS-SSIM}(x, y). \tag{11.3}$$

The total loss function for RGT estimation is a hybrid one defined as follows:

$$\mathcal{L}_{\text{RGT}} = (\mathcal{L}_{\text{MS-SSIM}} + \mathcal{L}_{\text{H}})/2. \tag{11.4}$$

(2) *Fault loss:* In the fault detection branch, we consider it is a binary segmentation problem. Binary cross-entropy (BCE) is a common loss function in binary segmentation problem, which can be written as follow:

$$\mathcal{L}_{\mathrm{BCE}} = -\sum_{i=1}^{i=N} y_i \log p_i - \sum_{i=1}^{i=N} (1 - y_i) \log(1 - p_i), \tag{11.5}$$

where N represents the number of points in the fault image. y_i represents the positive label and p_i represents the prediction possibility which is calculated by the sigmoid layer in the network.

BCE can not work well in the dataset with an imbalanced distribution of positive and negative samples. The positive (fault) samples are much less the negative (non-fault) samples in a fault image (Wu et al., 2019). When training with the common BCE loss, the network can easily make only negative predictions everywhere to minimize the training objective. Several loss functions have been proposed in recent years to solve this problem. The dice coefficient score (DCS) (Milletari et al., 2016) is a measurement used to evaluate the similarity between prediction and label, and is widely used in the task with an imbalanced training dataset. DCE can be written as follow:

$$\mathrm{DCS} = 2 \times \frac{|P \cap G|}{|P| + |G|}, \tag{11.6}$$

where P and G denote the prediction and ground truth, respectively. $|P \cap G|$ represents the intersection of P and G. $|P|$ and $|G|$ represent the pixel number of P and G. The range of DCS is from 0 to1, and the larger the value is, the more similar prediction and label are. We define the loss function for fault detection based on the DCS as follows:

$$\mathcal{L}_{\mathrm{Fault}} = 1 - \mathrm{DCS}. \tag{11.7}$$

Finally, the total loss function of training the MTL network can be written as follow:

$$\mathcal{L}_{\mathrm{Total}} = (\mathcal{L}_{\mathrm{RGT}} + \mathcal{L}_{\mathrm{Fault}})/2. \tag{11.8}$$

11.2.4 Evaluation Metric

We propose a metric to measure the RGT estimation which is similar to the one discussed in (Bi et al., 2021), and call this metric as Horizon Extraction Error by RGT Values (HEERV). In caclulating the metric, we first uniformly select several RGT values R_i and extract the corresponding contours from the RGT label. Then, we compute the average values $\tilde{R}_i$ of the RGT values estimated at the extracted contours, and further extract the corresponding contours of $\tilde{R}_i$ from the estimated RGT map. Finally, we compute the vertical distances between the contours extracted from the RGT label and predicted RGT map. The distances form our HEERV metric to evaluate the accuracy of the RGT estimation or horizon extraction.

11.3 Training and Validation

In this section, we first illustrate the workflow of the training data generation. Then, we introduce the detail of training and validation. Finally, we show the performance of our model on synthetic validation data.

11.3.1 Training Data

As a supervised learning method, our proposed MTL network needs a large number of samples (seismic images and horizon segments) and labels (RGT and fault images) to train a model with stability and robustness. The fault labels obtained from field seismic images by manually picking are time-consuming and highly subjective, and estimating RGT maps accurately from field seismic images are almost impossible.

Therefore, we use a workflow to automatically build a synthetic training dataset (Wu et al., 2020). First, we build an initial flat reflectivity model and add some folding and faulting controlled by some parameters. Next, we convolve the model with a Ricker wavelet to obtain a clean synthetic seismic image, and add some noise extracted from field data with a random SNR to make the synthetic seismic image (Fig. 11.5a) more similar to field data. The reflectivity, structure parameters, and the peak frequency of wavelet and SNR are all randomly chosen in a reasonable range. To obtain the corresponding fault labeling image, we select the points between the hanging wall or footwall sides as fault labels. The fault label points and other points, which are respectively assigned with ones and zeros, form the binary fault label image (Fig. 11.5b) corresponding to the seismic image.

To obtain the corresponding RGT label, an initial flat RGT model is built with vertically monotonic increasing values. Then, we use the same structure parameters which are chosen for the reflectivity model to simulate the same folding and faulting structures in this RGT model (Fig. 11.5c). By using this workflow, we randomly choose the parameters in the above description, and generate 10,000 pairs of 2D seismic images and corresponding fault and RGT labeling maps with the size of 128 × 256. Finally, we randomly extract several horizons (RGT contours) (Fig. 11.5d) from each RGT map in every training epoch and use them as part of the network inputs to provide prior constraints of known horizons. In defining a known horizon, we choose a partial or complete contour randomly, because we may not be able to obtain a whole horizon of a field seismic image (Bi et al., 2022). In this way, the model is trained to be able to deal with situations of various horizon segments as input.

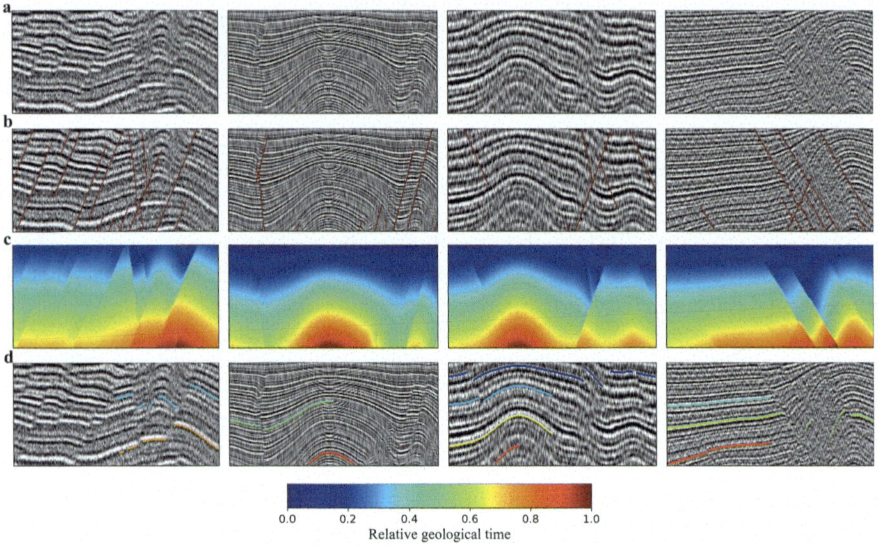

Fig. 11.5 Four of ten thousand synthetic training samples. The training dataset includes seismic images (**a**), the corresponding fault (**b**), and RGT (**c**) labels. Horizons segments (**d**) are randomly extracted from the RGT labels and are input together with seismic images into our neural network to provide constraints

11.3.2 Training Details

We train our MTL network by using 9600 pairs of synthetic seismic images and corresponding RGT and fault images, and another 400 pairs are used for validation. To relieve the effects of the seismic amplitude variations between field and synthetic seismic images, we use a normalization that seismic images are subtracted by their means and divided by their standard deviations. To define the values of horizon segments, every RGT map is normalized by its minimum and maximum, and the range of RGT values is from 0 to 1. We use the pretrained model which was trained by common RGB images with three channels. To be consistent, we define our inputs with three channels of [horizon segments, seismic, seismic]. We set the training epoch with 800 and batch size with 80. From our experience, a transformer network needs a large amount of training data and a relatively small learning rate (LR) compared to a CNN model to obtain an excellent result, and we use a StepLR with the initial value of 0.001 and decay the learning rate by 0.5 after every 100 epochs (Fig. 11.6c). Our proposed network is optimized with the Adams optimizer whose weight decay is set to 0.0001.

The RGT loss curves (Fig. 11.6a) and the fault loss curves (Fig. 11.6b) converge to less than 0.004 and 0.3 after 800 epochs, respectively. As shown in Fig. 11.7, the reflection features are highly contaminated by noise, and the fault structures are complex (Fig. 11.7a). However, the predicted RGT (Fig. 11.7d) and fault images

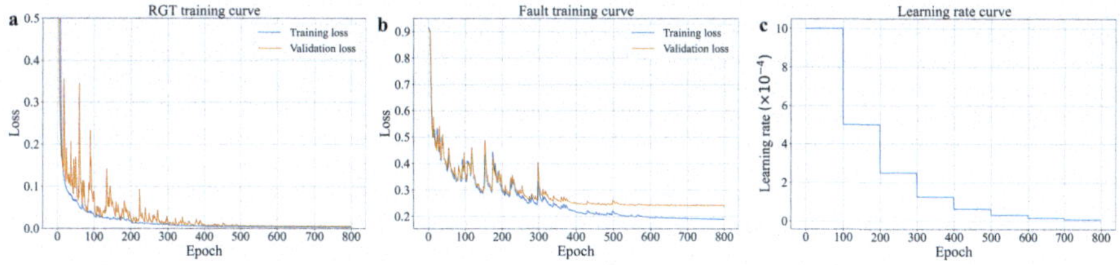

Fig. 11.6 **a** Training (*blue*) and validation (*orange*) loss curves for RGT estimation. **b** Training (*blue*) and validation (*orange*) loss curves for fault detection. These two training processes are simultaneously performed. **c** Learning rate curve is updated at every 100 epochs

(Fig. 11.7h) are almost the same as the labels (Fig. 11.7c, g). The contours (Fig. 11.7e) extracted from predicted RGT maps (Fig. 11.7d) represent horizons, which accurately follow seismic reflections, even across the faults (Fig. 11.7d). As shown in Fig. 11.7f, the extracted horizons can also fit the known horizon segments well. This indicates that our network predicts RGT maps confirming to the known horizon segments, which enables human interactions in complex field examples to compute reasonable RGT results.

11.4 Comparison Experiments

In this part, we first perform several field data tests to verify the superiority of our RGT estimation with transformer over the previous CNN-based method (Geng et al., 2019a, 2019b). We show that the transformer architecture and horizon constraints significantly improve the RGT estimation, and we also show the improvement by adding an integral layer before outputting the RGT map. Further, with 3D horizon constraints, we can decompose the 3D RGT estimation to 2D predictions slice by slice. Finally, we show the results containing predicted RGT and fault images from the whole MTL network.

11.4.1 CNN Versus Transformer

The previous CNN method (Geng et al., 2019a, 2019b) for 2D RGT estimation obtains an excellent performance, and we can use it to extract horizons easily from a seismic image. But this method also has some limitations: (1) The network in this method has high costs in computation. After four residual blocks in the encoder part, the number of the feature maps steeply rises to 2048. (2) The network in this method has a relatively small receptive field. Maintaining a large receptive field is important for the RGT estimation problem to capture global structural information.

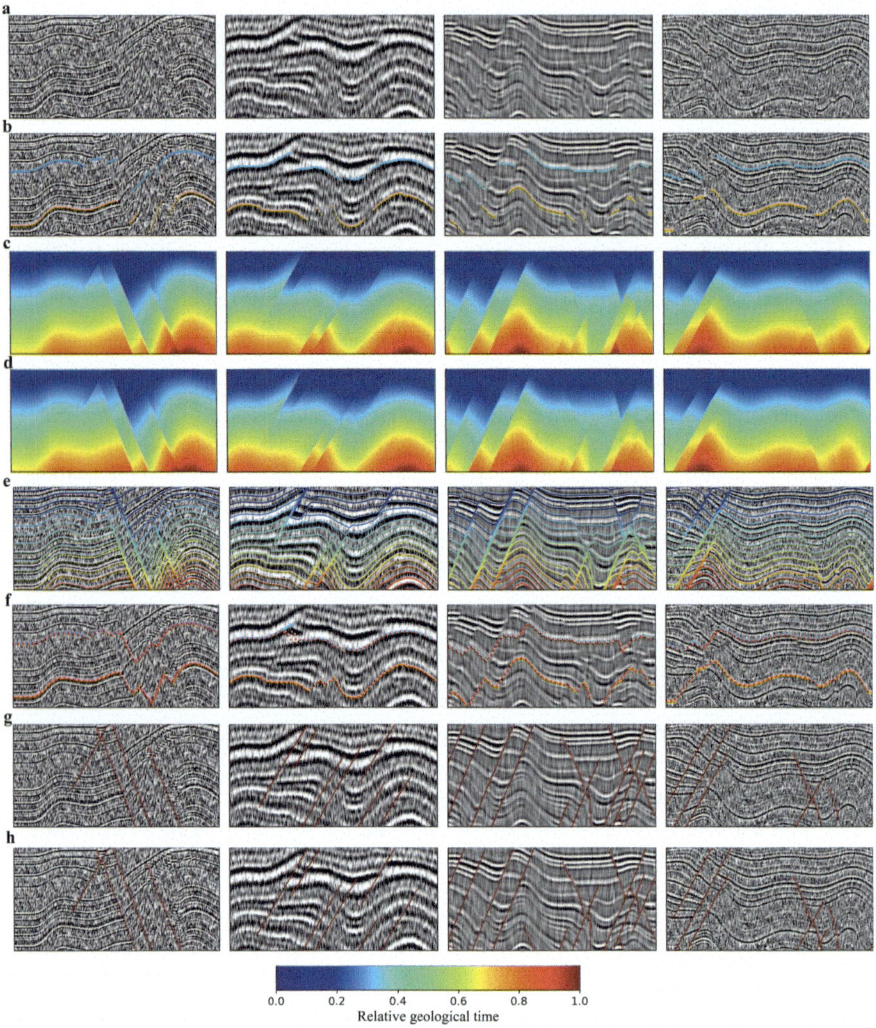

Fig. 11.7 We show several inputs from the validation dataset and the corresponding results, including seismic images (**a**), horizon segments (**b**), RGT labels (**c**), predicted RGT maps (**d**), and contours (**e**) extracted from (**d**). The contours (*purple dotted curves*) (**f**) extracted from (**d**) perfectly fit the input horizon segments (**b**). The fault images (**h**), predicted at the same time as the RGT estimation, also fit the fault labels (**g**) well

The hierarchical and efficient encoder in our RGT estimation branch helps solve these problems. Floating point operations (FLOPs) and trainable parameters indicate the efficiency and complexity of a network, respectively. As shown in (Table 11.1), our RGT estimation branch is much more efficient and lightweight than the one proposed in (Geng et al., 2019a, 2019b). Next, we further compare the receptive field and RGT estimation accuracy of our method with the CNN-based method.

Table 11.1 The comparison of different network in efficiency

Network	FLOPs ↓ (GB)	Trainable parameters ↓ (MB)
Geng et al., (2019a, 2019b)	129.739	214.780
Our RGT estimation branch	**7.148**	61.152

A unit in a feature map of a neural network depends on a region of the input image, and the region can be defined as the receptive field for this unit. With a bigger size of the receptive field, a network is able to better extract high-level features and fuse global information more effectively. A large receptive field is important for the task of RGT estimation that requires globally fitting structures in an input seismic image. The receptive field is theoretically dependent on the network architecture (e.g., convolutional kernal size, number of layers, downsampling) but the effective receptive field (ERF) can also be affected by the training process and the network inputs (Luo et al., 2016). Compared to theoretical receptive field (TRF), ERF is related to the features of the input image, and it can more directly show the ability of a trained network in extracting and fusing global information in practice. Figure 11.8a is a seismic image fed into CNN (Geng et al., 2019a, 2019b) and our Network with a transformer backbone, Fig. 11.8b, c are the ERFs of the center unit (red points in Fig. 11.8b, c) in these two networks. We can observe that the RGT estimation branch of our network shows a larger ERF than CNN. Moreover, the CNN receptive field (Fig. 11.8b) is isotropic in space while the one of our network (Fig. 11.8c) appears an anisotropic shape reasonably follow the layered structure (denoted by white curves in Fig. 11.8c) and even the dislocation across the fault (denoted by red line in Fig. 11.8c) in the input seismic image. We think the anisotropic receptive field, with a long-distance extension following seismic structures, would be essential for estimating an accurate RGT result from a seismic image.

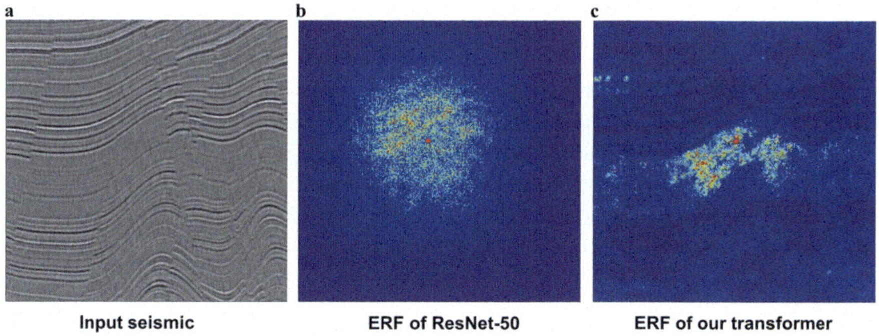

Fig. 11.8 With the same input seismic image (**a**), the previous CNN-based method yields an isotropic effective receptive field (ERF) centered at the target point (*denoted by the red dot*) (**b**), while our network generated a geologically more reasonable ERF which is spatially anisotropic and follows seismic reflection and fault structures (**c**)

We use a field example (128[vertical] × 256[inline] samples) in Fig. 11.9 to demonstrate the advantage of the RGT estimation branch of our network. The reflections in this field seismic image (Fig. 11.9a, b) appear significant lateral variations and are dislocated by several faults, which poses challenges to the RGT estimation from the seismic image. In this example, we manually interpret two horizons (solid curves in Fig. 11.9a, b) to validate the predictions by networks. The second row of Fig. 11.9 shows the RGT results predicted by the previous CNN (Fig. 11.9c) and our RGT estimation branch (Fig. 11.9d). They look quite similar but our result appears sharper discontinuities across the faults denoted by black arrows (Fig. 11.9d). To better evaluate the CNN-based and our RGT results, from which we extract two horizons (corresponding to the manual ones) and display them with dotted curves in Fig. 11.9a, b, respectively. The horizons extracted from our RGT results visually better fit the manual horizons than the ones extracted from the CNN-based RGT. The quantitative metric HEERV of our result (5.957) is also lower than the CNN result (7.712). The HEERV values indicate that our network is able to obtain a more reasonable horizon and fault interpretation result from a seismic image compared with the CNN-based method. Finally, we can easily extract any number of contours from the RGT results to obtain a set of dense horizons that follow reflections and dislocate across faults as shown in Fig. 11.9c, d.

11.4.2 Horizon Constraints

It is a common challenge to compute geologically reasonable horizons or faults from a seismic image with complex structures, artifacts, and low SNR for existing auto methods (Bi et al., 2021; Figueiredo et al., 2007; Geng et al., 2019a, 2019b; Wu & Janson, 2017; Wu & Zhong, 2012). In these cases, experts typically need to interpret with their prior knowledge of geological background instead of simply tracking seismic geometric features which may be unclear or even geologically untrue due to limited resolution and artifacts of seismic imaging. Similarly, it is necessary to incorporate prior expert geological knowledge into automatic seismic structural interpretation methods, including our MTL method, to deal with the complicated cases. Inputting manually interpreted horizons into our network is an effective way to incorporate prior constraints to help estimate reasonable RGT maps from seismic images with complex fault systems, poor data quality, and unconformities. Moreover, with the constraints of 3D horizons, we are able to decompose a 3D RGT estimation task into computationally cheap 2D predictions slice by slice in a 3D seismic volume but still maintain lateral consistency among the 2D RGT results.

(1) *Complex fault systems:* For the same example in Fig. 11.9, we input the two manually interpreted horizons together with the 2D seismic image into our network and obtain an RGT result and its associated horizons (contours) in the left column of Fig. 11.10. As shown in Fig. 11.10a, the horizons (dotted curves) extracted from the estimated RGT map fit well with the manual ones used as

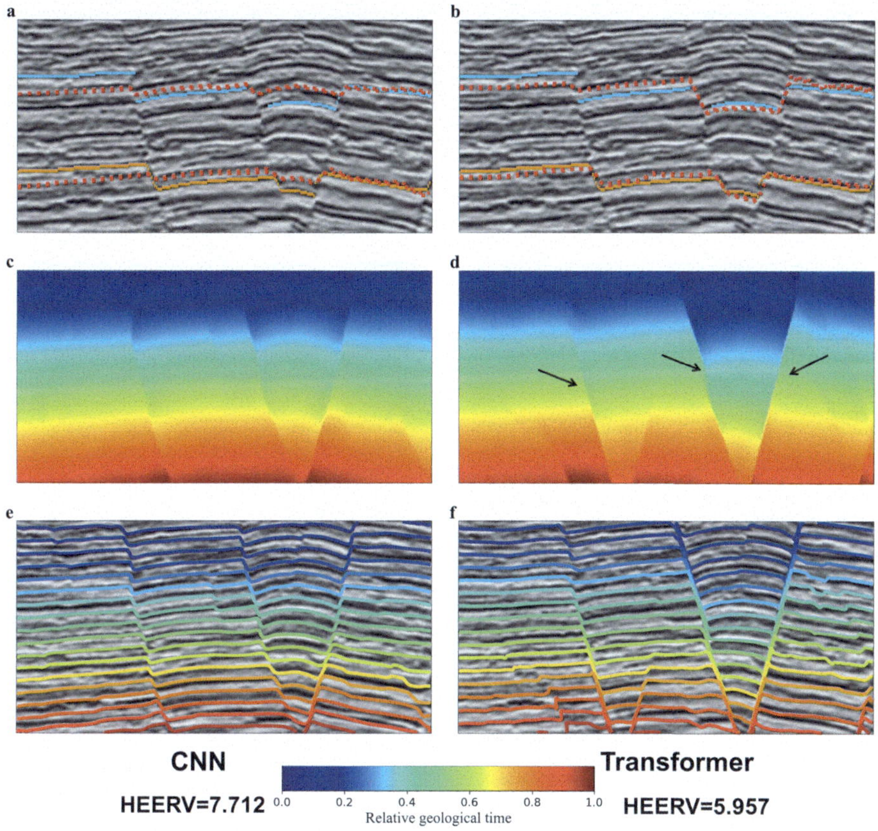

Fig. 11.9 Previously proposed CNN-based (Geng et al., 2019a, 2019b) (*left column*) and our transformer-based (*right column*) RGT estimation methods are applied to a field example with complex faults. Compared to the results of the CNN-based method, our RGT result appears sharper discontinuities across the faults as denoted by black arrows. In addition, the horizons (*dashed curves*) extracted from our RGT map better fit the manually interpreted horizons (*solid curves*) as shown in the first row

constraints, which indicates that the network performs RGT estimation honoring the input horizon constraints as what we expect. As shown in Fig. 11.10e, a set of dense horizons (contours) extracted from the RGT map follows seismic reflections, which indicates the network also tends to fit seismic structures in estimating the RGT map. Moreover, the input horizon constraints not only ensure accurate RGT estimation near the horizon positions but also improve and stabilize the RGT estimation in other areas, which we will illustrate in more examples in the following sections.

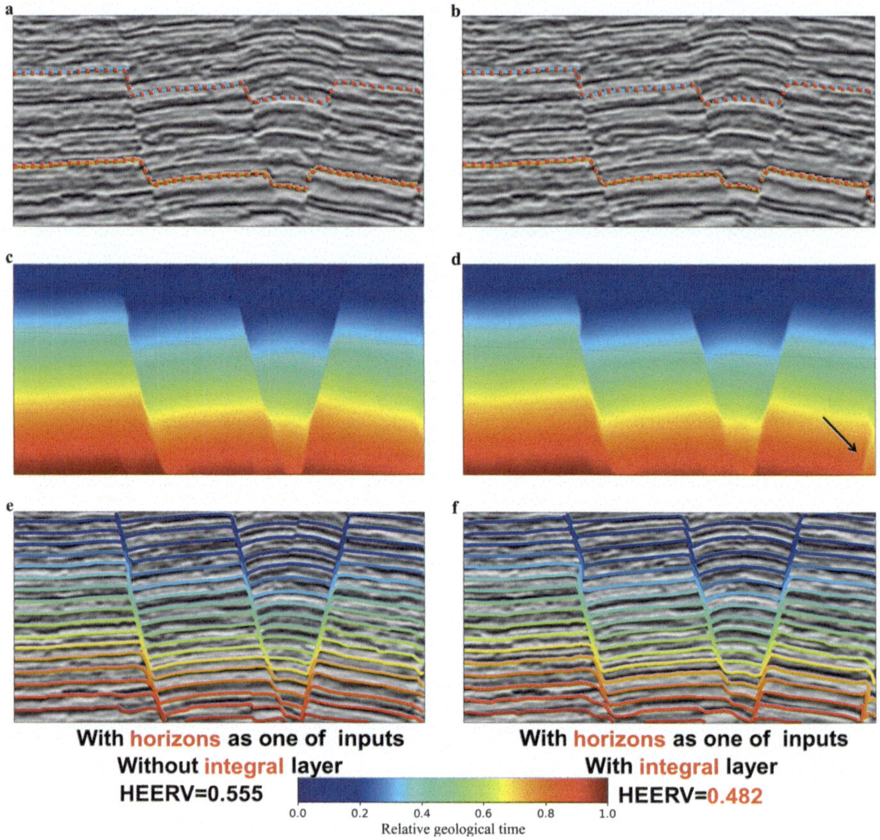

Fig. 11.10 For the same example in Fig. 11.9, we use the interpreted horizons (*solid curves in the first row*) as constraints to improve the RGT estimation (**c**), from which the extracted horizons or contours more accurately follow seismic reflections (**e**) and the manual interpretation (**a**) than those in Fig. 11.9. We add an integral layer (**b**, **d**), as discussed in Wu et al. (2023), at the end of the RGT estimation network to further improve the RGT estimation and obtain a lower HEERV (0.482) (**f**)

As discussed in (Wu et al., 2023), we can further improve the performance of our network by adding a layer of vertical integration operation before the output layer as denoted by the dark green arrow in the RGT estimation branch in Fig. 11.3. With this layer, we assume the network learns to estimate RGT derivatives from the inputs and the integration converts the derivatives to an RGT map. Such integration will accumulate prediction errors, if exist, and therefore emphasize the misfit between the predictions and labels during the training process, which would be helpful to better train the network. In addition, the integration is a global operator that is helpful for the network to fuse global features for the RGT estimation. Although the RGT results and their associated contours (horizons) of the network without (left column of Fig. 11.10) and with (right column) the integration layer are visually similar, the HEERV metric of the later is relatively smaller (0.482) than the former (0.555). In addition, the latter (Fig. 11.10f) can accurately produce a discontinuity across the fault (denoted by the black arrow in Fig. 11.10f) while the former missed that.

(2) *Poor data quality:* In addition to complex faults, horizon can also guide the network to track reflections in a seismic image with low SNR and artifacts. The first row of Fig. 11.11 shows a low SNR dataset (128[vertical] $\times$ 256[inline] samples) from the Soda Lake geothermal field, where some of the reflections are almost invisible and therefore are hard to track for an automatic horizon interpretation method. As shown in Fig. 11.11a, if we use only the seismic image as the input, the horizons computed by our network are quite messy and do not fit the manual ones (solid curves in Fig. 11.11a), which indicates that our network can not compute a reasonable RGT map from a seismic image with unclear reflections. Therefore, we gradually add more and more horizon segments as constraints to illustrate how the horizon constraints affect the RGT estimation from a seismic with low data quality. As denoted by the solid curves of the second column of Fig. 11.11a–f, we track eight horizons by manually picking. By inputting one horizon (solid curve in the first image of Fig. 11.11b) as constraints into the network, we are able to compute much more stable horizons (curves in the third image of Fig. 11.11b), compared to those in Fig. 11.11a. However, the computed horizons still do not fit well with the manual ones except near the constraint horizon as shown in the second image of Fig. 11.11b. As shown in Fig. 11.11c, we further add two more horizons as constraints and obtain more reasonable horizons that fit well with most of the interpreted horizons, even in areas far away from the constraint horizons. To further improve the result, we add one more horizon in the region with blurry reflections (the green one in the first image of Fig. 11.11d), which leads to geologically reasonable horizons everywhere in the seismic image.

In addition, we find that the result constrained by four horizons (Fig. 11.11d) is almost identical to the result constrained by all eight horizon segments (Fig. 11.11f). This indicates that the constraint of these four horizons is sufficient for the network to estimate an accurate RGT map from this seismic image with low data quality. In some situations, experts may not interpret a complete horizon, which means that only partially interpreted horizons or horizon segments are available to be used as constraints. We mimic this case in Fig. 11.11e, where we use only the left part of the green horizon as constraints. In this case, we still obtain accurate horizons (Fig. 11.11e) which are almost identical to those computed with the constraint of the entire green horizon in Fig. 11.11d. This test indicates that our method allows to use incompletely interpreted horizons or horizon segments as constraints to improve the RGT estimation, which facilitates its applications in practice.

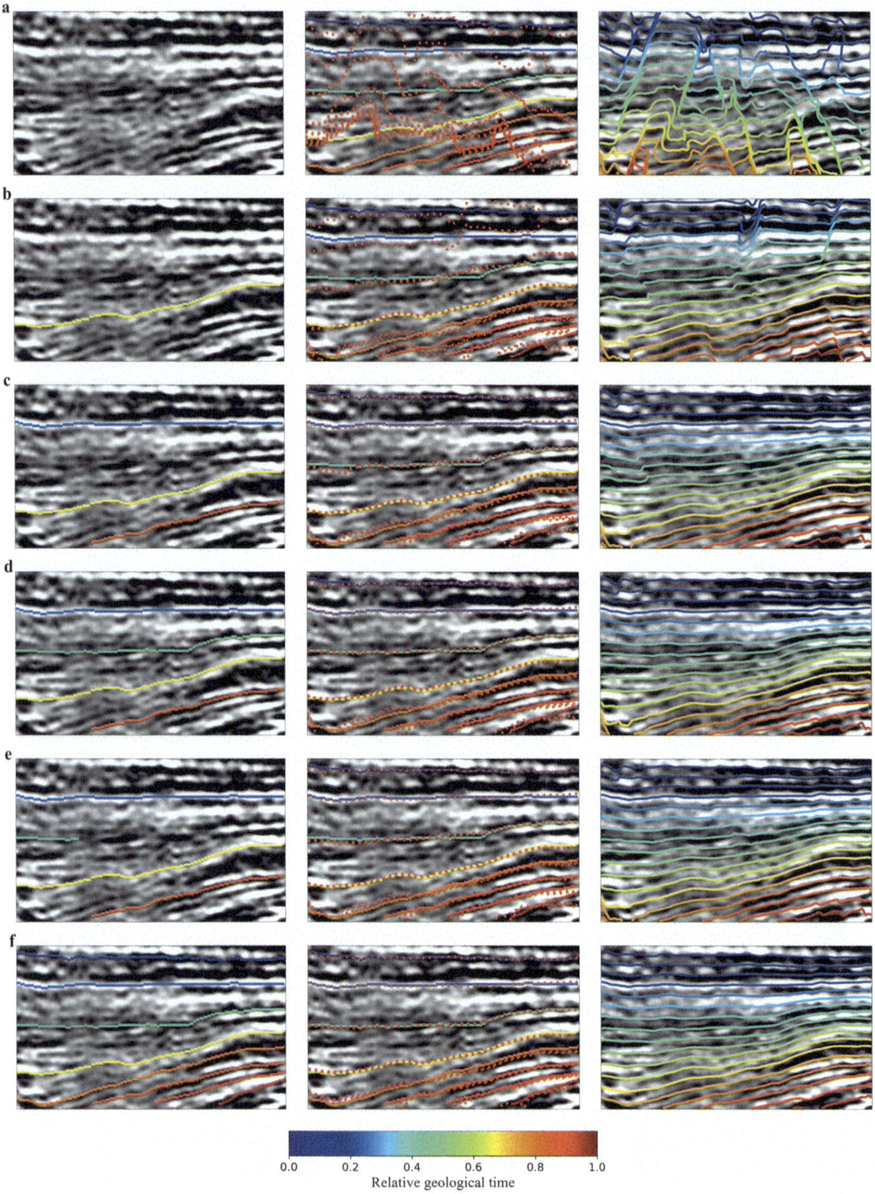

Fig. 11.11 A comparison study of using different numbers of horizons as constraints. In the first column, a seismic image with different number of horizons are input to our network for RGT estimation. In **a–d**, as we increase the number of constraint horizons from 0 to 4, the contours or horizons, extracted from the estimated RGT maps, become increasingly better fitted to the interpreted horizons (*solid curves in the second column*) and seismic reflections (*third column*). The contours (**d**) from the RGT map constrained by four horizons are similar to those (**f**) constrained by eight horizons. Compared to using constraints from all four complete horizons (**d**), we can still obtain a quite similar RGT map by making one [*the left part of the green horizon in* (**e**)] of the constraint horizons incomplete

In our experiments, we use two methods to determine the RGT values of horizon patches. In most cases, we calculate the relative locations of horizon patches within the vertical space, using values ranging from 0 to 1 as the RGT values. If the layer in the area is inclined or folded significantly, the RGT values may be unreasonable. In this instance, we may need to interpret complete horizons through the entire survey to decide reasonable RGT assignments. To solve this problem, we need to first manually determine the geologic time order of the constraint horizons or horizon patches based on the geologic background of the survey or the stratigraphic sequence appear in the seismic volume. We then assign some reasonable RGT values to the horizons in order. Our method is designed to automatically compute an RGT map with an adaptive range of values honoring the RGT assignments at the horizons, we only need to make sure the RGT assignments are in a geologically reasonable sequence, that is the assigned RGT values of vertically shallower horizons should be relatively smaller than the deeper ones. Note that here we assume no overturning layers appear in the seismic images.

(3) *Unconformities:* Unconformity (e.g., the region between dark blue and light green solid curves in Fig. 11.12a) can be regarded as a boundary where seismic reflectors terminate. It is indicated by vertical geologic time discontinuity in the corresponding RGT map. The reflector terminations or geologic time discontinuities near unconformities pose another type of challenges for task of automatic horizon interpretation in a seismic image. It is also a challenge for our network trained by a dataset without any unconformity features. By inputting the seismic images (128[vertical] $\times$ 256[inline] samples) (Fig. 11.12a) into our network, we obtain the corresponding RGT maps in Fig. 11.12b, where we do not observe any vertical discontinuities near the unconformities as expected. To be able to reasonably deal with the unconformities, one potential way is to retrain the network by seismic images with unconformities and the corresponding RGT labels with discontinuities at the unconformities. Here, however, we provide a more convenient way by incorporating some horizon constraints without the need of retraining the network.

We manually interpret four horizon segments (solid curves in Fig. 11.12a) and use them as constraints in our network for the RGT estimation. Note that the horizons used as constraints need to be geologically isochronous. Therefore, the light blue and cyan horizons are interpreted to be terminated at unconformities where the geologic time are varying. Figure 11.12c shows the RGT maps computed by using the horizon constraints, where we observe sharp vertical discontinuities near the unconformities as denoted by the yellow and red arrows. Figure 11.12d shows horizons (contours) extracted from the estimated RGT maps, from which we observe geologically reasonable horizon onlaps and downlaps at the unconformities denoted by the yellow and red arrows, respectively. This example shows that, using horizon constraints, our network is able to reasonably estimate the geologic time discontinuities corresponding to the onlap and downlap of the unconformity, although there are no structures of unconformities in our training dataset. It is hard to prepare a truly diverse training dataset that contains all kinds of geologic structures or features, which typically limits the

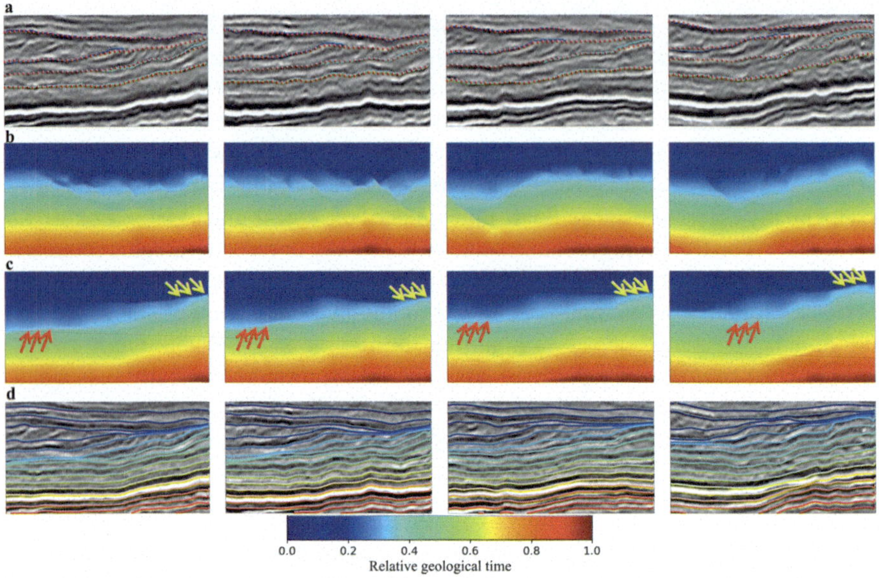

Fig. 11.12 RGT estimation in seismic images (**a**) with unconformities. The network does not work well to estimate RGT maps (**b**) from seismic images with unconformities because it is trained with synthetic datasets without unconformity features. However, we can input some horizons (*solid curves in the first row*) into the network as constraints to improve the RGT estimation (**c**) and horizon extraction (**d**), where we observe reasonable discontinuities near the unconformities as denoted by yellow and red arrows

generalization of a trained network in various field datasets. However, the idea of using horizon constraints provides a convenient way to involve human interactions into the inference step of a trained network and effectively improve its generalization in various structures that may not be included in the training datasets.

(4) *Obtain a 3D RGT volume from 2D predictions:* Field seismic data today are mostly in 3D and it is necessary for an algorithm to be able to deal with 3D volumes. However, training a 3D deep neural network with 3D convolutional kernels and feature maps is much more time-consuming than training a 2D network, even using relatively small volumes of training data. In addition, predicting an RGT volume from a large 3D seismic volume is also expensive in computational time and memory for a trained network, which limits its applications in practice. Bi et al. (2021) solve this problem by dividing a large 3D volume into smaller sub-volumes and make a prediction for each sub-volume, which requires a complex post-processing step to merge the predicted sub-volumes of RGT to obtain a large RGT volume without merging artifacts.

We propose a more efficient and practical way to compute a 3D RGT volume from a seismic volume by 2D predictions slice-by-slice in the inline or crossline directions. The main challenge of this way is to maintain lateral consistency among these independent 2D predictions. For example, Fig. 11.13b shows a 3D RGT volume (128[vertical] × 256[inline] × 150[crossline] samples) obtained by directly assembling 2D predictions, where we observe obvious unreasonable lateral jumps in the predicting direction. We propose to maintain lateral consistency among the slice-by-slice 2D predictions by introducing the lateral control of 3D horizons. Specifically, for each 3D horizon surface used for control, we assign the same RGT value to all points on the 3D surface as shown in Fig. 11.13c. This ensures the horizon constraints input into the network for each 2D prediction are consistent and therefore makes sure the RGT values predicated at the constraint horizons in all 2D predictions are identical. In this way, we are able to compute laterally consistent 2D RGT predictions that are directly assembled together in the predicting direction to form a reasonable 3D RGT result as shown in Fig. 11.13d. In addition to the lateral consistency, the horizon constraints also improve the RGT estimation near the unconformities where we observe obvious vertical RGT discontinuities. As shown in Fig. 11.13e, f, the 3D horizons extracted from the 3D RGT accurately follow seismic reflections and some of them reasonably terminate at the unconformity. This example demonstrates that we are able to decompose a 3D RGT prediction task into much more efficient 2D predictions but still maintain lateral consistency by introducing 3D horizon constraints.

11.4.3 Simultaneous RGT and Fault Prediction

In this section, we use the same seismic image (Fig. 11.14) as in Figs. 11.9 and 11.10 to demonstrate the effectiveness of the simultaneous RGT and fault prediction by the complete MTL network with two branches. The inputs for the branch of fault detection consist of the 2nd derivatives of the predicted RGT map and features concatenated from the transformer encoder. Because the RGT map is a scalar field globally fitting the seismic structures, the RGT derivatives provide mostly detailed seismic structural features to help obtain a noise free and continuous fault detection. The feature maps from the transformer encoder contain more detailed structure features to help detect small-scale faults. The fault detection branch network merges both the inputs and obtains a fault map with multi-scale faults as shown in Fig. 11.14d, where the fault features are clean and continuously trackable. During the training process, the gradients of the fault detection loss function are back propagated from the fault detection branch back to the RGT estimation branch, which helps the network to produce discontinuities across faults in the RGT result. As shown in the estimated RGT results (Fig. 11.14b) and their associated horizons or contours (Fig. 11.14c), we observe sharp RGT discontinuities and horizon dislocations across the detected fault positions.

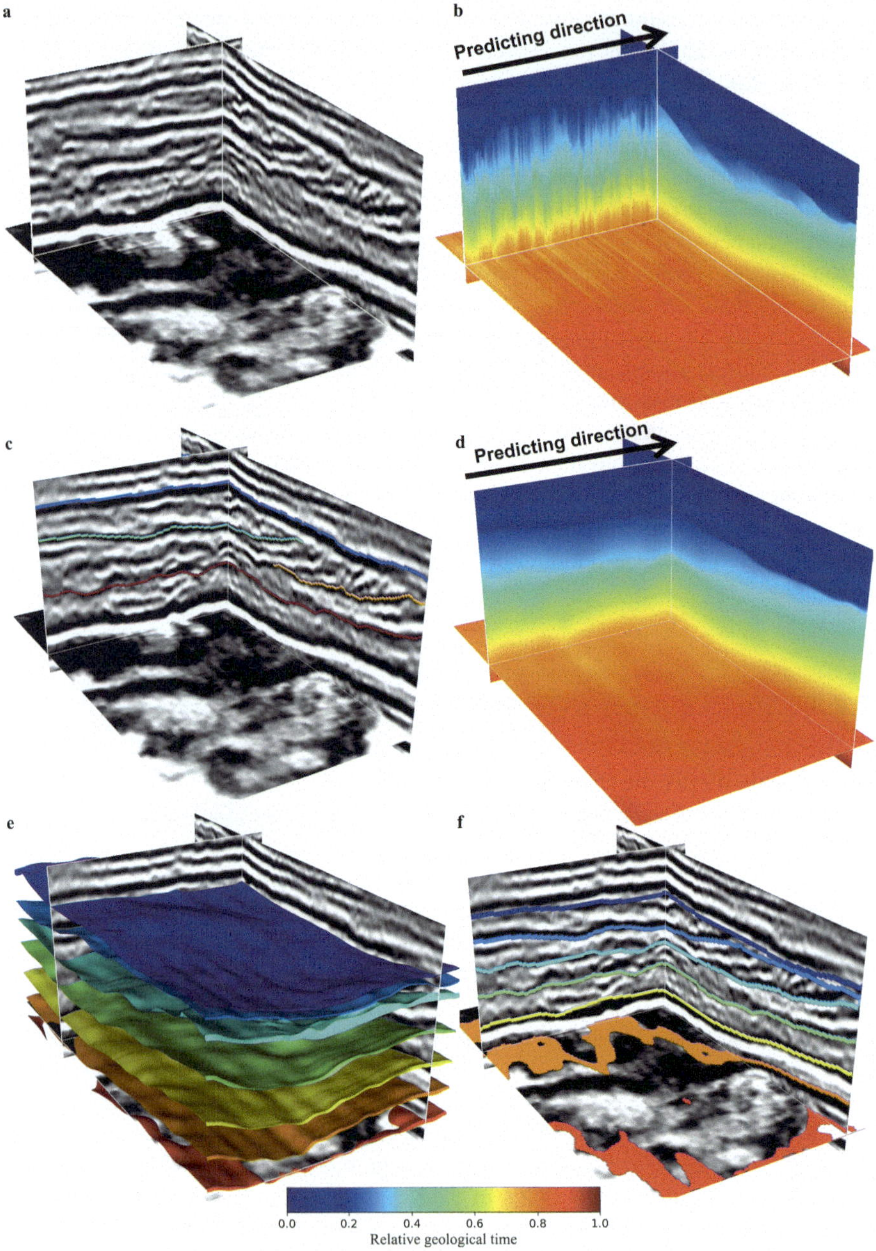

Fig. 11.13 Computing a 3D RGT volume from a 3D seismic image (**a**) by independent 2D predictions slice-by-slice yields unreasonable lateral jumps (**b**) in the predicting direction. We use 3D horizon surfaces (solid curves in **c**) as constraints to maintain lateral consistency of the 2D predictions to obtain a reasonable 3D RGT volume (**d**). The 3D horizons [(**e**) and (**f**)] extracted from the RGT volume (**d**) fit seismic reflections well and reasonably terminate at the unconformity

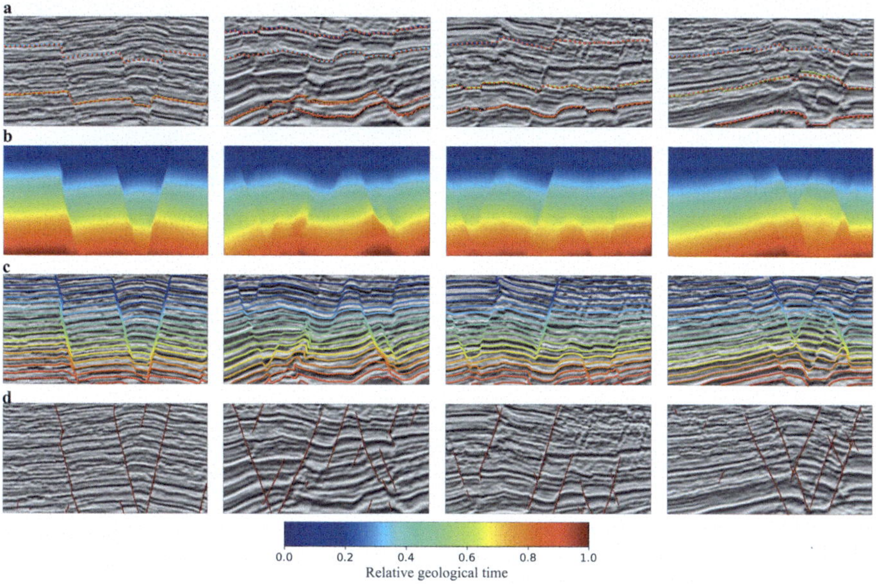

Fig. 11.14 By our MTL network, we simultaneously obtain reasonable RGT maps (**b**) and fault images (**d**) from seismic images (**a**) and horizon constraints (*solid curves in* ***a***). There are obvious sharp edges in the RGT maps (**b**), which correspond to the faults in fault predictions (**d**), and the contours (**c**) extracted from RGT maps (**b**) fit seismic reflectors well

11.5 Application

In this section, we apply optimal schemes, based on the above comparison experiments, to a 3D seismic dataset with faults. First, we use the horizon segments and seismic images (Fig. 11.15a) as inputs to obtain RGT maps (Fig. 11.15b) and fault images (Fig. 11.15c). The contours extracted from predicted RGT maps can fit interpreted horizons well (Fig. 11.15a), and these contours (Fig. 11.15c) accurately follow continuous seismic reflections and dislocations across faults. The fault detections in Fig. 11.15d are consistent with the sharp edges appear in the predicted RGT maps, and we think it is benefited from the multi-task learning scheme of simultaneous RGT estimation and fault detection.

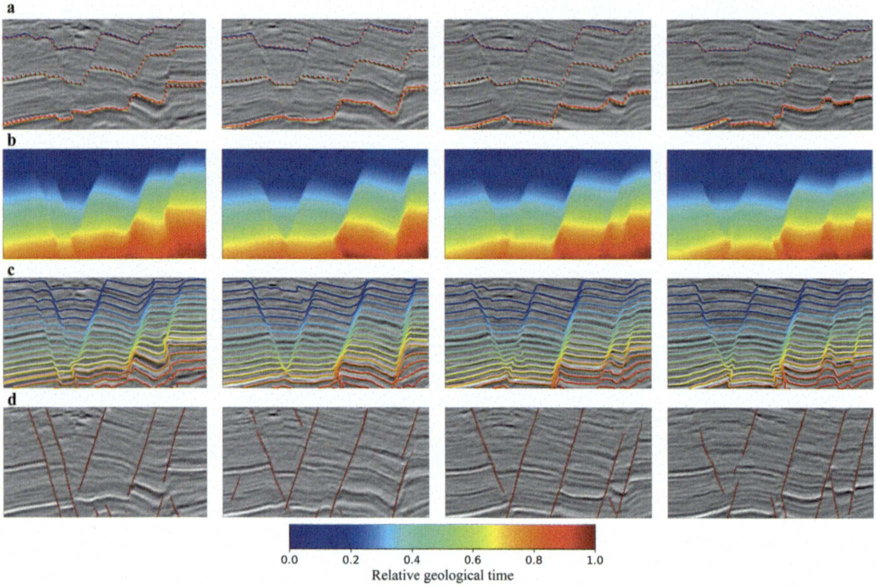

Fig. 11.15 By using our MTL network, we simultaneously compute reasonable predictions of RGT estimation and fault detection from the 2D seismic images and horizon constraints (**a**). In the estimated RGT maps (**b**), we observe sharp edges or discontinuities, which correspond to dislocations in the extracted horizons (contours) (**c**) and high fault probability in the fault images (**d**)

With the constraints of the three 3D horizons, the independently computed 2D predictions can be directly assembled in the predicting direction to obtain laterally consistent 3D volumes (128[vertical] $\times$ 256[inline] $\times$ 150[crossline] samples) of RGT values and fault probabilities. The 3D RGT volume (Fig. 11.16d) and the corresponding iso-surfaces (Fig. 11.16f) accurately fit the reflections in the seismic volume (Fig. 11.16a). The faults in the seismic volume are also accurately detected as shown in Fig. 11.16c. Finally, the iso-surfaces extracted from RGT volume (Fig. 11.16d) correspond to 3D horizons as shown in Fig. 11.16e.

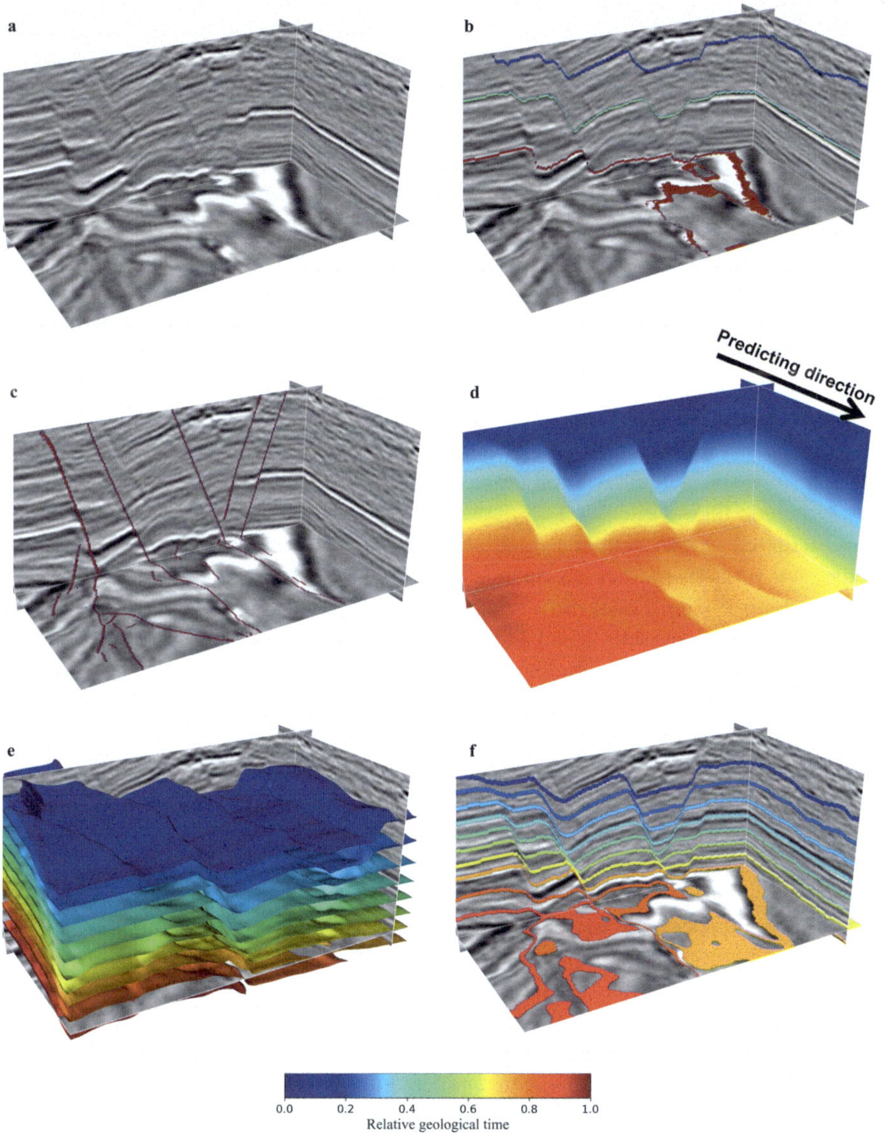

Fig. 11.16 With the 3D horizon constraints (**b**), we compute 2D fault images and RGT maps (Fig. 11.15) slice-by-slice in a 3D seismic volume (**a**) and directly assemble these predictions in the predicting direction to obtain 3D volumes of fault probabilities (**c**) and RGT values (**d**). Due to the constraints of 3D horizons in the 2D predictions, the assembled 3D RGT volume (**d**) maintains lateral consistency in the predicting direction, and the iso-surfaces [(**e**) and (**f**)] extracted from the RGT volume (**c**) fit seismic reflectors well. The horizon dislocations (**f**) correspond to the high fault probabilities in the fault volume (**c**)

11.6 Conclusion

We propose an MTL network for simultaneous RGT estimation and fault detection from a seismic image with the interpreted horizons as prior constraint information. The predicted RGT map can be further used to extract any horizon in the seismic image. The transformer encoder in the RGT estimation branch of the MTL network can capture global structural information from the inputs, and SFF can fuse the global and local features from the transformer encoder and CNN decoder. The simple end-to-end CNN uses the fault information from the 2nd gradients of the predicted RGT and the feature maps from the transformer encoder to detect faults. By sharing feature maps and network parameters in the training, these two tasks could provide controls for each other. Two loss functions constrained by prior horizon information are used in the RGT estimation branch, and the loss function to fit the imbalanced distribution of the positive and negative samples is used in the fault detection branch. We carry out comprehensive comparison studies to optimally design our schemes, which include the architectures of networks, the addition of prior horizon information, predicted results from 2 to 3D, and MTL. In multiple filed applications, our method can robustly predict reasonable RGT maps even in seismic images containing complex fault systems, unconformities, and blurry reflections, and at the same time compute a fault image with clean and continuous fault features. Finally, this method significantly improves the efficiency and saves the memory of 3D RGT estimation by computing and merging parallel 2D results slice by slice with the constraints of 3D horizons.

Some limitations still remain in our method. Firstly, our network model was trained on training samples with a size of 128 by 256 pixels. In practical data applications, it is often necessary to reshape seismic data into the same size through interpolation or downsampling to obtain optimal results. When the actual data size is much larger than 128 by 256, we need to perform block-wise predictions and require additional postprocessing to merge the block-wise results. Second, due to the large demand for memory and computational resources, we only tested the implementation of our method in two dimensions in our paper. This two-dimensional implementation can be applied to process a three-dimensional seismic volume and obtain a reasonable result with lateral consistency by introducing constraints of three-dimensional horizons. However, we believe that the implementation of our method in three dimensions is still valuable for better analyzing and fitting three-dimensional structures. The extension of our method from two dimensions to three dimensions is straightforward in algorithm implementation, but optimization of the network's computational efficiency and memory requirements is needed given the current performance of GPUs.

References

Al-Dossary, S., & Marfurt, K. J. (2006). 3D volumetric multispectral estimates of reflector curvature and rotation. *Geophysics, 71*(5), P41–P51. https://doi.org/10.1190/1.2242449

Arias, E. (2016). *Estimating seismic reflection slopes*. Dissertation, Colorado School of Mines.

Bader, S., Spikes, K., & Fomel, S. (2018). Missing well-log data prediction using Bayesian approach in the relative-geologic time domain. In *SEG international exposition and annual meeting* (pp. SEG-2018). SEG. https://doi.org/10.1190/segam2018-2997278.1

Bakker, P., van Vliet, L. J., & Verbeek, P. W. (1999). Edge preserving orientation adaptive filtering. In *Proceedings. 1999 IEEE computer society conference on computer vision and pattern recognition (Cat. No PR00149)* (Vol. 1, pp. 535–540). IEEE. https://doi.org/10.1109/CVPR.1999.786989

Bi, Z., Wu, X., Geng, Z., & Li, H. (2021). Deep relative geologic time: A deep learning method for simultaneously interpreting 3-D seismic horizons and faults. *Journal of Geophysical Research: Solid Earth, 126*(9), e2021JB021882. https://doi.org/10.1029/2021JB021882

Bi, Z., Wu, X., Li, Z., Chang, D., & Yong, X. (2022). DeepISMNet: Three-dimensional implicit structural modeling with convolutional neural network. *Geoscientific Model Development, 15*(17), 6841–6861. https://doi.org/10.5194/gmd-15-6841-2022

Caumon, G., Collon-Drouaillet, P. L. C. D., Le Carlier de Veslud, C., Viseur, S., & Sausse, J. (2009). Surface-based 3D modeling of geological structures. *Mathematical Geosciences, 41*, 927–945. https://doi.org/10.1007/s11004-009-9244-2

de Bruin, G., Hemstra, N., & Pouwel, A. (2007). Stratigraphic surfaces in the depositional and chronostratigraphic (Wheeler-transformed) domain. *The Leading Edge, 26*(7), 883–886. https://doi.org/10.1190/1.2756868

Di, H., Li, Z., Maniar, H., & Abubakar, A. (2020). Seismic stratigraphy interpretation by deep convolutional neural networks: A semisupervised workflow. *Geophysics, 85*(4), WA77–WA86. https://doi.org/10.1190/geo2019-0433.1

Figueiredo, A. M., Gattass, M., & Szenberg, F. (2007). Seismic horizon mapping across faults with growing neural gas. In *10th International Congress of the Brazilian Geophysical Society* (pp. cp–172). European Association of Geoscientists & Engineers. https://doi.org/10.3997/2214-4609-pdb.172.SBGF0266_07

Figueiredo, A. M., Silva, F. B., Silva, P. M., Milidiú, R. L., & Gattass, M. (2014). A seismic facies analysis approach to map 3D seismic horizons. In *SEG international exposition and annual meeting* (pp. 1501–1505.SEG-2014). SEG

Figueiredo, A. M., Silva, F. B., Silva, P. M., de O. Martins, L., L. Milidiú, R., & Gattass, M. (2015). A clustering-based approach to map 3D seismic horizons. In *14th International Congress of the Brazilian Geophysical Society & EXPOGEF* (pp. 1166–1170), Rio de Janeiro, Brazil, August 3–6, 2015. Brazilian Geophysical Society. https://doi.org/10.1190/sbgf2015-233

Fomel, S. (2002). Applications of plane-wave destruction filters. *Geophysics, 67*(6), 1946–1960. https://doi.org/10.1190/1.1527095

Fomel, S. (2010). Predictive painting of 3D seismic volumes. *Geophysics, 75*(4), A25–A30. https://doi.org/10.1190/1.3453847

Gao, H., Wu, X., & Liu, G. (2021). ChannelSeg3D: Channel simulation and deep learning for channel interpretation in 3D seismic images. *Geophysics, 86*(4), IM73–IM83. https://doi.org/10.1190/geo2020-0572.1

Geng, Z., Wu, X., Shi, Y., & Fomel, S. (2019b). Relative geologic time estimation using a deep convolutional neural network. In *89th Annual international meeting, SEG, Expanded abstracts* (pp. 2238–2242). Abstract

Geng, Z., Wu, X., Shi, Y., & Fomel, S. (2019a). Deep learning for relative geologic time and seismic horizons. *Geophysics*, p. in revision.

Geng, Z., Zhao, Z., Shi, Y., Wu, X., Fomel, S., & Sen, M. (2022). Deep learning for velocity model building with common-image gather volumes. *Geophysical Journal International, 228*(2), 1054–1070. https://doi.org/10.1093/gji/ggab385

Grose, L., Laurent, G., Aillères, L., Armit, R., Jessell, M., & Caumon, G. (2017). Structural data constraints for implicit modeling of folds. *Journal of Structural Geology, 104*, 80–92. https://doi.org/10.1016/j.jsg.2017.09.013

Hale, D. (2013). Methods to compute fault images, extract fault surfaces, and estimate fault throws from 3D seismic images. *Geophysics, 78*(2), O33–O43. https://doi.org/10.1190/geo2012-0331.1

Kusznir, N. J., & Karner, G. D. (2007). Continental lithospheric thinning and breakup in response to upwelling divergent mantle flow: Application to the Woodlark, Newfoundland and Iberia margins. *Geological Society, London, Special Publications, 282*(1), 389–419. https://doi.org/10.1144/SP282.16

Li, J., Wu, X., & Hu, Z. (2021). Deep learning for simultaneous seismic image super-resolution and denoising. *IEEE Transactions on Geoscience and Remote Sensing, 60*, 1–11. https://doi.org/10.1109/TGRS.2021.3057857

Lou, Y., Zhang, B., Fang, H., Cao, D., Wang, K., & Huo, Z. (2021a). Simulating the procedure of manual seismic horizon picking. *Geophysics, 86*(1), O1–O12. https://doi.org/10.1190/geo2020-0021.1

Lou, Y., Zhang, B., Wang, R., Lin, T., & Cao, D. (2019). Seismic fault attribute estimation using a local fault model. *Geophysics, 84*(4), O73–O80. https://doi.org/10.1190/geo2018-0678.1

Lou, Y., Zhang, B., Yong, P., Fang, H., Zhang, Y., & Cao, D. (2021b). Semiautomatic fault-surface generation and interpretation using topological metrics. *Geophysics, 86*(3), O13–O27. https://doi.org/10.1190/geo2020-0038.1

Luo, W., Li, Y., Urtasun, R., & Zemel, R. (2016). Understanding the effective receptive field in deep convolutional neural networks. In *Advances in neural information processing systems* (Vol. 29).

Luo, S., & Hale, D. (2013). Unfaulting and unfolding 3D seismic images. *Geophysics, 78*(4), O45–O56. https://doi.org/10.1190/geo2012-0350.1

Marfurt, K. J. (2006). Robust estimates of 3D reflector dip and azimuth. *Geophysics, 71*(4), P29–P40. https://doi.org/10.1190/1.2213049

Marfurt, K. J., Kirlin, R. L., Farmer, S. L., & Bahorich, M. S. (1998). 3-D seismic attributes using a semblance-based coherency algorithm. *Geophysics, 63*(4), 1150–1165. https://doi.org/10.1190/1.1444415

Marfurt, K. J., Sudhaker, V., Gersztenkorn, A., Crawford, K. D., & Nissen, S. E. (1999). Coherency calculations in the presence of structural dip. *Geophysics, 64*(1), 104–111. https://doi.org/10.1190/1.1444508

Milletari, F., Navab, N., & Ahmadi, S. A. (2016, October). V-net: Fully convolutional neural networks for volumetric medical image segmentation. In *2016 fourth international conference on 3D vision (3DV)* (pp. 565–571). IEEE. https://doi.org/10.1109/3DV.2016.79

Mousavi, S. M., Ellsworth, W. L., Zhu, W., Chuang, L. Y., & Beroza, G. C. (2020). Earthquake transformer—an attentive deep-learning model for simultaneous earthquake detection and phase picking. *Nature Communications, 11*(1), 3952. https://doi.org/10.1038/s41467-020-17591-w

Patel, M. D., & McMechan, G. A. (2003). Building 2-D stratigraphic and structure models from well log data and control horizons. *Computers & Geosciences, 29*(5), 557–567. https://doi.org/10.1016/S0098-3004(03)00039-6

Pedersen, S. I., Skov, T., Hetlelid, A., Fayemendy, P., Randen, T., & Sønneland, L. (2003). New paradigm of fault interpretation. In *SEG international exposition and annual meeting* (pp. 350–353. SEG-2003). SEG

Posamentier, H. W., Davies, R. J., Cartwright, J. A., & Wood, L. (2007). Seismic geomorphology–an overview. https://doi.org/10.1144/GSL.SP.2007.277.01.01

Qayyum, F., Betzler, C., & Catuneanu, O. (2017). The Wheeler diagram, flattening theory, and time. *Marine and Petroleum Geology, 86*, 1417–1430. https://doi.org/10.1016/j.marpetgeo.2017.07.034

Rivenæs, J. C., Otterlei, C., Zachariassen, E., Dart, C., & Sjøholm, J. (2005). A 3D stochastic model integrating depth, fault and property uncertainty for planning robust wells, Njord Field, offshore Norway. *Petroleum Geoscience, 11*(1), 57–65. https://doi.org/10.1144/1354-079303-612

Saad, O. M., Chen, Y., Savvaidis, A., Fomel, S., & Chen, Y. (2022). Real-time earthquake detection and magnitude estimation using vision transformer. *Journal of Geophysical Research: Solid Earth, 127*(5), e2021JB023657. https://doi.org/10.1029/2021JB023657

Shi, Y., Wu, X., & Fomel, S. (2019). SaltSeg: Automatic 3D salt segmentation using a deep convolutional neural network. *Interpretation, 7*(3), SE113–SE122. https://doi.org/10.1190/INT-2018-0235.1

Stark, T. J. (2003). Unwrapping instantaneous phase to generate a relative geologic time volume. In *SEG international exposition and annual meeting* (pp. 1707–1710. SEG-2003). SEG.

Stark, T. J. (2004). Relative geologic time (age) volumes—Relating every seismic sample to a geologically reasonable horizon. *The Leading Edge, 23*(9), 928–932. https://doi.org/10.1190/1.1803505

Vail, P. R., Todd, R. G., & Sangree, J. B. (1977). Seismic stratigraphy and global changes of sea level: Part 5. Chronostratigraphic significance of seismic reflections: Section 2. Application of seismic reflection configuration to stratigraphic interpretation. In *M 26: Seismic stratigraphy–applications to hydrocarbon exploration* (Vol. 26, pp. 99–116). AAPG Memoirs.

Van Bemmel, P. P., & Pepper, R. E. (2000). U.S. Patent No. 6,151,555. U.S. Patent and Trademark Office, Washington, DC.

Vasiljevic, I., Kolkin, N., Zhang, S., Luo, R., Wang, H., Dai, F. Z., Daniele, A.F., Mostajabi, M., Basart, S., Walter, M. R., & Shakhnarovich, G. (2019). *Diode: A dense indoor and outdoor depth dataset.* arXiv preprint arXiv:1908.00463. https://doi.org/10.48550/arXiv.1908.00463

Vaswani, A., Shazeer, N., Parmar, N., Uszkoreit, J., Jones, L., Gomez, A. N., Kaiser, Ł., & Polosukhin, I. (2017). Attention is all you need. In *Advances in neural information processing systems* (Vol. 30).

Wang, Z., Simoncelli, E. P., & Bovik, A. C. (2003). Multiscale structural similarity for image quality assessment. In *The Thirty-seventh Asilomar conference on signals, systems & computers* (Vol. 2, pp. 1398–1402). IEEE. https://doi.org/10.1109/ACSSC.2003.1292216

Wu, X., & Hale, D. (2015). Horizon volumes with interpreted constraints. *Geophysics, 80*(2), IM21–IM33. https://doi.org/10.1190/geo2014-0212.1

Wu, X., Luo, S., & Hale, D. (2016). Moving faults while unfaulting 3D seismic images. *Geophysics, 81*(2), IM25–IM33. https://doi.org/10.1190/geo2015-0381.1

Wu, X., & Hale, D. (2016). 3D seismic image processing for faults. *Geophysics, 81*(2), IM1–IM11. https://doi.org/10.1190/geo2015-0380.1

Wu, X., & Fomel, S. (2018b). Least-squares horizons with local slopes and multigrid correlations. *Geophysics, 83*(4), IM29–IM40. https://doi.org/10.1190/geo2017-0830.1

Wu, X., Liang, L., Shi, Y., & Fomel, S. (2019). FaultSeg3D: Using synthetic data sets to train an end-to-end convolutional neural network for 3D seismic fault segmentation. *Geophysics, 84*(3), IM35–IM45. https://doi.org/10.1190/geo2018-0646.1

Wu, X., Geng, Z., Shi, Y., Pham, N., Fomel, S., & Caumon, G. (2020). Building realistic structure models to train convolutional neural networks for seismic structural interpretation. *Geophysics, 85*(4), WA27–WA39. https://doi.org/10.1190/geo2019-0375.1

Wu, X., Ma, J., Si, X., Bi, Z., Yang, J., Gao, H., Xie, D., Guo, Z., & Zhang, J. (2023). Sensing prior constraints in deep neural networks for solving exploration geophysical problems. *Proceedings of the National Academy of Sciences, 120*(23), e2219573120. https://doi.org/10.1073/pnas.2219573120

Wu, X., & Fomel, S. (2018a). Automatic fault interpretation with optimal surface voting. *Geophysics, 83*(5), O67–O82. https://doi.org/10.1190/geo2018-0115.1

Wu, X., & Janson, X. (2017). Directional structure tensors in estimating seismic structural and stratigraphic orientations. *Geophysical Journal International, 210*(1), 534–548. https://doi.org/10.1093/gji/ggx194

Wu, X., & Zhong, G. (2012). Generating a relative geologic time volume by 3D graph-cut phase unwrapping method with horizon and unconformity constraints. *Geophysics, 77*(4), O21–O34. https://doi.org/10.1190/geo2011-0351.1

Yang, L., Fomel, S., Wang, S., Chen, X., Chen, W., Saad, O. M., & Chen, Y. (2023). Porosity and permeability prediction using a transformer and periodic long short-term network. *Geophysics, 88*(1), WA293–WA308. https://doi.org/10.1190/geo2022-0150.1

Yang, F., & Ma, J. (2019). Deep-learning inversion: A next-generation seismic velocity model building method. *Geophysics, 84*(4), R583–R599. https://doi.org/10.1190/geo2018-0249.1

Yu, Y., Kelley, C., & Mardanova, I. (2013). Volumetric seismic dip and azimuth estimation with 2D log-Gabor filter array. In *SEG international exposition and annual meeting* (pp. 1357–1362.SEG-2013). SEG

Yu, S., Ma, J., & Wang, W. (2019). Deep learning for denoising. *Geophysics, 84*(6), V333–V350. https://doi.org/10.1190/geo2018-0668.1

Zeng, H., Backus, M. M., Barrow, K. T., & Tyler, N. (1998). Stratal slicing; Part 1, realistic 3-D seismic model. *Geophysics, 63*(2), 502–513. https://doi.org/10.1190/1.1444351

Zeng, Q., Liu, J., Chu, S., Wang, Y., Sun, Y., Duan, X., & Zhou, L. (2012). Mesozoic molybdenum deposits in the East Xingmeng orogenic belt, northeast China: characteristics and tectonic setting. *International Geology Review, 54*(16), 1843–1869. https://doi.org/10.1080/00206814.2012.677498

Zhao, H., Gallo, O., Frosio, I., & Kautz, J. (2016). Loss functions for image restoration with neural networks. *IEEE Transactions on Computational Imaging, 3*(1), 47–57. https://doi.org/10.1109/TCI.2016.2644865

Zu, S., Cao, J., Qu, S., & Chen, Y. (2020). Iterative deblending for simultaneous source data using the deep neural network. *Geophysics, 85*(2), V131–V141. https://doi.org/10.1190/geo2019-0319.1

Zu, S., Ke, C., Hou, C., Cao, J., & Zhang, H. (2022). End-to-end deblending of simultaneous source data using transformer. *IEEE Geoscience and Remote Sensing Letters, 19*, 1–5. https://doi.org/10.1109/LGRS.2022.3174106

Chapter 12
New Mechanisms for Earthquakes Distribution and Fault Patterns

Abstract In this chapter, we present evidence for damage stratigraphy in which a stratigraphic unit evolves dynamically into a weak decollement zone. The analysis is focused on the internal structure of a deep, tectonically active, decollement zone in the Sichuan Basin, China. The analyzed zone is about a 1 km thick sequence of Triassic rocks that extends across most of the Sichuan Basin at a depth range of 5.8–6.8 km. This sequence is composed of evaporites (mostly anhydrite) and carbonates (dolomite and limestone). We used three-dimensional seismic surveys, borehole observations, and rock-mechanics experiments to document pervasive internal damage revealed by fractures and flow features within this unit. As this damaged unit is defined by its stratigraphic position, we term it a Damaged Stratigraphic Unit (DSU). The damage-induced weakness of the DSU controls two active tectonic features in the western Sichuan Basin. First, large, intra-basin, frontal thrust faults terminate against the DSU from above and from below. Second, regional earthquakes display a profound decrease in frequency and seismic moment at the depth interval of 5.8–6.8 km that coincides with the DSU.

12.1 Introduction

12.1.1 Strength of the Layered Crust

The large scale strength and rheological variations of the crust are dominated by depth variations of pressure and temperature (Brace & Kohlstedt, 1980; Ranalli & Murphy, 1987). The crust can be divided into two main parts: an upper brittle part, in which the strength is bounded by frictional slip along discontinuities, and a lower ductile part, in which the strength is determined by nonlinear viscoelasticity that depends on temperature, lithology and strain-rate. It is generally assumed that in tectonically active regions, the in-situ stress depth profile mimics the strength profile, and that the strength profiles can be applied to earthquake distribution (Konstantinou, 2010; Scholz, 1998), the earthquakes cycle (Reches et al., 1994), and large scale decollements associated with plate collisions (Avouac et al., 1993). However, the strength of

Z. Liao, *Applied Seismic Geostructures*, https://doi.org/10.1007/978-981-97-7079-3_12

a rock unit does not necessarily stay constant as the damage processes that are active during tectonic deformation could lead to rock-units weakening causing deformation beyond failure (e.g., Busetti et al., 2012; Kachanov, 1994; Katz & Reches, 2004; Lyakhovsky & Ben-Zion, 2009). The evolving weakening of a stratigraphic unit by ongoing damage is in contrast to relatively static strength associated with thermal, pressure and lithological properties that were applied by Brace and Kohlstedt (1980).

12.1.2 Tectonic Setting of the Present Analysis

We analyze here tectonic effects of the damage weakening of the Triassic age sequence in Sichuan basin, China. The study area is located in the western part of the basin, tens of km east of the seismically active Longmen Shan mountain range (Fig. 12.1a) (Burchfiel et al., 2008; Hubbard & Shaw, 2009; Jia et al., 2010; Lu et al., 2019; Wallis et al., 2003). The Longmen Shan lies to the the eastern Tibetan Plateau and defines a marginal topographic transition from the Plateau to Sichuan Basin. It is characterized by four major fault zones: Beichuan (BCF), Pengxian (PXF): Guankou (GKF), and Longquan (LQF) (Fig. 12.1). Figure 12.1b illustrates the interpreted 3D view of structural architecture for Longmen Shan range and western Sichuan basin, which is likely formed by a Late Triassic compressional episode and a Cenozoic deformation episode during the India–Asia collision (Burchfiel et al., 2008; Delcaillau et al., 2022; Jia et al., 2010; Izquierdo-Llavall et al., 2018; Qin et al., 2022). The uplift due to crustal shortening created the 2008 Wenchuan earthquake and is continuously activating this region and shaping the local topography (Hubbard & Shaw, 2009).

The general stratigraphy of this region consists of a Proterozoic crystalline basement of the Yangtze craton, that is overlain by a sequence of Cambrian dolomite and shales, Permian and Triassic shallow-marine carbonates and evaporites, and Triassic and Jurassic non-marine sandstones and shales (Fig. 12.2) (Jia et al., 2010). The Cenozoic tectonism, which is dominated by the India-Asia collision (Burchfiel et al., 2008), generated a series of north-east trending folds and thrusts.

We focus on the internal damage of the ~1 km thick sequence of Triassic carbonates and evaporites at depth interval of 5.8–6.8 km. The damage of this sequence was directly recognized in borehole logs, core samples, and 3D seismic survey. We show that this unit serves as a thick decollement zone that controls the depth distribution of earthquakes and faults. As the damage is bounded to a stratigraphic unit, we refer to it as 'damage stratigraphy', and the present analysis focuses on the *Damaged Stratigraphic Unit* (DSU) of the Triassic sequence in Sichuan basin (Figs. 12.1 and 12.2).

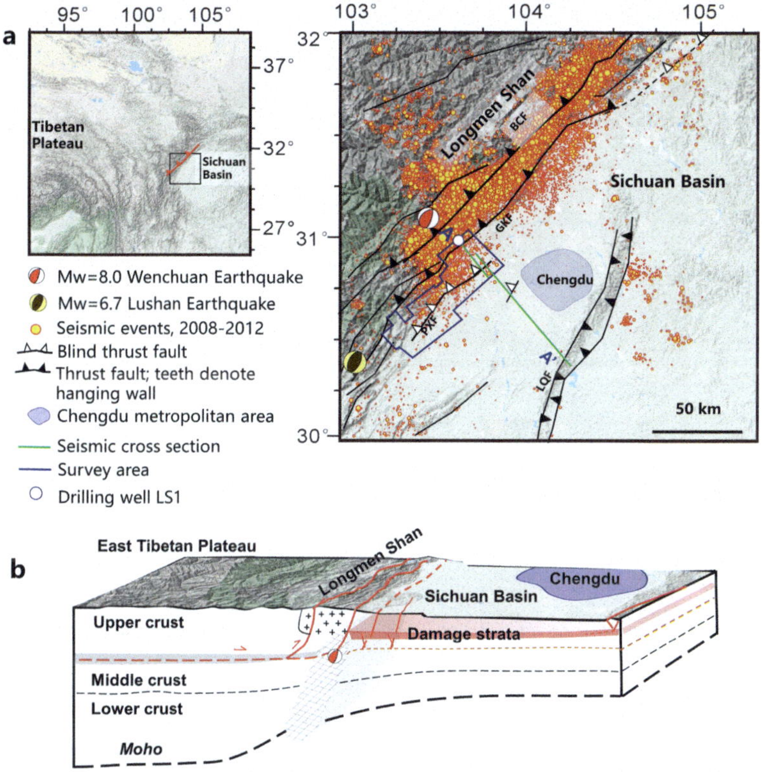

Fig. 12.1 Tectonic setting of the study area adjacent to the Longmen Shan range and Tibetan Plateau. **a** Left: General location with inset of study region. Right: Tectonic map of western Sichuan basin that displays: (1) Region of the analyzed seismic survey covering 1370 km^2 (blue polygon close to southwest corner); (2) Earthquakes epicenters for 2008–2012 period, including the 2008 Wenchuan Mw = 8.0 earthquake, and the 2015 Mw = 6.7 Lushan earthquake; (3) Site of the LS1 well; and (4) The major fault zones, Beichuan (BCF), Pengxian (PXF): Guankou (GKF), and Longquan (LQF). **b** Interpreted 3D view of structural architecture for Longmen Shan range and western Sichuan basin (after Zhang et al., 2009), including the hypocenter of 2018 Wenchuan earthquake and the analyzed Triassic damage unit (*light red*)

12.1.3 Methods of Subsurface Analyses

For the subsurface damage analysis, we utilized the 3D seismic attribute methodology, including the attributes of amplitude, variance, and curvature (Chopra & Marfurt, 2005) (Supporting Information). It was previously found that subsurface damage is best recognized by the variance attribute that represents the scattering response of acoustic waves by a rock body (Chopra & Marfurt, 2005; Iacopini & Butler, 2011; Liao et al., 2019). High values of the variance attribute indicate intensely damaged and fractured body (Iacopini & Butler, 2011), and it was used to map subsurface fault-zones (Liao et al., 2019).

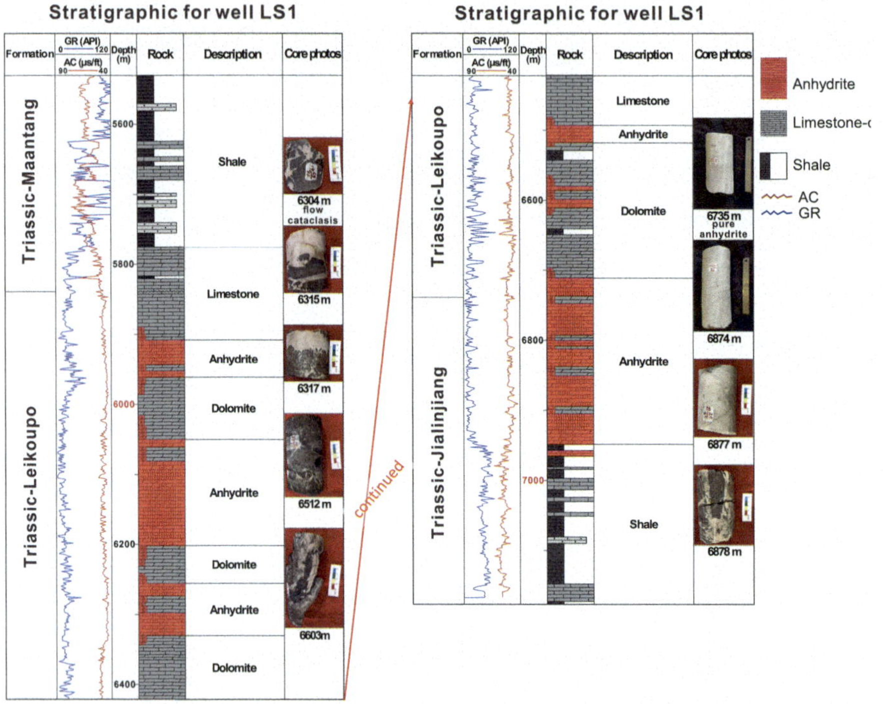

Fig. 12.2 Stratigraphic section of the Triassic damage unit at depth interval of 5.8–7 km as determine in Longshen 1 borehole. The unit's lithology is dominated by anhydrite (*red*), dolomite and limestone, and core images appear in right column. Close-up of selected core images are displayed in Fig. 12.5

12.2 Observations: Structure, Damage, and Earthquakes

12.2.1 Subsurface Structure

We investigated the subsurface structure and internal damage of the Triassic sequence DSU by analyzing the three-dimensional seismic surveys within an area of 1370 km^2 (Fig. 12.1b); these seismic data are part of the basin-scale petroleum exploration (Lu et al., 2019). The analyzed area is located in the frontal region of the Longmen Shan range, as close as 20 km from the Mw8.0 Wenchuan earthquake hypocenter, and it borders many of the aftershocks of this earthquake (Fig. 12.1a).

The structure of the study area is presented by three subsurface maps (Fig. 12.3a–c) and three complementary vertical seismic profiles (Fig. 12.3d). The maps display the variance and curvature attributes at three depth levels: the Upper Triassic dolomite the is positioned above the damage strata, the Middle Triassic sequence that we identify as the damage stratigraphic unit (DSU), and the units below the DSU. These maps show consistent higher variance intensity within the DSU (Fig. 12.3b), and along the

Pengxian fault-zone (PXF in Fig. 12.3). The vertical seismic profiles (Fig. 12.3d) display the DSU at depth of 5.8–6.8 km (light red) where high values of the variance attribute indicate high damage density related to intense fracturing. These apparent displacements of the Pengxian fault (Fig. 12.3b, d) indicate layer-parallel shear within the DSU that ranges from ~8 km shear at the southwest that gradually decreases and vanishes at the northeast after a horizontal distance of ~50 km. The relatively uniform distribution of the variance attribute within the DSU suggests that the shear was accommodated quasi-uniformly as show schematically in Fig. 12.3e.

The Pengxian fault, which is the dominant fault in the study area (PXF in Figs. 12.1b and 12.3), is regarded as a frontal blind thrust that propagated from the Longmen Shan range (Lu et al., 2019). The Pengxian fault displays some striking relationships with the DSU. First, this fault is terminated and displaced by the DSU (red arrows in Fig. 12.3d): In the southwest profile (left profile, Fig. 12.3d), the upper segment of Pengxian fault is displaced eastward by about 8 km relatively to the lower segment, in the middle profile, the fault is displaced only ~2 km, and in the northeast profile (right profile, Fig. 12.3d), the Pengxian fault is not displaced. Further, at the northeast profile, the DSU displays notice thickening at the intersection with the Pengxian fault. Second, the damage zones of the Pengxian fault reveals contrasting structures above and below the DSU. Above the DSU (Fig. 12.3a), the Pengxian fault appears as a relatively linear zone that extends for 55 km by a fold-fault system. Below the DSU (Fig. 12.3c), the fault zone is bent and twisted with curved subsidiary segments, and it extends for greater length than above the DSU. We deduce that the presence of the DSU led to separation of the Pengxian fault into these two contrasting structural styles; the controlling mechanism will be discussed later.

12.2.2 Depth Distribution of Subsurface Damage

We presented above maps and profiles of the variance attribute as indicator of subsurface damage intensity (Fig. 12.3), and now examine the depth distribution of this variance. The seismic profile (Fig. 12.4a) and the attribute depth profile (Fig. 12.4b) reveals three zones of high damage intensity: (1) A shallow zone at <3 km depth, with the most-intensive variance, 0.3–0.5 of intense damage. We relate this damage to intense brittle fracturing under the relatively low confining pressure; (2) The zone at 5.8–6.8 km depth of variance of 0.1–0.2 of the Triassic DSU described above. The mechanisms of the brittle-ductile damage associated with the lithology in this stratigraphic unit as discussed later; and (3) A zone at 8–10 km depth with variance of 0.1–0.2. We relate this damage to deformation localized at the contact between the sedimentary sequence and the crystalline basement.

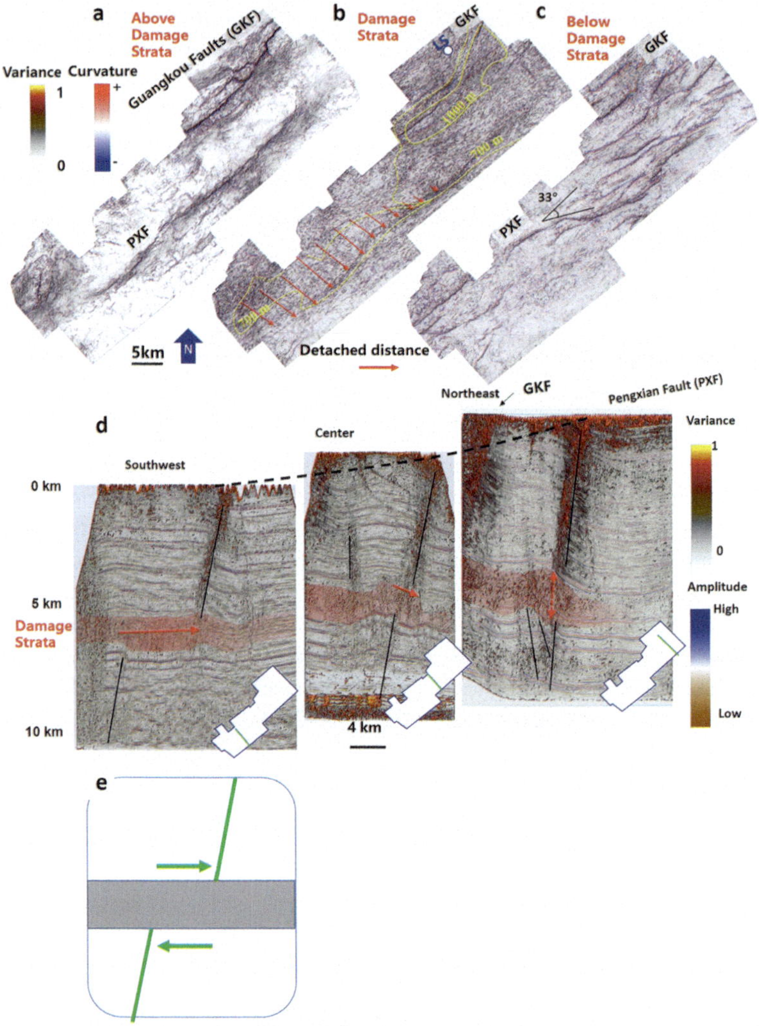

Fig. 12.3 Horizontal and vertical sections of the damage-stratigraphic-unit (DSU) that are displayed by the seismic attribute of variance co-rendered with curvatures recognized in the three-dimensional seismic survey (Methods, Supporting material). **a** Horizontal section of the dolomite layers above the DSU. Note the linear trend of the Pengxian Fault (PXF). **b** Horizontal section within the DSU with thickness contours (*yellow dashed curves*). The red arrows indicate the displacement of the Pengxian Fault by the DSU as recognized by comparing PXF positions in (**a**) and (**c**) (test). **c** Horizontal section of dolomite and limestone layers below the DSU. **d** Vertical seismic sections across the DSU; the location of each section is marked by a green line at the lower inserts; the sections display the amplitude attribute co-rendered with variance attribute, and the Pengxian Fault (*black lines*). **e** A schematic presentation of the apparent displacement of the PXF by ~8 km within the DSU as interpreted in (**d**). Note the DSU (*red*), and the red arrows indicating displacement direction of the PXF by the DSU

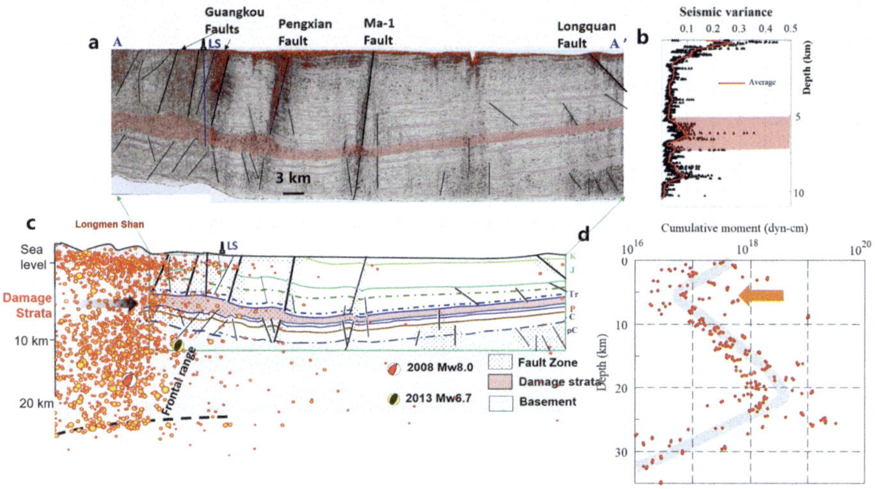

Fig. 12.4 Damage stratigraphy, fault patterns, and earthquake distributions. **a** Extended vertical seismic section in NW–SE direction (*green line A-A' in* Fig. 12.1a); displays amplitude co-rendered with variance attribute. The section displays the DSU (transparent red) and fault systems (*black lines*). Horizontal scale: vertical scale = 1:2. **b** Depth profile of the average values of seismic variance for the stratum surrounding the Pengxian Fault, but excluding the fault damage zones. Note three zones of higher variance values that correspond to intense damage (text). **c** Interpreted geological cross-section of (4**a**), with projected hypocenters of the seismic events. **d** Depth profile of cumulative moments of the earthquakes of the 2008–2012 period (4**c**) calculated at 0.25 km depth intervals. Note a relative increase of cumulative seismic moment at shallow depth (*blue arrow*) and a relative decrease of the cumulative seismic moment at 5–8 km depth of the DSU (text)

12.2.3 Regional Seismic Activity

The study area is located at the margins of the Longmen Shan range, close to the hypocenter of the 2008 Mw8.0 Wenchuan earthquake. To explore the effects of a ~1 km thick DSU on the regional seismic activity, we examined the catalog of more than 40,000 earthquakes in the upper crust (depth < 25 km) for this area during the 2008–2012 period. The spatial distribution of the epicenters is displayed in Fig. 12.1b, and their depth distribution is displayed on a tectonic cross-section across with earthquake hypocenters projected to the section (Fig. 12.4c). The depth distribution of the cumulative seismic moments was calculated in depth intervals of 0.1 km, which is the depth resolution of the earthquakes in the catalog, and the smoothen data in 0.5 km depth intervals is displayed in Fig. 12.4d.

The outstanding feature in Fig. 12.4 is the correlation of the depth distribution of the earthquake cumulative moments (Fig. 12.4d) with the damage zone as observed in Figs. 12.3 and 12.4b. This correlation is noted by the decrease of earthquake moments at hypocenter depths of 4–8 km that correlates with the 5.8–6.8 depth of the damage stratigraphic unit. We argue below that this decrease of seismic activity surrounding DSU can be attributed to the relaxation of tectonic loading by weak DSU.

12.3 Analysis: Rheology and Strength of the DSU in Sichuan Basin

The above descriptions indicate that the Triassic sequence (DSU) in the western Sichuan basin is pervasively damaged, and that its presence affects the fault pattern (Fig. 12.3) and earthquake distribution (Fig. 12.4). We now analyze possible mechanisms for these observations.

The lithological composition of the DSU in the study area was documented in the LS1 borehole that was drilled to a depth of 7,100 m (location in Figs. 12.1b, and 12.4a, c). The borehole logs (Fig. 12.2) indicate that the 5.8–6.8 km depth interval includes ~400 m of anhydrite, ~200 m anhydrite mixed with dolomite breccia, and ~600 m of fractured dolomites and limestones. Rock samples collected during drilling provide direct evidence of pervasive damage within the DSU (Figs. 12.5 and 12.6). These samples reveal cataclasis at scales of centimeter to decimeter of the brittle dolomite samples, and at the millimeter scale in thin sections. For example, Fig. 12.5 shows fractures in the dolomite matrix that are filled with anhydrite that display similar style to the experimental fracturing of a dolomite core embedded inside ductile limestone (Fig. 12.5e). The thin-sections display abundant elongated anhydrite crystal grains that surround the dolomite fragments (Fig. 12.6). The brittle dolomite blocks rotated within the ductile anhydrite zones, that together composed brittle-ductile deformation within the DSU.

12.3.1 Strength and Ductility of Triassic Rocks in Sichuan Basin

We envision that the evaporite/carbonate composition of the Middle and Upper Triassic in Sichuan basin controls the rheology and strength by integration of thermally controlled ductility of the anhydrite and damage cataclasis flow of the brittle carbonates. We evaluate below the strength of the DSU in three steps: experimental tests of samples within the DSU, thermal ductility of anhydrite, and evaluation of damage weakening. Recently, the mechanical properties of Triassic rocks from Sichuan basin were analyzed in rock-mechanics experiments (Supporting Information). The tested samples, which belong to the same stratigraphic unit of the DSU, were collected at a depth range of 4.7–4.8 km in oil production wells located ~100 km away from our study area (Fig. 12.2) (Kibikas, 2021).

We present the experimental results of three samples that represent the lithology range of the DSU (Fig. 12.7): pure anhydrite (sample EV1), calcitic dolomite (sample

Fig. 12.5 **a–d** Images of core samples collected at Longshen 1 borehole (Figs. 12.1a and 12.2) displaying centimeter-scale flow of anhydrite (*bright zones*) that fills the spaces between fractured and fragmented dolomite (*dark gray*). **e** Cross-section of an experimental sample deformed under high confining pressure that displays a core of fractured brittle dolomite embedded within ductile limestone (Griggs & Handin, 1960)

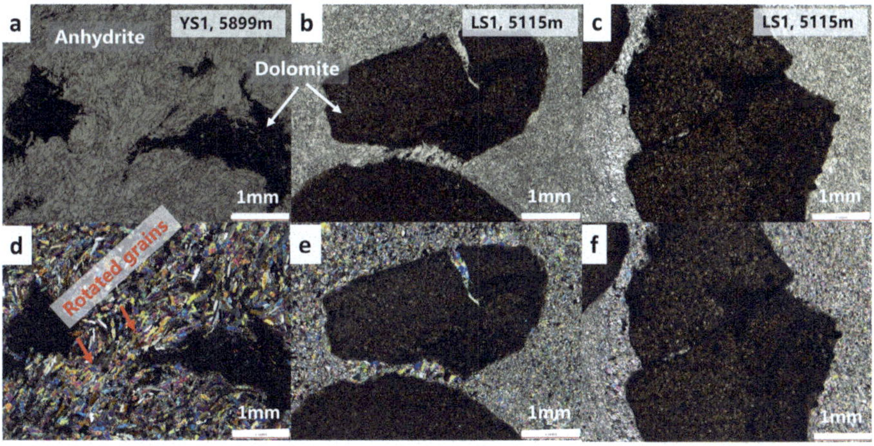

Fig. 12.6 Photomicrographs of thin-sections of core samples from the damage unit displaying anhydrite and dolomite relationships; sample collection depth in upper right. **a–c** plane-light views of dolomite fragments (black) floating in anhydrite (dark gray). **d–f** cross-polarized views of (**a–c**); note elongated anhydrite grains in (**d**)

DS1), and dolomitic-limestone (sample LS2) (Kibikas, 2021). The triaxial testing was conducted on water-saturated samples that were loaded to failure and beyond under 22.5 MPa confining pressure and temperatures of 23 and 50 °C. The samples reached the ultimate strength at an axial strain of 0.5–0.7%. The stress–strain data of these experiments show that the anhydrite (EV1) deformed in a ductile style even at room temperature with only gentle strength reduction with increasing strain (Fig. 12.7). The carbonate samples (LS2 and DS1) failed in a brittle-ductile style at both temperatures, but continued to flow in a quasi-ductile style after axial strain > 1% under approximately constant differential stress of ~130 MPa (LS1) and ~160 MPa (DS1). We envision that under in-situ conditions of 5.8–6.8 km depth of the DSU will flow at even lower differential stresses.

12.3.2 Thermally Controlled Ductility

The large scale strength of the crust is evaluated for a brittle upper part, in which the strength is bounded by frictional slip, and a ductile lower part, in which the strength is controlled by viscoelasticity (Brace & Kohlstedt, 1980). We apply this approach to the study area by assuming that the DSU strength is governed by the flow of its anhydrite component. We used the experimental analysis of Moller and Briegel (1978) who tested Riburg anhydrite samples from a borehole in the Jura tectonic region, Switzerland. These samples were deformed at 150 MPa confining pressure, temperatures up to 450 °C, and strain-rates as low as 10^{-7} s^{-1}. Moller and Briegel (1978) found that at temperature of ~200 °C and under tectonic strain-rate of 10^{-14} s^{-1}, which are the approximate conditions at the DSU depth, the tested anhydrite is stronger than halite and weaker than Solenhofen limestone (Fig. 12.7 of Moller & Briegel, 1978). We regard these experimental results as a suitable representative of the pure anhydrite layers in Sichuan basin. The strength profile for Sichuan basin was calculated (Supporting Information) by application of anhydrite rheology to the rock sequence at the DSU depth (Fig. 12.8a). The profile indicates that the shear strength of this unit is 30–60% of the strength of the layers above and below it.

12.3.3 Damage Induced Weakening

Experimental analyses revealed that brittle rocks are continuously damaged by microfracturing from loading initiation to failure and beyond (e.g., Lockner et al., 1991). The microfracturing damage leads to global weakening and reduction of Young modulus, as demonstrated in experiments (Katz & Reches, 2004; Lockner et al., 1991), predicted in theoretical modeling (Kachanov, 1994; Lyakhovsky & Ben-Zion, 2009), and simulated in numerical calculations (Busetti et al, 2012) (Fig. 12.7b). These analyses indicate that after failure, damaged brittle rocks may

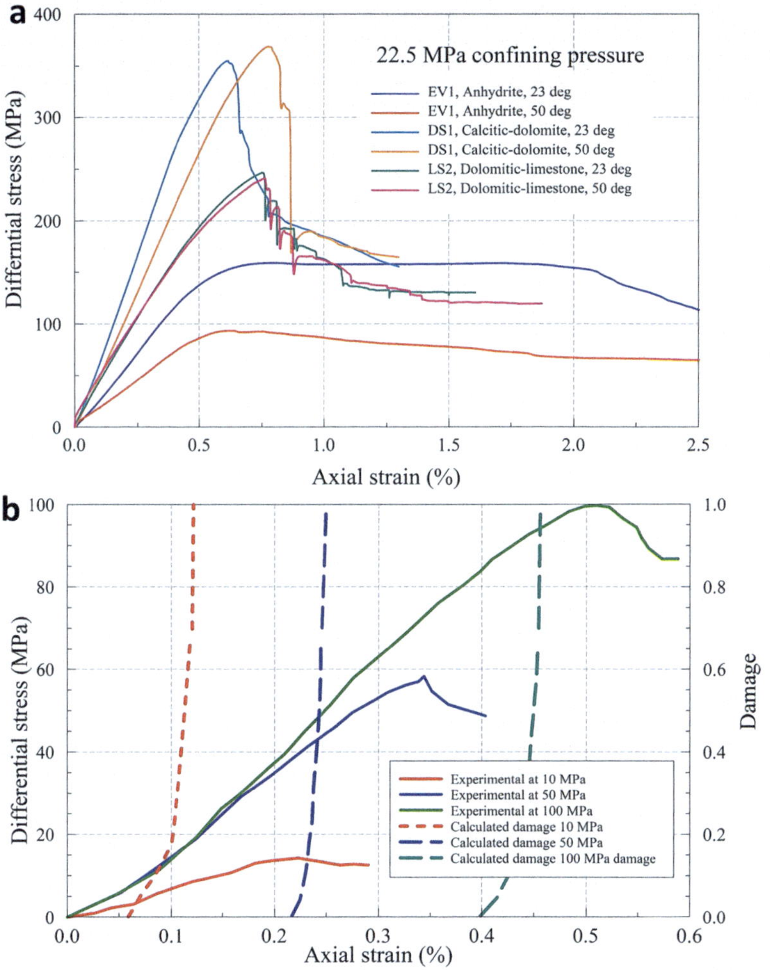

Fig. 12.7 Stress–strain relationships in rock-mechanics experiments of anhydrite and carbonates samples. **a** Differential stress (axial stress—confining pressure) as function of axial strain of three rock types from the Triassic damage unit in Sichuan basin; composition and experimental details appear in text, after Kibikas (2021). The plot shows results of saturated samples at 22.5 MPa effective confining pressure that were loaded to failure and beyond at temperatures of 23 and 50 °C. **b** Differential stress curves of three experiments of Berea sandstone loaded at three confining pressures (*solid curves*); the numerically simulated damage intensity for each test in marked by dashed curves. Modified after Fig. 11.6 of Busetti et al. (2012)

continuously deform under a quasi-constant, lower differential stress by cataclastic flow (Fig. 12.7b) (Busetti et al., 2012; Vajdova et al., 2010; Wong et al., 1997).

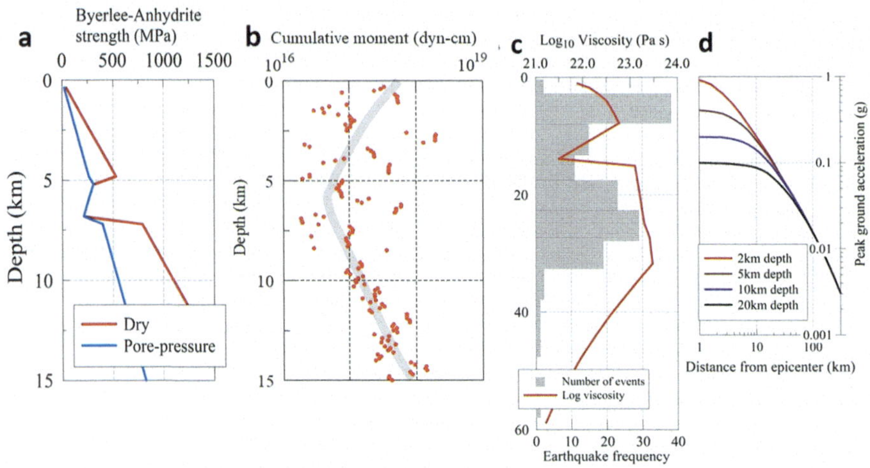

Fig. 12.8 Crustal strength and earthquake depth distribution. **a** The crust strength in Sichuan basin as calculated to two rheological components: frictional strength of the brittle part following Byerlee law, and ductile strength of the evaporite-carbonate sequence by using the experimental rheological results of anhydrite (text and Supporting Information). **b** Depth profile of cumulative moments of the earthquakes of the 2008–2012 period calculated at 0.25 km depth intervals (same as Fig. 12.4d, text). **c** The relationships between the crustal strength, shown by effective viscosity (*red curve*), and depth distribution of 142 earthquakes (*gray columns*) in the southern Aegean region (modified after Konstantinou, 2010, see text). Note the positive correlation between the strength reduction and reduction of earthquake occurrence at the 12–16 km depth. **d** Expected intensity of peak horizontal acceleration for a moderate earthquake (Ms 5.5) for the displayed range of hypocentral depths and distance from the earthquake epicenter (after Bommer et al., 2001)

12.4 Discussion: Damage Stratigraphy and Earthquakes Distribution Decollement

12.4.1 DSU Control of Fault Patterns

The pervasive damage in the DSU (Figs. 12.3 and 12.4b), the evaporite flow (Fig. 12.5), and rock-mechanics observations (Fig. 12.7), all indicate that the Middle-Upper Triassic sequence in Sichuan basin is relatively weak. This weakening evolved by a combination of ductile flow of the evaporites layers coupled with cataclastic flow of the damaged brittle carbonates. This ~1 km thick DSU has the suitable properties to become a decollement zone: relative weakness (Figs. 12.8a and 12.9a), sub-horizontal position, and regional extension across the entire Sichuan basin. The effects of this Triassic decollement is first examined with respect to the termination and shifting of crustal-scale faults (Figs. 12.3d and 12.4a).

It was shown above that the Pengxian fault terminates toward the DSU and shifted by it (Figs. 12.3d and 12.4a). In a brittle medium, the front zone of a fault is characterized by intense stress amplification that facilitates the fault propagation (e.g.,

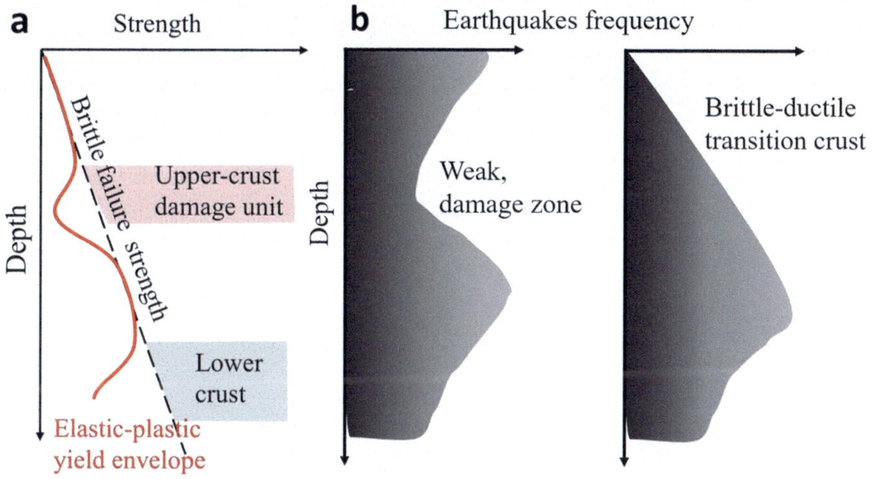

Fig. 12.9 Schematic model of crustal strength and earthquake depth distribution. **a** Schematic strength profile for sedimentary sequences with a weak, damage unit. **b** Expected earthquake depth distribution at the presence of a weak zone due to damage-induced thermal control or lithology (see text and Fig. 12.8)

Reches & Lockner, 1994). However, the inverse is expected as a fault approaches a weak region in which the failure/flow relaxes and diffuses the stress amplification. As a result, the fault propagation as a distinct discontinuity is blunted, and the localized shear is converted to distributed deformation. For example, experimental simulations with viscous layering as evaporite analogue revealed blunting of brittle faults (Brun et al., 1994; Koyi & Skelton, 2001). Similarly, the blunting of fault propagation by ductile layer was documented in rifting off-shore Norway where salt layers of Middle-Upper Triassic disconnected between the faults in the brittle layers above a below. We envision that the thick weak-ductile layers of the present DSU, act as agent of rupture blunting (Begley et al., 1977) that leads to wide deformation zone (Fig. 12.3d), and prevents fault propagation across it.

12.4.2 DSU Control of Earthquake Distribution

Distinct reduction of earthquake occurrence with depth (Fig. 12.3c) was observed at the proximity of weak, ductile/viscous zones elsewhere (e.g., Bonner et al., 2003). For example, Sibson (1982) reviewed the depth distribution of earthquakes in the continental US, and concluded that "…the depth of seismic activity is controlled by the passage from a pressure-sensitive, dominantly frictional regime to strongly temperature-dependent, quasi-plastic mylonitization at greenschist and higher grades…". Konstantinou (2010) analyzed the relationships between the crustal strength, earthquake distribution, and the 1956 Amorgos earthquake in the southern

Aegean region (Fig. 12.8c). Their heat flow and lithology calculations revealed that the interval of 7–15 km depth has the lowest effective viscosity, namely the weakest layer, in the upper 30 km of the crust (red curve in Fig. 12.8c). Konstantinou (2010) further showed that the depth distribution of 142 relocated earthquakes in these regions display profound reduction at the lower-crust depth (bars in Fig. 12.8c). Similarly, we attribute the reduction of earthquake occurrence in our analysis (Fig. 12.4d, Fig. 12.8b) to the damage-controlled weakness that relaxes the buildup of in-situ stresses during interseismic periods by continuous damage and cataclastic flow.

12.4.3 Potential DSU Effect on Geohazard

Our analysis also reveals a profound increase of the occurrence of earthquakes in the upper two kilometers of the study area (blue arrow in Fig. 12.4d). This observation may pose increased geohazard to Chengdu region. The regional geohazard estimations focused on the moderate-to-large earthquakes of the Longmen Shan range (Burchfiel et al., 2008; Hubbard et al., 2010; Jia et al., 2010; Lu et al., 2019; Wallis et al., 2003). However, small-to-moderate magnitude earthquakes that occur at shallow depth and near-field can pose major local hazard. Bommer et al. (2001) described a few cases of moderate earthquakes with catastrophic outcome, and stated: "…there have been many cases where earthquakes of moderate and even small magnitude have caused very significant destruction when they have coincided with population centers. Even though the area of intense ground shaking caused by such events is generally small, the epicentral motions can be severe enough to cause damage even in well-engineered structures." Further, Bommer et al. (2001), emphasized the combined effect of epicenter distance and earthquakes depth, for example, the predicted peak ground acceleration for an M5.5 earthquake at 1 km epicenter distance and 20 km depth is 0.1 g, but the same event at 2 km depth could generate ~1 g of peak acceleration (Fig. 12.8d). Based on our observations (Fig. 12.4c–d) and the analysis of Bommer et al. (2001), we argue that Chengdu area may suffer significant ground motion during small-to-moderate shallow earthquakes (Fig. 12.9b) if they occur in the near-field.

12.5 Conclusions

The present analysis demonstrates that the pervasive damage and flow within a thick stratigraphic unit would generate a weakened zone during tectonic deformation. This unit will serve as a perfect, thick decollement zone within the upper crust that controls the style of faulting and earthquake distribution. While our investigation focused on the western Sichuan basin, the lithological characteristics on the studied Triassic unit are common in sedimentary basins, and we argue that similar damage decollement zones are active elsewhere.

References

Avouac, J.-P., Tapponnier, P., Bai, M., You, H., & Wang, G. (1993). Active thrusting and folding along the northern Tien-Shan and Late Cenozoic rotation of the Tarim relative to Dzungaria and Kazakhstan. *Journal of Geophysical Research, 98*(B4), 6755–6804. https://doi.org/10.1029/92JB01963

Begley, J. A., Logsdon, W. A., & Landes, J. D (1977) Ductile rupture blunt-notch fracture criterion. In *Flaw growth and fracture*. ASTM Int

Bommer, J. J., Georgallides, G., & Tromans, I. J. (2001). Is there a near-field for small-to-moderate magnitude earthquakes? *Bulletin of Earthquake Engineering, 5*, 395–423.

Bonner, J. L., Blackwell, D. D., & Herrin, E. T. (2003). Thermal constraints on earthquake depths in California. *Bulletin of the Seismological Society of America, 93*, 2333–2354. https://doi.org/10.1785/0120030041

Brace, W. F., & Kohlstedt, D. L. (1980). Limits on lithospheric stress imposed by laboratory experiments. *Journal of Geophysical Research, 85*, 6248–6252. https://doi.org/10.1016/0148-9062(82)90774-4

Brun, J. P., Sokoutis, D., & Van Den Driessche, J. (1994). Analogue modeling of detachment fault systems and core complexes. *Geology, 22*, 319–322. https://doi.org/10.1130/0091-7613(1994)022%3c0319:AMODFS%3e2.3.CO;2

Burchfiel, B. C., Royden, L. H., Van der Hilst, R. D., & Hager, B. H. (2008). A geological and geophysical context for the Wenchuan earthquake of 12 May 2008, Sichuan, PRC. *GSA Today, 18*, 4–11. https://doi.org/10.17615/yw9w-jm47

Busetti, S., Mish, K., & Reches, Z. (2012). Damage and plastic deformation of reservoir rocks: Part 1. Damage fracturing. *AAPG Bulletin, 96*, 1687–1709. https://doi.org/10.1306/02011211010

Chopra, S., & Marfurt, K. J (2005) Seismic attributes—A historical perspective. *Geophysics, 70*, 3SO–28SO. https://doi.org/10.1190/1.2098670

Delcaillau, B., Graveleau, F., Saint Carlier, D., Rao, G., Le Béon, M., Charreau, J., & Nexer, M. (2022). Geomorphic analysis of active fold growth and landscape evolution in the central Qiulitage fold belt, southern Tian Shan, China. *Geomorphology, 398*, 108063. https://doi.org/10.1016/j.geomorph.2021.108063

Griggs, D., & Handin, J. (1960). Observations on fracture and a hypothesis of earthquakes. In: Griggs, & Handin (Eds.), *"Rock deformation" GSA Memoir* (Vol. 79, Ch. 13).

Hubbard, J., & Shaw, J. H. (2009). Uplift of the LMS and Tibetan Plateau, and the 2008 Wenchuan (M=7.9) earthquake. *Nature, 458*, 194–197.

Hubbard, J., Shaw, J. H., & Klinger, Y. (2010). Structural setting of the 2008 M w 7.9 Wenchuan, China, earthquake. *Bulletin of the Seismological Society of America, 100*, 2713–2735. https://doi.org/10.1785/0120090341

Iacopini, D., & Butler, R. W. (2011). Imaging deformation in submarine thrust belts using seismic attributes. *Earth and Planetary Science Letters, 302*, 414–422. https://doi.org/10.1016/j.epsl.2010.12.041

Izquierdo-Llavall, E., Roca, E., Xie, H., Pla, O., Muñoz, J. A., Rowan, M. G., Yuban, S., & Huang, S. (2018). Influence of overlapping décollements, syntectonic sedimentation, and structural inheritance in the evolution of a contractional system: The central Kuqa fold-and-thrust belt (Tian Shan Mountains, NW China). *Tectonics, 37*(8), 2608–2632. https://doi.org/10.1029/2017TC004928

Jia, D., Li, Y., Lin, A., Wang, M., Chen, W., Wu, X., Ren, Z., Zhao, Y., & Luo, Y. (2010). Structural model of 2008 Mw 7.9 Wenchuan earthquake in the rejuvenated Longmen Shan thrust belt, China. *Tectonophysics, 491*, 174–184. https://doi.org/10.1016/j.tecto.2009.08.040

Kachanov, M. (1994). On the concept of damage in creep and in the brittle-elastic range. *International Journal of Damage Mechanics, 3*, 329–337. https://doi.org/10.1177/105678959400300402

Katz, O., & Reches, Z. (2004). Microfracturing, damage, and failure of brittle granites. *Journal of Geophysical Research, 109*. https://doi.org/10.1029/2002JB001961

Kibikas, W. (2021). *Characterizing rock properties and their impact on the mechanical behavior of crystalline basement and caprocks*. Dissertation, University of Oklahoma.

Konstantinou, K. I. (2010). Crustal rheology of the Santorini-Amorgos zone: Implications for the nucleation depth and rupture extent of the 9 July 1956 Amorgos earthquake, southern Aegean. *Journal of Geodynamics, 50*, 400–409. https://doi.org/10.1016/j.jog.2010.05.002

Koyi, H. A., & Skelton, A. (2001). Centrifuge modelling of the evolution of low-angle decollement faults from high-angle normal faults. *Journal of Structural Geology, 23*, 1179–1185. https://doi.org/10.1016/S0191-8141(00)00185-1

Liao, Z., Liu, H., Carpenter, B. M., Marfurt, K. J., & Reches, Z. (2019). Analysis of fault damage zones using three-dimensional seismic coherence in the Anadarko Basin, Oklahoma. *AAPG Bulletin, 103*, 1771–1785. https://doi.org/10.1306/1219181413417207

Lockner, D., Byerlee, J. D., Kuksenko, V., Ponomarev, A., & Sidorin, A. (1991). Quasi-static fault growth and shear fracture energy in granite. *Nature, 350*, 39–42. https://doi.org/10.1038/350039a0

Lu, R., He, D., Xu, X., Tan, X., Li, Y., Cai, M., & Wang, Z. (2019). Geometry and kinematics of buried structures in the piedmont of the central Longmen Shan: Implication for the growth of the Eastern Tibetan Plateau. *Journal of the Geological Society, 176*, 323–333. https://doi.org/10.1144/jgs2018-015

Lyakhovsky, V., & Ben-Zion, Y. (2009). Evolving geometrical and material properties of fault zones in a damage rheology model. Geochemistry, Geophysics, *Geosystems, 10*(11). https://doi.org/10.1029/2009GC002543

Moller, W. H., & Briegel, U. (1978). The rheological behavior of polycrystalline anhydrite. *Eclogae Geologicae Helvetiae, 71*, 397–407.

Qin, X., Chen, X., Shao, Z., Zhang, Y., Wang, Y., & Li, B. (2022). Cenozoic multi-phase intra-continental deformation of the Tianshan Range (NW China): Constraints from detrital zircon provenance and syn-tectonic sedimentation of the Kuqa Depression. *Journal of Asian Earth Sciences, 232*, 105183. https://doi.org/10.1016/j.jseaes.2022.105183

Ranalli, G., & Murphy, D. C. (1987). Rheological stratification of the lithosphere. *Tectonophysics, 132*, 281–295. https://doi.org/10.1016/0040-1951(87)90348-9

Reches, Z., & Lockner, D. A. (1994). Nucleation and growth of faults in brittle rocks. *Journal of Geophysical Research, 99*(B9), 18159–18173. https://doi.org/10.1029/94JB00115

Reches, Z. E., Schubert, G., & Anderson, C. (1994). Modeling of periodic great earthquakes on the San Andreas fault: Effects of nonlinear crustal rheology. *Journal of Geophysical Research: Solid Earth, 99*(B11), 21983–22000. https://doi.org/10.1029/94JB00334

Scholz, C. H. (1998). Earthquakes and friction laws. *Nature, 391*, 37–42. https://doi.org/10.1038/34097

Sibson, R. H. (1982). Fault zone models, heat flow, and the depth distribution of earthquakes in the continental crust of the United States. *Bulletin of the Seismological Society of America, 72*, 151–163. https://doi.org/10.1785/BSSA0720010151

Vajdova, V., Zhu, W., Chen, T. M. N., & Wong, T. F. (2010). Micromechanics of brittle faulting and cataclastic flow in Tavel limestone. *Journal of Structural Geology, 32*, 1158–1169. https://doi.org/10.1016/j.jsg.2010.07.007

Wallis, S., Tsujimori, T., Aoya, M., Kawakami, T., Terada, K., Suzuki, K., & Hyodo, H. (2003). Cenozoic and Mesozoic metamorphism in the Longmenshan orogeny: Implications for geodynamic models of eastern Tibet. *Geology, 31*, 745–748. https://doi.org/10.1130/G19562.1

Wong, T. F., David, C., & Zhu, W. (1997). The transition from brittle faulting to cataclastic flow in porous sandstones: Mechanical deformation. *Journal of Geophysical Research, 102*, 3009–3025. https://doi.org/10.1029/96JB03281

Zhang, Y., Dong, S., & Yang, N. (2009). Active faulting pattern, present-day tectonic stress field and block kinematics in the east Tibetan Plateau. *Acta Geologica Sinica, 83*, 694–712. https://doi.org/10.1111/j.1755-6724.2009.00093.x

Appendix A
Simulation Tests

As shown in Fig. A1, we need to do three simulation tests for calculating the elements of an equivalent stiffness tensor of a 2D model of interest according to the 2D stress–strain correlation. Let $D = (0, L_x) \times \left(0, L_y\right)$ be an area that represents the interest model. Performing the first test (Fig. A1a), a uniform vertical displacement is imposed on the bottom and top boundaries, with two lateral boundaries confined and no shear stress applied. As a result, the x- and y-directions' boundary conditions are as follows

$$\begin{cases} \mathbf{U} \cdot \nu = -\Delta U, & (x, y) \in \Gamma^b \cup \Gamma^t, \\ \mathbf{U} \cdot \nu = 0, & (x, y) \in \Gamma^l \cup \Gamma^r, \\ (\sigma \nu) \cdot \chi = 0, & (x, y) \in \Gamma. \end{cases} \tag{A1}$$

where $\mathbf{U}$ is the solid displacement, ΔU is the displacement shift, $\boldsymbol{\sigma}$ represents the total stress tensor, Γ is the model boundary, Γ^b, Γ^t, Γ^l, and Γ^r are the bottom, top, left, and right boundary, respectively, χ is a unit tangent and ν is the unit outer normal so that $\{\nu, \chi\}$ is an orthonormal system on Γ.

Figure A1b illustrates the second test analogous to the first with a uniform horizontal displacement to the right and left boundaries and the two other boundaries confined. Thus, the corresponding boundary conditions are expressed as

$$\begin{cases} \mathbf{U} \cdot \nu = -\Delta U, & (x, y) \in \Gamma^l \cup \Gamma^r, \\ \mathbf{U} \cdot \nu = 0, & (x, y) \in \Gamma^b \cup \Gamma^t, \\ (\sigma \nu) \cdot \chi = 0, & (x, y) \in \Gamma. \end{cases} \tag{A2}$$

For the third test (Fig. A1c), a shear test is applied to the interested model, where the top and bottom boundaries are loaded with displacement. The corresponding boundary conditions are

Z. Liao, *Applied Seismic Geostructures*, https://doi.org/10.1007/978-981-97-7079-3

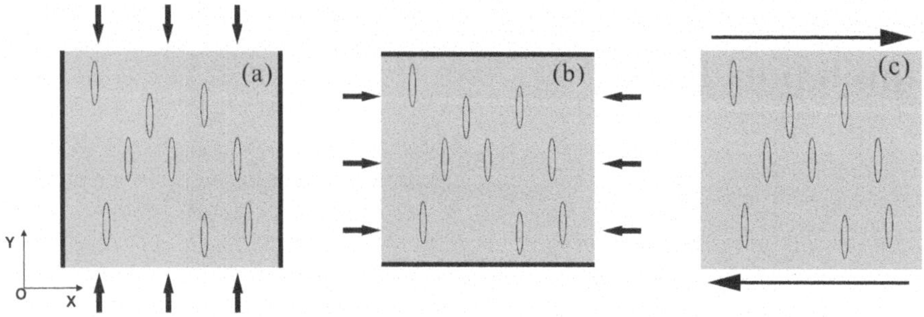

Fig. A1 Sketch map of simulation tests to determine 2D stiffness tensors. The black arrows denote the different orientations of solid displacements. Let the solid displacement vertical to the boundary on the black edge lines be zero

$$\begin{cases} \mathbf{U} \cdot \chi = -\Delta U, & (x, y) \in \Gamma^b \cup \Gamma^t, \\ \mathbf{U} \cdot \chi = 0, & (x, y) \in \Gamma^l \cup \Gamma^r, \\ (\sigma \nu) \cdot \nu = 0, & (x, y) \in \Gamma^l. \end{cases} \tag{A3}$$

Appendix B
Least-Squares Procedure for 2D Equivalent Stiffness Tensors

The least-squares procedure to minimize the following cost function $F(\mathbf{M})$ for computing equivalent stiffness matrix M_{ij} is described as below.

With the goal, the partial derivatives of the cost function related to the elements M_{ij} are set to be 0, resulting in the linear equations:

$$\frac{\partial F(\mathbf{M})}{\partial M_{11}} = 0$$

$$\sum_{s=1}^{3} \overline{\sigma_{xx}^{s} e_{xx}^{s}} = M_{11} \sum_{s=1}^{3} \overline{e_{xx}^{s} e_{xx}^{s}} + M_{12} \sum_{s=1}^{3} \overline{e_{yy}^{s} e_{xx}^{s}} + 2M_{16} \sum_{s=1}^{3} \overline{e_{xy}^{s} e_{xx}^{s}}, \tag{B1}$$

$$\frac{\partial F(\mathbf{M})}{\partial M_{12}} = 0$$

$$\begin{aligned} \sum_{s=1}^{3} \left(\overline{\sigma_{xx}^{s} e_{yy}^{s}} + \overline{\sigma_{yy}^{s} e_{xx}^{s}} \right) &= M_{11} \sum_{s=1}^{3} \overline{e_{xx}^{s} e_{yy}^{s}} + M_{12} \sum_{s=1}^{3} \left(\overline{e_{xx}^{s} e_{xx}^{s}} + \overline{e_{yy}^{s} e_{yy}^{s}} \right) \\ &+ 2M_{16} \sum_{s=1}^{3} \overline{e_{xy}^{s} e_{yy}^{s}} + 2M_{26} \sum_{s=1}^{3} \overline{e_{xy}^{s} e_{xx}^{s}} + M_{22} \sum_{s=1}^{3} \overline{e_{xx}^{s} e_{yy}^{s}}, \end{aligned} \tag{B2}$$

$$\frac{\partial F(\mathbf{M})}{\partial M_{16}} = 0$$

$$\begin{aligned} \sum_{s=1}^{3} \left(2\overline{\sigma_{xx}^{s} e_{xy}^{s}} + \overline{\sigma_{xy}^{s} e_{xx}^{s}} \right) &= 2M_{11} \sum_{s=1}^{3} \overline{e_{xx}^{s} e_{yy}^{s}} + 2M_{12} \sum_{s=1}^{3} \left(\overline{e_{yy}^{s} e_{xy}^{s}} \right) \\ &+ M_{16} \sum_{s=1}^{3} \left(4\overline{e_{xy}^{s} e_{xy}^{s}} + \overline{e_{xx}^{s} e_{xx}^{s}} \right) + M_{26} \sum_{s=1}^{3} \overline{e_{yy}^{s} e_{xx}^{s}} + 2M_{66} \sum_{s=1}^{3} \overline{e_{xx}^{s} e_{xy}^{s}} \end{aligned} \tag{B3}$$

$$\frac{\partial F(\mathbf{M})}{\partial M_{22}} = 0$$

Z. Liao, *Applied Seismic Geostructures*, https://doi.org/10.1007/978-981-97-7079-3

$$\sum_{s=1}^{3} \overline{\sigma_{yy}^{s} e_{yy}^{s}} = M_{12} \sum_{s=1}^{3} \overline{e_{xx}^{s} e_{yy}^{s}} + M_{22} \sum_{s=1}^{3} \overline{e_{yy}^{s} e_{yy}^{s}} + 2M_{26} \sum_{s=1}^{3} \overline{e_{xy}^{s} e_{yy}^{s}}, \tag{B4}$$

$$\frac{\partial F(\mathbf{M})}{\partial M_{26}} = 0$$

$$\sum_{s=1}^{3} \left(2\overline{\sigma_{yy}^{s} e_{xy}^{s}} + \overline{\sigma_{xy}^{s} e_{yy}^{s}} \right) = 2M_{12} \sum_{s=1}^{3} \overline{e_{xx}^{s} e_{xy}^{s}} + 2M_{22} \sum_{s=1}^{3} \left(\overline{e_{yy}^{s} e_{xy}^{s}} \right)$$

$$+ 2M_{26} \sum_{s=1}^{3} \left(4\overline{e_{xy}^{s} e_{xy}^{s}} + \overline{e_{yy}^{s} e_{yy}^{s}} \right) + M_{16} \sum_{s=1}^{3} \overline{e_{yy}^{s} e_{xx}^{s}} + 2M_{66} \sum_{s=1}^{3} \overline{e_{xy}^{s} e_{yy}^{s}} \tag{B5}$$

And

$$\frac{\partial F(\mathbf{M})}{\partial M_{66}} = 0$$

$$\sum_{s=1}^{3} \overline{\sigma_{xy}^{s} e_{xy}^{s}} = M_{16} \sum_{s=1}^{3} \overline{e_{xx}^{s} e_{xy}^{s}} + M_{26} \sum_{s=1}^{3} \overline{e_{yy}^{s} e_{xy}^{s}} + 2M_{66} \sum_{s=1}^{3} \overline{e_{xy}^{s} e_{xy}^{s}}. \tag{B6}$$

SPRINGER NATURE

GPSR Compliance

The European Union's (EU) General Product Safety Regulation (GPSR) is a set of rules that requires consumer products to be safe and our obligations to ensure this.

If you have any concerns about our products, you can contact us on ProductSafety@springernature.com

In case Publisher is established outside the EU, the EU authorized representative is:

Springer Nature Customer Service Center GmbH
Europaplatz 3
69115 Heidelberg, Germany

The manufacturer's authorised representative in the EU is Springer
Nature Customer Service Centre GmbH, Europaplatz 3, 69115 Heidelberg,
Germany. If you have any concerns regarding our products, please
contact ProductSafety@springernature.com

Printed and bound by CPI Group (UK) Ltd, Croydon, CR0 4YY
01/10/2026
02225969-0006